ALTERNATIVE SYSTEMS FOR POULTRY – HEALTH, WELFARE AND PRODUCTIVITY

Poultry Science Symposium Series

Executive Editor (Volumes 1–18): B.M. Freeman

*Out of print
Volumes 1–24 were not published by CAB International. Those still in print may be ordered from:

Carfax Publishing Company
PO Box 25, Abingdon, Oxfordshire OX14 3UE, UK

Alternative Systems for Poultry – Health, Welfare and Productivity

Poultry Science Symposium Series
Volume Thirty

Edited by

Victoria Sandilands

Avian Science Research Centre, SAC Auchincruive, Ayr, UK

and

Paul M. Hocking

Division of Genetics and Genomics, The Roslin Institute and Royal (Dick) School of Veterinary Studies, University of Edinburgh, UK

www.cabi.org

CABI is a trading name of CAB International

CABI
Nosworthy Way
Wallingford
Oxfordshire OX10 8DE
UK
Tel: +44 (0)1491 832111
Fax: +44 (0)1491 833508
E-mail: cabi@cabi.org
Website: www.cabi.org

CABI
875 Massachusetts Avenue
7th Floor
Cambridge, MA 02139
USA
Tel: +1 617 395 4056
Fax: +1 617 354 6875
E-mail: cabi-nao@cabi.org

A catalogue record for this book is available from the British Library, London, UK.

Library of Congress Cataloging-in-Publication Data

Alternative systems for poultry : health, welfare and productivity / edited by Victoria Sandilands and Paul M. Hocking.
       p. cm. -- (Poultry science symposium series ; v. 30)
   ISBN 978-1-84593-824-6 (alk. paper)
   1. Poultry--Housing. 2. Animal welfare. I. Sandilands, Victoria. II. Hocking, P. M. (Paul M.) III. Series:

Poultry science symposium ; no. 30.
   SF494.5.A48 2012
   636.55--dc23

                                2011037994

ISBN-13: 978 1 84593 824 6

Commissioning editor: Rachel Cutts
Editorial assistant: Alexandra Lainsbury
Production editor: Simon Hill

Typeset by Columns Design XML Ltd, Reading.
Printed and bound in the UK by CPI Group (UK) Ltd, Croydon, CR0 4YY.

CONTENTS

Part III Village and Backyard Poultry

Part IV Waterfowl and Game Birds

Part V Laying Hens

Part VI Meat Birds

CONTRIBUTORS

Robyn G. Alders, *International Rural Poultry Centre, KYEEMA Foundation, Brisbane, Australia, and Environment and Population Health, International Veterinary Medicine Section, Tufts University, 200 Westboro Road, North Grafton, MA 01536, USA; e-mail: Robyn. Alders@gmail.com*

Michael C. Appleby, *World Society for the Protection of Animals, London WC1X 8HB, UK; e-mail: michaelappleby@wspa-international.org*

James S. Bentley, *Consultant, Woodbank, John Street, Utkinton, Tarporly, Cheshire, CW6 0LU, UK; e-mail: Jamess.Bentley@virgin.net*

Jutta Berk, *Institut für Tierschutz und Tierhaltung (FLI), Dörnbergstrasse 25/27, D-29223 Celle, Germany; e-mail: Jutta.Berk@fli.bund.de*

Badi Besbes, *Animal Production and Health Division, FAO, Viale Delle Terme Di Caracalla, I-00153 Rome, Italy; e-mail: Badi.Besbes@fao.org*

Charles Deeming, *Department of Biological Sciences, University of Lincoln, Riseholme Park, Lincoln LN2 2LG, UK; e-mail: Charlie@ Deemingdc.freeserve.co.uk*

Ingrid C. de Jong, *Wageningen UR Livestock Research, PO Box 65, 8200 AB Lelystad, The Netherlands; e-mail: Ingrid.deJong@wur.nl*

Koen De Reu, *Institute for Agricultural and Fisheries Research, Technology and Food Unit, Product Quality and Food Safety, Brusselsesteenweg 370, B-9090 Melle, Belgium; e-mail: Koen.Dereu@ilvo.vlaanderen.be*

Jeroen Dewulf, *Veterinary Epidemiology Unit, Department of Reproduction, Obstetrics and Herd Health, Faculty of Veterinary Medicine, Ghent University, Sallsburylaan 133, B-9820 Merelbeke, Belgium; e-mail: Jeroen.Dewulf@ugent.be*

H. Arnold Elson, *ADAS Gleadthorpe, Meden Vale, Mansfield, Nottingham NG20 9PF, UK; e-mail: Arnold.Elson@adas.co.uk*

Ernst K.F. Fröhlich, *Centre for Proper Housing: Poultry and Rabbits (Zentrum für tiergerechte Haltung, Geflügel und Kaninchen, ZTHZ), Swiss Federal Veterinary Office, Burgerweg 22, CH-3052 Zollikofen, Switzerland; e-mail: Ernst.Froehlich@bvet.admin.ch*

Daniel Guémené, *INRA, UR83 Recherches Avicoles, F-37380 Nouzilly, France, and SYSAAF, Centre INRA de Tours-Nouzilly, URA, F-37380 Nouzilly, France; e-mail: Daniel.Guemene@tours.inra.fr*

E. Fallou Guèye, *Regional Animal Health Centre for Western and Central Africa, FAO, BP 1820, Bamako, Mali; e-mail: EFGueye@gmail.com*

Gérard Guy, *INRA, UE89 Palmipèdes à foie gras, Domaine d'Artiguères,*

F-40280 Benquet, France, and INRA, UMR 1289, Tissus Animaux, Nutrition, Digestion, Ecosystèmes et Métabolisme, F-31326 Castanet-Tolosan, France; e-mail: Gerard.Guy@bordeaux.inra.fr

Paul M. Hocking, The Roslin Institute and Royal (Dick) School of Veterinary Studies, University of Edinburgh, Easter Bush, Midlothian, EH25 9RG, UK; e-mail: Paul.Hocking@roslin.ed.ac.uk

Wiebke Icken, Lohmann Tierzucht GmbH, Am Seedeich 9–11, PO Box 460, D-27454 Cuxhaven, Germany; e-mail: Icken@ltz.de

Tracey A. Jones, Compassion in World Farming, River Court, Mill Lane, Godalming, Surrey, GU7 1EZ, UK; e-mail: Tracey.Jones@CIWF.org

Frieda Jorgensen, Health Protection Agency, Microbiology Services, Porton Down, Salisbury SP4 0JG, UK; e-mail: Frieda.Jorgensen@hpa.org.uk

Stephen Lister, Crowshall Veterinary Services, 1 Crows Hall Lane, Attleborough, Norfolk, NR17 1AD, UK; e-mail: salister@crowshall.co.uk

Murdo G. MacLeod, The Roslin Institute and Royal (Dick) School of Veterinary Studies, University of Edinburgh, Midlothian, EH25 9RG, UK; e-mail: Murdo.MacLeod@roslin.ed.ac.uk

Marion McMillan, SAC Consulting Veterinary Services, Ayr Disease Surveillance Centre, Auchincruive, Ayr KA6 5AE, UK; e-mail: Marion.McMillan@sac.co.uk

Knut Niebuhr, Institute of Animal Husbandry and Animal Welfare, Department of Farm Animals and Veterinary Public Health, University of Veterinary Medicine, Veterinärplatz 1, A-1210 Wien, Austria; e-mail: Knut.Niebuhr@vetmeduni.ac.at

Hans Oester, Centre for Proper Housing: Poultry and Rabbits (Zentrum für tiergerechte Haltung, Geflügel und Kaninchen, ZTHZ), Swiss Federal Veterinary Office, Burgerweg 22, CH-3052 Zollikofen, Switzerland; e-mail: Hans.Oester@bvet.admin.ch

Tom Pennycott, SAC Consulting Veterinary Services, Ayr Disease Surveillance Centre, Auchincruive, Ayr KA6 5AE, UK: e-mail: Tom.Pennycott@sac.co.uk

Rudi Preisinger, Lohmann Tierzucht GmbH, Am Seedeich 9–11, PO Box 460, D-27454 Cuxhaven, Germany; e-mail: Preisinger@ltz.de

David G. Pritchard, Formerly Senior Veterinary Consultant, Animal Welfare, DEFRA, London, UK, and Animal Welfare Science and Prentice, Argyll House, London SW18 1EP, UK; e-mail: DavidGeorgePritchard@gmail.com

Robert A.E. Pym, School of Veterinary Science, University of Queensland, Gatton , QLD 4343, Australia; e-mail: R.Pym@uq.edu.au

T. Bas Rodenburg, Animal Breeding and Genomics Centre, Wageningen University, PO Box 338, 6700 AH Wageningen, The Netherlands; e-mail: Bas.Rodenburg@wur.nl

Antonio Rota, International Fund for Agricultural Development, Via Paolo Di Dono 44, I-00142 Rome, Italy; e-mail: A.Rota@ifad.org

Victoria Sandilands, Avian Science Research Centre, SAC Auchincruive, Ayr KA6 5HW, UK; e-mail: Vicky.Sandilands@sac.ac.uk

Matthias Schmutz, Lohmann Tierzucht GmbH, Am Seedeich 9–11, PO Box 460, D-27454 Cuxhaven, Germany; e-mail: Schmutz@ltz.de

Lars Schrader, Institute of Animal Welfare and Animal Husbandry, Friedrich-Loeffler-Institut, Dörnbergerstrasse 25/27, D-29223 Celle, Germany; e-mail: Lars.Schrader@fli.bund.de

Z. Dan Shi, Department of Animal Sciences, South China Agricultural University, Guangzhou 510642, People's Republic of China; e-mail: ZDShi@scau.edu.cn

Magnus Swalander, Aviagen Turkeys Ltd, Chowley Oak Business Park, Tattenhall, Cheshire, CH3 9GA, UK; e-mail: MSwalander@aviagen.com

Hans-Heinrich Thiele, Lohmann Tierzucht GmbH, Am Seedeich 9–11, PO Box 460, D-27454 Cuxhaven, Germany; e-mail: Thiele@ltz.de

Ragnar Tauson, Department of Animal Nutrition and Management, Swedish University of Agricultural Sciences, Kungsängens forskningscentrum, 753 23 Uppsala, Sweden; e-mail: Ragnar.Tauson@slu.se

Olaf Thieme, Animal Production and Health Division, FAO, Viale Delle Terme Di Caracalla, I-00153 Rome, Italy; e-mail: Olaf.Thieme@fao.org

Frank A.M. Tuyttens, Institute for Agricultural and Fisheries Research, Animal Sciences Unit, Animal Husbandry and Welfare, Scheldeweg 68, B-9090 Melle, Belgium: e-mail Frank.Tuyttens@ilvo.vlaanderen.be

Sebastiaan Van Hoorebeke, Veterinary Epidemiology Unit, Department of Reproduction, Obstetrics and Herd Health, Faculty of Veterinary Medicine, Ghent University, Salisburylaan 133, B-9820 Merelbeke, Belgium; e-mail: Sebastiaan.VanHoorebeke@UGent.be

Filip Van Immerseel, Department of Pathology, Bacteriology and Avian Diseases, Faculty of Veterinary Medicine, Ghent University, Salisburylaan 133, B-9820 Merelbeke, Belgium; e-mail: Filip.VanImmerseel@UGent.be

Bert van Nijhuis, Verbeek's Broederij en Opfokbedrijven BV, Postbus 11, 6741 AA Lunteren, The Netherlands; e-mail: Nijhuis@verbeek.nl

PREFACE

A symposium such as this, collated from the work of many authors and the efforts of the many contributors who compiled them into a sensible chapter and meeting paper, does not come about easily. The process for the 30th Poultry Science Symposium, which was held at the University of Strathclyde in Glasgow, started at a UK branch council meeting of the World's Poultry Science Association (WPSA) in 2009, where it was agreed that it was time to host another Symposium in 2011, since the last (Biology of Breeding Poultry) was held four years previously. Paul had the initial topic idea but wished to share the workload; Vicky, knowing no better, agreed. We are grateful for the support of the UK branch's council and also that of WPSA.

We would like to thank the generous donations of time that the authors, scientific committee and independent referees gave. All of the processes that result in a successful meeting and book take longer than one expects, but without any one of them this Symposium would not have happened. The authors you will be familiar with, leafing through this book. The scientific committee and referees were made up of: Anna Bassett, James Bentley, Alice Clark, Laura Dixon, Arnold Elson, Patrick Garland, Andrew Joret, Steve Lister, Kelvin McCracken, Dan Pearson, Tom Pennycott and Claire Weeks. In addition, we were greatly assisted by Liz Archibald who maintained the web site. Kelvin McCracken deserves a second mention as the Symposium treasurer, and maintained his role even after his official retirement from his post as the UK branch treasurer.

The poultry industry continues to grow, change and modernize, and so we are grateful to our sponsors who, particularly in these difficult economic times, found the generosity to support this meeting. A full list of sponsors can be seen on page xiii.

Conferences are not easily arranged by fully committed scientific staff, nor would they run smoothly or cover all the necessary arrangements. Therefore we must thank the team at Congrex UK Ltd, and in particular Sharon MacIntyre and Kristina Milicevic who looked after us so ably.

At the end of a long chain of events is CABI Publishing and in particular Rachel Cutts, who would occasionally ask us when we might be turning in our manuscript. To her, a large apology for the delay and a big thank you for being so patient.

Finally, we must not forget you, the conference-goers, who make a considerable effort to attend, participate, and take home new information to your organizations and countries. We realize times are tough and funding for such meetings is shrinking. Thank you for choosing to attend the 30th Poultry Science Symposium.

Vicky Sandilands and Paul Hocking
July 2011

ACKNOWLEDGEMENTS

Support for this symposium is gratefully acknowledged from:

Principal Sponsor:
P.D. Hook (Hatcheries) Ltd

Session and Speaker Sponsors:
BOCM Pauls
British Poultry Science
Cherry Valley Farms
Huvepharma
Jansen Cilag
Lohmann Tierzucht
Pfizer Animal Health
St David's Poultry Team
Scottish Government

Sponsors:
Alpharma
DSM Nutritional Products Ltd
Hy-Line
Positive Action Publications Ltd
Taylor and Francis

CHAPTER 1

What are Alternative Systems for Poultry?

E.K.F. Fröhlich, K. Niebuhr, L. Schrader and H. Oester

ABSTRACT

By the late 1960s, poultry production had developed from a small-scale rural enterprise to an economically important branch of agriculture. Flock sizes increased and production systems, for hygienic and economic reasons, became more intensive. Rearing and housing of laying hens took place in conventional cages. At the same time, public concern for intensively housed birds began to increase, particularly following publications such as *Animal Machines* written by Ruth Harrison in 1964. New animal protection laws came into force and agriculture was forced to adapt to the welfare concerns of consumers. Alternative systems for housing laying hens that provided greater freedom of movement and facilities for natural behaviour including the use of the third dimension (perching, nesting) were developed. Production systems for meat birds were introduced that, in addition to higher space allowances, specified maximum rates of growth and feed ingredients. Time alone will show which type of system for poultry egg or meat production will survive the evolving social and economic pressures on producers and consumers.

INTRODUCTION

The term 'alternative system' is not specific and has changed its meaning during the last hundred years. This development reflects both economic and technical progress and the shift of cultural attitudes in the western, in particular European society. Animal protection has been a public issue for many years but the meaning of 'alternative system' changed dramatically in the last four decades. Therefore, it is necessary to give a definition of 'alternative system'. We then provide a short summary of the development of housing systems for laying hens and meat poultry since the beginning of intensive egg production.

Directive 1999/74/EC of the European Council (CEC, 1999) defined three categories of production systems for laying hens: 'unenriched cages', 'enriched cages' and 'alternative systems'. 'Alternative' in the meaning of the

EC directive is defined as a non-cage system 'which is operated with the human keepers entering it' (AHAW, 2005). Thus, the distinction between cage and non-cage systems was based on the human, not the animal perspective. In this book 'alternative' will be defined as any system that is not a barren cage for laying hens or an equally barren deep litter house for meat birds. The definition, applied to cages or deep litter, does not give us any information as to whether the system meets important requirements, for example, in meeting the behavioural needs of laying hens or broilers. For instance, it might become possible to improve cages further, e.g. by adding a scratching area and providing litter, or increasing height and space per hen, as has been done recently in Germany (Schrader, 2010).

ALTERNATIVE PRODUCTION SYSTEMS FOR LAYING HENS

The deadline (2012) fixed in Directive 1999/74/EC for the ban of conventional cages is approaching and will cause a European Union (EU)-wide change in the way laying hens are housed in the future. However, the provisions of this directive leave several options open. Each laying hen keeper has the difficult task of deciding which housing system will be the most sustainable for his farm. Is it the 'indoor path' with newly designed furnished cages or with aviaries, with the possibility to add outside runs later, or should he switch completely to the 'outdoor path'? The large number of new housing systems on the market makes the decision even more difficult. Furthermore, we can expect that the market for eggs will be increasingly specialized for specific markets including eggs for direct marketing, retailing or the egg processing industry. The example of Switzerland demonstrates the impact of the decision by the major retailers to sell nothing but free range and preferably organic eggs. Recently, several comprehensive reports on alternative systems have been published (e.g. Pickett, 2007) and therefore we only present a short overview of 'alternative systems' actually available in the second part of this chapter. It is not possible to give advice on the most sustainable or suitable housing system for the future as this depends on the country, the farm, the stockman and his marketing of eggs. Nevertheless, we try to demonstrate which solutions we think are the most promising for both the laying hens and the stockmen.

History of the development of alternative housing for laying hens

Until the beginning of the 20th century, nearly all the poultry in the world were kept in small backyard flocks providing eggs and culled birds for household consumption or relatively small-scale deep litter houses with access to an outdoor run. Housing, nutrition and care were often not ideal, because poultry keeping was not an economic enterprise and substantial investment was uneconomic. This changed dramatically in the early decades of the 20th century. Egg production became an economically viable enterprise as a consequence of increased urbanization, the development of new products containing processed eggs, and the focus provided by the military needs of two world wars. To cope with these changes, large-scale egg and poultry meat production was necessary that could exploit improved breeds, nutrition and disease prevention strategies (van de Poel, 1998). This was the challenge a hundred years ago and the answers included intensive deep litter systems and, in parallel, the development of conventional cage systems. Both were 'alternative systems' at the time and the conventional cage system was the most successful.

Van de Poel (1998) reported that J. Hulpin, Professor of the University of Wisconsin, built probably the first cages to house laying hens in 1911 and identified the trials at the Ohio Agricultural Experimental Station around 1924 as the first experimental study with cages arranged in batteries. The concept of keeping chickens in battery cages therefore celebrated its 100-year anniversary in 2011! However, the use of conventional cages under commercial conditions only started in the 1930s (Arndt, 1931) mainly in California. It should be pointed out that, at this time, conventional cages were considered as an alternative to overcome most of the problems of large-scale poultry farms using deep litter systems:

> This form of battery is coming into widespread use throughout the country and apparently is solving a number of the troubles encountered with laying hens in the regular laying house on the floor. (Arndt, 1931.)

These troubles mainly were diseases like coccidiosis and Marek's disease but also other parasitic infections, cannibalism and the lack of economic efficiency. Ebbell (1959) introduced what is probably the first large fully metal conventional cage system for layers in Europe on the Ovaltine® Farm at Niederwangen, Switzerland in 1935 (Fig. 1.1) and considered the much higher efficiency, in terms of work load and stocking density, as the main advantage of this system. This development towards intensive indoor conventional cage systems is widely considered as the outcome of the 'ideology of efficiency' (van de Poel, 1998) that governed economic theory from the beginning of the 20th century. However, we should be aware that at that time the threats of diseases were very high because of the nearly complete absence of treatments and vaccines (Hofrogge, 2000). This also means that with today's possibilities and scientific knowledge the conventional cages would probably not have spread in the way they did.

The dominance of this ideology ended in Switzerland some 56 years later in 1991 and will also be phased out in the EU by 2012 because of Directive 1999/74/EC of the European Council (CEC, 1999) and even in California by 2015 (California Legislature, 2009). Governor Schwarzenegger signed the Bill after a successful state-wide ballot in 2008 that brought into law the Prevention of Farm Animal Cruelty Act, also known as Proposition 2. The reason for this change was the increasing awareness of animals, and farm animals in particular, as sentient beings rather than machines or simple means of production (Harrison, 1964). In 1965 the Brambell committee, a panel of experts on livestock production, scientists and members of animal welfare organizations, proposed a set of five basic 'freedoms' for (farmed) animals (Brambell, 1965).

Fig. 1.1. The first conventional cages for pullets with heating and washing of the manure belt in Switzerland in 1935.

Among them, the freedom to express normal behaviour has had the greatest impact on housing conditions. Consequently, public attitudes to housing systems changed and the primacy of economics, the 'ideology of efficiency', was no longer the only relevant paradigm! The 'needs' of the animals and animal welfare considerations also became an important issue. Hughes (1973) stated:

> The last three findings described identify three ways in which cages are inadequate. They constrain various behaviour patterns, either through space limitation or through lack of a suitable substrate like litter, they fail to provide suitable stimuli to release nesting behaviour and thereby lead to frustration and they produce an environment which potentiates the expression of undesirable behaviour like feather pecking.

During the last 35 years, several European countries have implemented animal welfare legislation. Directive 1999/74/EC of the European Council (CEC, 1999) is probably the most important single piece of legislation to affect poultry keeping. This directive bans conventional cages from January 2012. However, the complete ban on cages in Switzerland and Austria and the withdrawal of all cage eggs from the shelves in German supermarkets suggest that a substantial proportion of European consumers do not regard furnished cages as an acceptable 'alternative' to conventional cages. Therefore, we would, in line with other publications (e.g. AHAW, 2005), suggest avoiding the term 'alternative systems' and distinguish between cage and non-cage systems for laying hens.

Types of husbandry systems for laying hens

There are different ways to categorize housing systems (LAYWEL, 2006b). For the purpose of this chapter we distinguish two main categories, cage and non-cage systems. Regarding cage systems, we concentrate on furnished cages (enriched cages, modified cages) and for non-cage systems, we consider single-level systems (in some countries called deep litter, in others barn) and multi-level systems (aviaries; multi-tier systems). Of more historical interest are fully slatted floor systems (Pennsylvania systems). All non-cage systems may be combined with outside runs (Fig. 1.2).

Furnished cages

An approach to improve the cage environment is to furnish them with relevant resources like nests, perches and a litter area. All furnished cages maintain important characteristics of the conventional cage such as wire mesh flooring

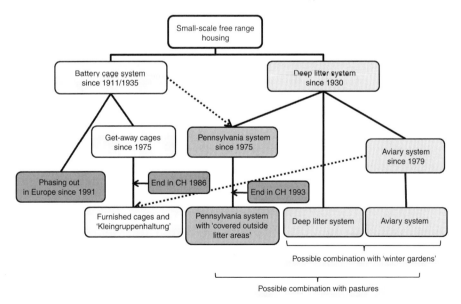

Fig. 1.2. Development of housing systems in the last 100 years in Europe. Dotted lines indicate major influences.

to separate the birds from their manure to reduce the risk of disease. In addition, the group size should be kept small to reduce aggressive and cannibalistic behaviours (Abrahamsson and Tauson, 1995). Furthermore, it has been argued that cages facilitate good management because of the possibility that the producer can control the behaviour of the laying hens (Appleby, 1998). For example, in conventional cages eggs roll away on a collecting belt immediately after laying. Therefore egg eating or dirty eggs are not a problem. On the other hand, the furnished cage should overcome some inherent disadvantages of conventional cages (EC, 1996), where the birds are unable to perform most natural behaviours (Appleby *et al.*, 1993) and may suffer skeletal fragility (Knowles and Broom, 1990) because of limited physical space allowance and the lack of resources to exercise.

Elson (1976) was probably the first to propose 'get-away' cages. His idea was to develop an alternative to the conventional cage that combined economy, animal welfare and more 'freedom' for the hens to move and conduct more behaviours. Furthermore, it should be possible to pile these cages up and the increase in fixed costs should be minimized. The original cage (Fig. 1.3) for 60 laying hens had a flat floor, littered group nests and one or two raised perches to enable birds to get away from the others. That is the reason for its name: 'get-away cage'. Later additional versions (Fig. 1.4) for smaller groups of ten or 20 birds had wire mesh flooring, additional perches and a dust bath facility (Oester, 1985).

Research projects on the 'get-away' cage in the UK (Elson, 1981), the Netherlands (Brantas *et al.*, 1978), Switzerland (Oester, 1985) and Germany (Wegner, 1981) ended with controversial results. There was some evidence of an improvement in the welfare of the birds if a dust bath was available, but also for severe problems of aggression, cannibalism and feather pecking. In addition,

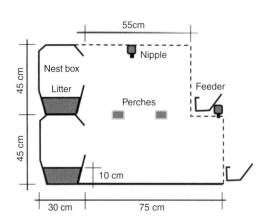

Fig. 1.3. Get-away cage: side elevation plan of the experimental cage (Bareham, 1976).

Fig. 1.4. Get-away cage at Zollikofen with dust bath and nest box (Oester, 1985).

the inspection and catching of birds was difficult (Appleby, 1998). In Switzerland, the 'get-away' cage was tested with a negative result at the Swiss Poultry Husbandry School between 1978 and 1980 (Oester, 1985). Despite these results, further efforts to improve cage design went on (e.g. Elson, 1988; Appleby, 1998). In the 1980s several Swiss manufacturers developed furnished cages with perches and nest boxes for either 42 or 58 hens and a space allowance of 800 cm^2 per bird; however, all failed to pass the Swiss testing procedure. The main criticism was that an illumination of the inside of the cages with a light intensity of 5 lux, which is required by the Swiss Animal Welfare Ordinance as the minimum light intensity, caused extremely high mortalities of up to 30% due to cannibalism (Fröhlich and Oester, 1989, 2001).

Recent examples of furnished cages are based on the considerable work of many scientists (e.g. Elson, 1990; Nicol, 1990; Appleby and Hughes, 1995; Tauson, 1998) and on the later provisions of Directive 1999/74/EC (CEC, 1999). The EU directive stipulates a minimum space allowance of at least 750 cm^2 area per hen of which 600 cm^2 shall have 45 cm free height above the area. In addition the cage shall have a nest with no direct contact to any wire mesh floor, 15 cm of perches per hen, a feed trough of 12 cm per hen, claw shortening devices as proposed by Tauson (1984) and 'friable material enabling the hens to satisfy their ethological needs for pecking and scratching'. Cages fulfilling these provisions have been developed throughout the EU. The LAYWEL (2006b) study defined three categories of furnished cages based on the number of hens housed: large, medium and small cages. In the large ones, groups of up to 115 hens are kept, whereas in the medium cages 15 to 30 hens and in the small ones (Fig. 1.5) up to 15 hens are kept.

In Germany, the legal requirements for furnished cages are higher than in the EU directive. The floor space per hen must be at least 800 cm^2 including 90 cm^2 litter area per hen. In addition, a nest area of 90 cm^2 per hen should be provided, resulting in a total area of 890 cm^2 per hen. Perches have to be at different heights and the minimum cage height has to be 50 cm and 60 cm at the feed troughs. Furthermore, the minimum total area of each cage must be 2.5 m^2 (Fig. 1.6). This enlarged furnished cage is called 'Kleingruppenhaltung' ('small group housing').

Furnished cages have been intensively investigated (e.g. AHAW, 2005; FAL, 2005; LAYWEL, 2006a) because they are meant to be the standard system within the EU in the future. According to the LAYWEL study the overall productivity in furnished cages was found to be as good as in conventional cages. The percentage of eggs laid in the nest boxes as an indicator of the use of the nests varied between 88% and 99% in different houses with an average of 92% to 95%. An explanation for these differences was identified in differences in the management of the nest opening and the types of flooring used in the nest boxes. However, the hens may also use the nests as places to withdraw, which may cause hygiene problems (FAL, 2005) and suggests that no other adequate place to separate from the group was available. The perch use during night-time varied between 65% in larger and 80% to 87% in smaller furnished cages and during the daytime between 15% and 40%, respectively. Other studies reported similar results (Appleby *et al.*, 1993; Tauson, 2002;

(a)

(b)

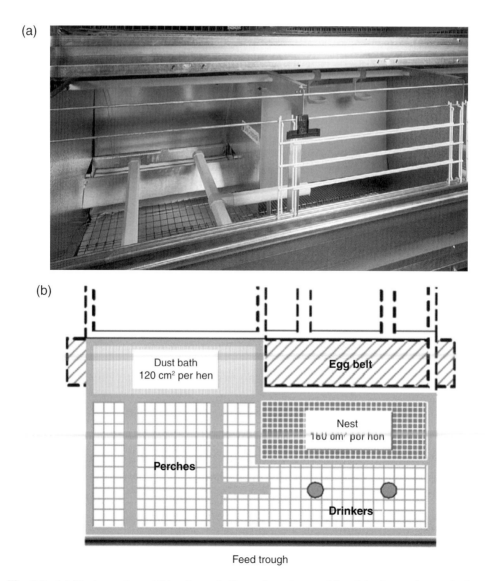

Fig. 1.5. (a) Photograph and (b) schematic floor plan of a small furnished cage for ten laying hens (Big Dutchman Inc.).

Sewerin, 2002; FAL, 2005; Rönchen, 2007). However, the factors influencing the use of these facilities are not only the cage size but also the breed, the nest material and the shape of the perches (Schrader, 2008). A particular problem in furnished cages is the litter or scratching area. Most often, the area is small and covered by a plastic mat without any surrounding boards. Therefore, the litter material is likely to disappear immediately by the hen's scratching activity. To ensure the proper use of this area for foraging and dust bathing, frequent administration of additional material, preferably several times a day, is needed

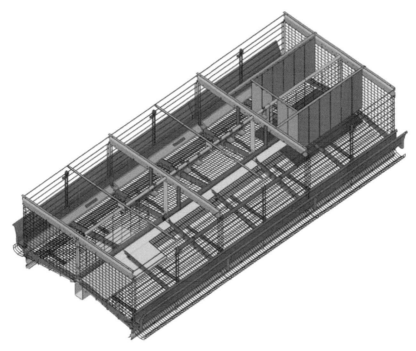

Fig. 1.6. Schematic drawing of a 'Kleingruppenhaltung' furnished cage (Big Dutchman Inc.).

(Schrader, 2008). Furthermore, in medium and large furnished cages only 1 to 2% of the hens have been observed dust bathing during the main activity time for this behaviour (FAL, 2005), while at the same time an equal number of hens or even more perform sham dust bathing on the wire mesh floor surrounding the litter area (Lindberg and Nicol, 1997). This observation implies that the area is too small to allow all motivated hens to dust bathe at the same time or the substrate provided does not meet the hens' preferences (LAYWEL, 2006b). In the German FAL study, the mortality and feather damage in medium and large furnished cages were low. However, the light intensities within the cages were also low and could explain this result.

Taken together, the results of different studies show that laying hens use the additional resources in furnished cages. However additional research has been proposed to improve the cage design, in particular with respect to the dimension of the foraging and dust bathing facilities (FAL, 2005), the administration and quality of the 'litter' substrate (FAL, 2005; LAYWEL, 2006b), the group sizes (LAYWEL, 2006b) and the light intensity within the cages (FAL, 2005). Modern furnished cages are therefore an improvement compared with conventional cages. However, they offer less space and less adequate resources than non-cage systems. Furnished cages will probably become the basic system in many countries of the EU except in central and northern European countries where there is some doubt about the long-term acceptability of this system.

Non-cage systems

Single-level systems. The single-level systems like the traditional deep litter system (Fig. 1.7) have been in use since large-scale poultry housing commenced. They are a more intensive variation of the traditional small-scale poultry house (Brade, 1999) and are often combined with covered outside runs ('winter gardens') and pasture. This system is still widespread in laying hen and parent stock systems, but less frequently in new houses, especially in Central Europe. The stocking density usually varies from 6 to 9 birds m^{-2} ground floor or usable area, depending on the breed used and on the provisions of organic legislation or of label programmes.

Single-level systems are often equipped with raised perches (aerial perches) above the slatted areas depending on the interpretation of the requirement of Directive 1999/74/EC for a minimum amount of perch space per hen. In relation to the design and position of the perches within the poultry house both the prevalence and severity of keel bone damage may increase (Sandilands *et al.*, 2009), but this is the case in any system with perches. The costs of building this system are low and it may therefore be an economically viable investment, despite the low stocking densities. This system is therefore often restricted to organic production and for housing birds in free range production systems.

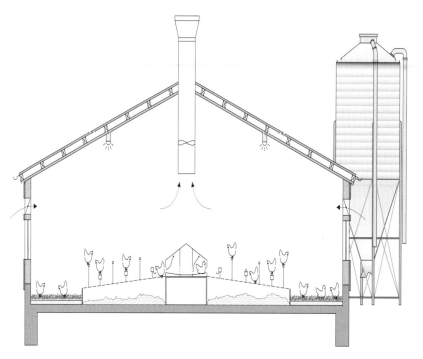

Fig. 1.7. Single-level system (Achilles *et al.*, 2006) with group nests in the middle, a droppings pit covered by wire mesh or plastic slats, and a litter area towards the walls of the pen. Feed and water troughs are over the droppings pits as are perches, if provided.

FULLY SLATTED FLOOR SYSTEMS (PENNSYLVANIA SYSTEMS). Fully slatted systems are also known as 'Pennsylvania' systems because of their origin in the USA (Fig. 1.8). These systems were built as alternatives to conventional deep litter pens as early as the 1970s to combine the advantages of cages – high stocking density and good hygiene – with the improved freedom of movement in deep litter systems (Scholtyssek, 1987). However, Prip (1976) already reported severe problems with hysteria in this system and Scholtyssek (1987) concluded that the system failed because of the high stocking densities of 16 birds m^{-2}. In Switzerland the Pennsylvania system got a second chance but at a lower stocking density of 12.5 birds m^{-2}. However, the Pennsylvania system failed to pass the obligatory authorization and testing procedure of Switzerland (Wechsler *et al.*, 1997; Oester and Troxler, 1999) even with the reduced stocking density and a raised scratching area. Pennsylvania systems were banned from the Swiss market in 1993 but are sometimes used in other European countries in combination with a covered run (winter garden) as a scratching area.

MULTI-TIER OR AVIARY SYSTEMS. The development of aviary or multi-tier systems started in Switzerland in the late 1970s, parallel to the development of the 'get-away' cages. The first prototype was created at the Swiss Federal Institute of Technology, Zürich in 1979 (Fölsch *et al.*, 1983). The multi-tier system attempted to improve on the conventional single-level system by providing elevated perches and additional floors on different levels. The aim is to make the third dimension of a barn accessible to the hens and to divide the environment into four different functional areas (Fröhlich and Oester, 1989).

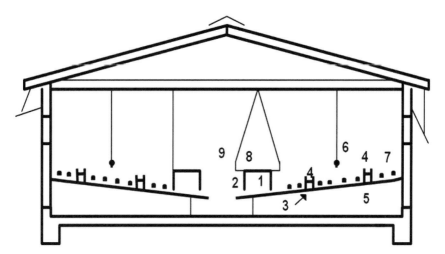

Fig. 1.8. Diagrammatic representation of a Pennsylvania system. A classical variation without litter and raised perches is shown on the left and the system with a litter area is shown on the right: 1, laying nests; 2, egg collecting; 3, wire mesh floor; 4, feed troughs; 5, droppings pit; 6, nipple drinkers; 7, perches; 8, litter area, closable; 9, gangway. (Image by E. Fröhlich.)

An area for resting and withdrawal equipped with perches to roost was placed at the highest level of the system. The lower tiers with wire or plastic mesh flooring and feed and water facilities serve as the feeding area. The floor, the lowest level of the barn, was either covered entirely or at least between the tiered rows, with litter material for the daytime activities like exploration, foraging, scratching or dust bathing and the fourth area was equipped with individual or group nests.

The EU directive stipulated some additional provisions for this type of system, e.g. the maximum stocking density is restricted to 9 birds m^{-2} (1111 cm^2 per bird) usable space, the maximum number of elevated floors is restricted to three and the feed trough length has to be at least 10 cm per bird. Furthermore, the floors have to be equipped with manure belts to prevent the birds defecating on each other. Apart from European legislation, in many European countries accreditation programmes, e.g. Verein für Kontrollierte Alternative Tierhaltungsformen eV (KAT, 2009), had a substantial impact on the design of multi-tier systems. KAT for instance allows only two additional elevated tiers (three levels including the floor). Therefore, aviary systems from many manufacturers are built to satisfy this provision (compare, as an example, the right and left row in Fig. 1.10). It should be pointed out that in many European countries the English word 'aviary' (German: *Voliere*; French, Dutch: *volière*; Spanish: *sistemas aviarios*; Italian: *sistema a voliera*) is used only in connection with multi-tier systems.

There is a large number of different designs of multi-tier systems. In principle, three types can be distinguished. The earlier systems (Fig. 1.9), mainly used in Switzerland, have nest boxes outside the system, either at the wall of the barn or in the middle between two rows. In these early systems only the area at the sides of the rows was littered. In newer designs (Fig. 1.10), the nests are integrated into the rows between two elevated floors, or placed at equal distance on the raised tiers, and the entire floor is littered. These systems are also called 'row' systems, as several rows are often placed beside each other in the barn. The caretaker usually does not enter the multi-tier system itself during stock inspection.

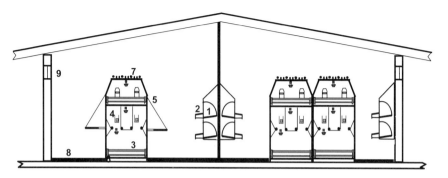

Fig. 1.9. Multi-tier system (Natura-450; Big Dutchman Inc., Inauen): 1, group nests; 2, egg-collecting facility; 3, plastic mesh floor; 4, raised feed troughs; 5, manure belt; 6, nipple drinkers; 7, perches; 8, litter area; 9, windows. (Image by E. Fröhlich.)

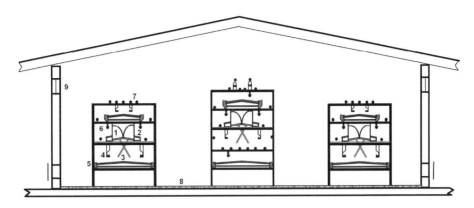

Fig. 1.10. Aviary system (Bolegg Terrace; Vencomatic BV, Rihs-Agro AG) with integrated group nests; a Swiss variation with raised feed troughs and additional drinkers in the centre. 1, Group nest; 2, egg-collecting facility; 3, plastic mesh floor; 4, feed troughs; 5, manure belt; 6, nipple drinkers; 7, perches; 8, litter area; 9, windows. (Image by E. Fröhlich.)

The third type is the so-called 'portal' system (Fig. 1.11). In this system the caretaker walks below and in the system during flock inspection.

After farmers gained experience with many different systems in several European countries, some portal systems became less popular and some manufacturers have stopped producing them. Misplaced eggs on the slats have been a major problem in recent years (Niebuhr *et al.*, 2009) and some manufacturers solved this by placing egg belts (and nests) along both tiers, allowing eggs mislayed on the slats to roll on to the egg belts.

The group size in aviaries varies considerably from a few hundred birds to several tens of thousands but is restricted to 2000–6000 in some European countries. The main reasons for this are the special provisions of accreditation programmes, legislation for organic agriculture or national upper limits (e.g. in

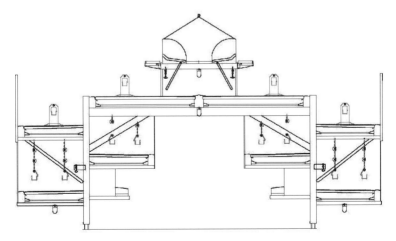

Fig. 1.11. Multi-tier 'portal' system (Red-L®; Vencomatic BV). (Image by Janker Stalltechnik, Kilb, Austria.)

Germany). Today, nearly 90% of all Swiss poultry houses are equipped with aviaries. However, aviaries are also becoming more popular in other European countries such as Germany, the Netherlands and Austria. In Austria approximately 50% of laying hens are currently housed in aviaries. Aviaries are also used for layer breeders (see Thiele, Chapter 10, this volume).

It is important to underline the fact that in all multi-tier systems only birds reared under similar conditions should be kept. An influence of early access to litter on feather pecking, productivity and mortality has been reported by several authors (e.g. Johnsen *et al.*, 1998; Huber-Eicher and Sebö, 2001). It has also been reported that early perching results in a lower prevalence of cloacal cannibalism (Gunnarsson *et al.*, 1999). However, the early access to elevated perches and floors might be even more important for the proper use of the aviary system in the layer house. Chickens with early perching experience are better flyers in adulthood (Fröhlich and Oester, 1989) and may reach raised perches, food troughs or nests more easily than those that have had no or much later access to them (Faure and Jones, 1982a,b; Appleby *et al.*, 1988).

Although the potential for good welfare of laying hens is greatest in multi-tier or aviary systems especially when equipped with outdoor runs (free range system), recent studies have revealed important risks, in particular with respect to increased prevalence for keel bone distortions and fractures (Nicol *et al.*, 2009; Käppeli *et al.*, 2011) and cannibalism (Nicol *et al.*, 2009).

An important issue for this category of system is the qualification of the stockman and of the management personnel. Practical experience in Switzerland showed clearly that multi-tier systems need good stockmanship and conscientious labour. The importance of stockmanship and management is critically important if hens are kept with intact beaks. In order to reduce feather pecking and cannibalism, fine-tuning of the whole chain including genetics, feeding, rearing conditions and husbandry in the layer house is important as management directly or indirectly influences many of these factors (Niebuhr *et al.*, 2006).

ALTERNATIVE SYSTEMS FOR MEAT POULTRY

A large number of different poultry species are farmed for meat production. All species are most often kept in an intensive way in barren barn systems with litter or wire mesh flooring or even in cages in the case of quail. However, all of these species are also kept in alternative housing systems. We focus in this chapter on broiler chickens and turkeys because they represent the large majority of birds in European poultry meat production. In contrast with alternatives in laying hen systems, alternative systems for meat birds encompass changes to housing, husbandry procedures and the use of different breeds. Reduced stocking density, lighting schedules with an uninterrupted dark period, natural daylight, the use of low input feed, and improved air and litter quality are important aspects differentiating alternative systems for meat poultry. However, the main welfare issues in conventional broiler and turkey breeds remain the fast growth rate with high susceptibility to metabolic and gait disorders (Knowles *et al.*, 2008) as well as low locomotor activity (Bessei, 2006). The main characteristic of an alternative production of broiler chickens and turkeys is a comparatively low growth rate. In fact, this is the pre-condition to enable the birds to benefit from alternative production systems. The European Commission was clearly aware of this fact when laying down the rules for marketing standards for poultry meat (CEC, 2008) since increased minimum age at slaughter was implemented for different alternative production systems. In reality, this can only be achieved with slow growing breeds. All major breeding companies now offer slow growing meat chickens, but also regional breeders supplying traditional breeds like Sulmthaler, New Hampshire or Plymouth Rock are realistic alternatives.

In addition to the choice of breed the welfare of all broilers can be improved, even for fast growing birds, by the broiler house design. As in laying hens, an alternative broiler house for standard breeds should be structured in different functional areas. Litter, food and water are always available, but the possibility of having access to the third dimension is normally missing. In the first few days of life the broiler chicken will explore environmental components such as perches, ramps or elevated surfaces (Fig. 1.12), but of course it will, depending on the hybrid, take around 12 to 30 days until the birds start to rest during the night on elevated structures (Oester and Wiedmer, 2005).

A covered outside run (winter garden) is a potential enrichment to a conventional broiler or turkey house. It adds space, a scratching area, fresh air and daylight and, if furnished, additional components to encourage the activity

Fig. 1.12. Swiss standard broiler house with ramps for fast growing broiler breeds. (Image courtesy of H. Oester.)

Fig. 1.13. Covered outside run for commercial breeds. (Image courtesy of H. Oester.)

of the birds (Fig. 1.13). Winter gardens for fast growing conventional hybrids are widely used in Switzerland and research showed that, in 2009, 88.4% of broilers had access to a covered outside area that was used intensively by the birds.

The stocking density in alternative systems for meat poultry should be considerably lower than in conventional production. More active birds need more space and additional facilities like perches are also space consuming. The EU stipulated maximum figures for various alternative systems (CEC, 2008), but it must be emphasized that these are upper limits and not the recommended stocking density in alternative systems for meat chickens. Examples of alternative meat chicken systems are numerous and most depend on specific requirement of local or organic labelling systems such as Label Rouge, Poulet de Bresse, Freedom Food and Freiland.

CONSUMER PERCEPTIONS AND PREFERENCES

During the last few decades, European citizens have completely changed their view of what are good housing conditions for livestock and in particular for laying hens. There are several reasons for this, the most important being that people have developed a very different attitude towards animal integrity and welfare. Farm animals in particular are now considered to be sentient beings with their own needs and value. In Switzerland, even the integrity or dignity of all creatures is protected following the revision of the constitution in 1999 (Schweizerische Bundesverfassung, 1999). Today, the majority of European consumers are very concerned about the way laying hens are kept (CEC,

2005). Fifty-eight per cent of the citizens in the EU think that the welfare of laying hens is bad or even very bad, whereas the housing of pigs and cattle is judged much more positively (44% and 25% bad and very bad, respectively). Only 33% consider the welfare of laying hens as good. Most criticism comes from the Dutch followed by the Danes and Germans. In contrast to these attitudes, only about four citizens in ten (38%) state that they prefer buying eggs from animal welfare friendly production systems such as free range or outdoor production systems. This could mean that another 20% of Europeans do not ask for free range or organic eggs but prefer that hens are kept in better indoor housing. Finally, the third group of about 42% do not care (9%) or do not see any problem in the actual housing conditions of laying hens. These results suggest that every housing system is suitable for some European consumers but it would be wrong to assume that the development in public attitudes would stop at the current level. Instead, we should be prepared for a continuing development of animal welfare provisions. A large range of different housing systems for laying hens has been developed in recent years and they all are alternatives to conventional cages. However, the initial question 'what are alternative systems?' could also be understood differently as 'which system or class of systems will be the most accepted in the future?'

BEHAVIOURAL AND WELFARE NEEDS AND PREFERENCES

In recent decades, an epic discussion among scientists as well as the public took place about the question of the 'needs' of laying hens, and, in particular, which needs should be considered essential for the welfare of the animals. De Mol *et al.* (2006) developed a computer model to assess the welfare of laying hens in 19 different production systems in the Netherlands based on the available scientific evidence. The authors identified five attributes: (i) feeding level, (ii) space per hen, (iii) perches, (iv) water availability and (v) nests, as the most important factors contributing to good laying hen welfare, whereas 'free range' was of minor importance. However, this approach did not include the attribute 'litter' or, more generally, the 'material to perform foraging and dust bathing behaviour' that has been considered by other authors as a basic attribute of alternative housing systems (de Jong *et al.*, 2007).

The question about essential needs implies that animals have important and less important needs. This approach may have its origin in the 'pyramid of needs of humans', proposed by Maslow (1943). However, this pyramid has five levels of needs and the fulfilment of the lower one is a prerequisite for the next higher level. Once the lower level is fulfilled, the motivation for the upper level becomes relevant. In other words, if the basic or essential needs are fulfilled, the needs of the next upper level become the goals to struggle for. In this perspective, the discussion about essential or luxury needs is therefore superfluous: there are only fulfilled or not fulfilled needs. In the case of the poultry, the basic needs (e.g. freedom of hunger and thirst) are necessary to keep them alive. Fulfilling these needs means that the next level of needs, like safety from predators, shelter from inclement weather and freedom from fear, starts to become essential. If these needs are also met, then the birds will try to satisfy the needs of the next level, and so on. To conclude: we should not discuss whether a resource is necessary to fulfil the needs of an animal to improve its welfare, but about the way to provide a well-structured environment with different functional areas and the availability of all the resources that laying hens, pullets or broiler chickens could possibly require to satisfy their needs.

CONCLUSIONS

Hughes stated as early as 1973 that:

> The real choice lies not between an ideal, natural system and an unsatisfactory, intensive system, but between several systems, all artificial to a greater or lesser extent and all with various imperfections.

Alternative systems for laying hens, pullets and meat chickens should provide an environment where the hen can choose and find resources she is motivated to seek. The challenge is to arrange the facilities in a poultry house in a way that minimizes conflicts between hens with different motivations and that still optimizes management. In our opinion only multi-tier systems or aviaries with covered outside runs have the potential to meet this requirement. However, the systems that are actually available are not perfect but will need constant endeavours to improve them.

REFERENCES

Abrahamsson, P. and Tauson, R. (1995) Aviary systems and conventional cages for laying hens. Effects on production, egg quality, health and bird location in three hybrids. *Acta Agriculturae Scandinavica Section A: Animal Science* 45, 91–203.

Achilles, W., Eurich-Menden, B., Grimm, E. and Schrader, L. (eds) (2006) *Nationaler Bewertungsrahmen Tierhaltungsverfahren*. KTBL-Schrift 446. Kuratorium für Technik und Bauwesen in der Landwirtschaft eV (KTBL), Darmstadt, Germany, p. 597.

AHAW (2005) Opinion of the Scientific Panel on Animal Health and Welfare (AHAW) on a request from the Commission related to the welfare aspects of various systems of keeping laying hens. *The EFSA Journal* 197, 1–23.

Appleby, M.C. (1998) Modification of laying hen cages to improve behaviour. *Poultry Science* 77, 1828–1832.

Appleby, M.C. and Hughes, B.O. (1995) The Edinburgh Modified Cage for laying hens. *British Poultry Science* 36, 707–718.

Appleby, M.C., Duncan, I.J.H. and McRae, H.E. (1988) Perching and floor laying by domestic hens: experimental results and their commercial application. *British Poultry Science* 29, 351–357.

Appleby, M.C., Smith, S.F. and Hughes, B.O. (1993) Nesting, bathing and perching by laying hens in cages: effects of design on behaviour and welfare. *British Poultry Science* 34, 835–847.

Arndt, M. (1931) *Battery Brooding*, 2nd edn. Orange Judd Publishing Co., New York, pp. 308–312.

Bareham, J.R. (1976) A comparison of the behaviour and production of laying hens in experimental and conventional battery cages. *Applied Animal Ethology* 2, 291–303.

Bessei, W. (2006) Welfare of broilers: a review. *World's Poultry Science Journal* 62, 455–466.

Brade, W. (1999) Haltung von Legehennen, Eiqualität und Verbraucher (Teil 1 und 2). *Tierärztliche Umschau* 54, 270–274.

Brambell, F.W.R. (1965) *Command Paper 2836*. Her Majesty's Stationery Office, London.

Brantas, G.C., Wennrich, G. and De Vos-Reesink, K. (1978) Behavioural observations on laying hens in get-away-cages (Ethologische Beobachtungen an Legehennen in Get-away-Kaefigen). *Archiv für Geflügelkunde* 42, 129–132.

California Legislature (2009) Assembly Bill No. 1437. ftp://leginfo.public.ca.gov/pub/09-10/bill/asm/ab_1401-1450/ab_1437_bill_20090702_amended_sen_v98.pdf (accessed 28 January 2011).

CEC (Commission of the European Communities) (1999) Council Directive 1999/74/EC laying down minimum standards for the protection of laying hens. *Official Journal of the European Communities* L203, 53–57.

CEC (Commission of the European Communities) (2005) Attitudes of consumers towards the welfare of farmed animals. Eurobarometer 229, Wave 63.2. http://www.ec.europa.eu/public_opinion/archives/ebs/ebs_229_en.pdf (accessed 28 January 2011).

CEC (Commission of the European Communities) (2008) Council Directive 2008/543/EC laying down detailed rules for the application of Council Regulation (EC) No 1234/2007 as regards the marketing standards for poultry meat. http://eur-lex.europa.eu/LexUriServ/LexUriServ.do?uri=OJ:L:2008:157:0046:0087:EN:PDF (accessed 28 January 2011).

De Jong, I.C., Wolthuis-Fillerup, M. and van Reenen, C.G. (2007) Strength of preference for dustbathing and foraging substrates in laying hens. *Applied Animal Behaviour Science* 104, 24–36.

De Mol, R.M., Schouten, W.G.P., Evers, E., Drost, H., Houwers, H.W.J. and Smits, A.C. (2006) A computer model for welfare assessment of poultry production systems for laying hens. *NJAS – Wageningen Journal of Life Sciences* 54, 157–168.

Ebbell, H. (1959) Intensiv- und Käfighaltung. *Geflügelzucht-Bücherei*, Heft 12. Eugen Ulmer Verlag GmbH, Stuttgart, Germany, pp. 1–94.

EC (European Commission) (1996) Report on the welfare of laying hens. Scientific Veterinary Committee, Animal Welfare Section. http://www.ec.europa.eu/food/fs/sc/oldcomm4/out33_en.pdf (accessed 28 January 2011).

Elson, H.A. (1976) New ideas on laying cage design – the 'get-away' cage. In: *Proceedings of the 5th European Poultry Conference*, Malta, 5–11 September 1976. World's Poultry Science Association (Malta Branch), vol. 2, pp. 1030–1041.

Elson, H.A. (1981) Modified cages for layers. In: *Alternatives to Intensive Husbandry Systems*. Universities Federation for Animal Welfare, Potters Bar, UK, pp. 47–50.

Elson, H.A. (1988) Making the best cage decisions. In: *Cages for the Future, Proceedings of the Cambridge Poultry Conference*. Agricultural Development and Advisory Service, Cambridge, UK, pp. 70–76.

Elson, H.A. (1990) Recent developments in laying cages designed to improve bird welfare. *World's Poultry Science Journal* 46, 34–37.

FAL (2005) *Modellvorhaben ausgestaltete Käfige: Produktion, Verhalten, Hygiene und Ökonomie in ausgestalteten Käfigen von 4 Herstellern in 6 Legehennenbetrieben*. FAL, Braunschweig, Germany.

Faure, J.M. and Jones, R.B. (1982a) Effects of sex, strain and type of perch on perching behaviour in the domestic fowl. *Applied Animal Ethology* 8, 281–293.

Faure, J.M. and Jones, R.B. (1982b) Effects of age, access and time of day on perching behaviour in the domestic fowl. *Applied Animal Ethology* 8, 357–364.

Fölsch, D.W., Rist, M., Munz, G. and Teygeler, H. (1983) Entwicklung eines tiergerechten Legehennen-Haltungssystemes: Die Volierenhaltung. *Landtechnik* 6, 255–257.

Fröhlich, E.K.F. and Oester, H. (1989) Application of ethological knowledge in the examination of proper keeping conditions of housing systems for laying hens. In: *Aktuelle Arbeiten zur artgemäßen Tierhaltung 1988*. KTBL-Schrift 336. Kuratorium für Technik und Bauwesen in der Landwirtschaft eV (KTBL), Darmstadt, Germany, pp. 273–284.

Fröhlich, E.K.F. and Oester, H. (2001) From battery cages to aviaries: 20 years of Swiss experiences. In: Oester, H. and Wyss, Chr. (eds) *Proceedings of the 6th European Poultry Conference*. World's Poultry Science Association, Zollikofen, Switzerland, pp. 51–59.

Gunnarsson, S., Keeling, L.J. and Svedberg, J. (1999) Effect of rearing factors on the prevalence of floor eggs, cloacal cannibalism and feather pecking in commercial flocks of loose housed laying hens. *British Poultry Science* 40, 12–18.

Harrison, R. (1964) *Animal Machines*. Vincent Stuart Ltd, London.

Huber-Eicher, B. and Sebö, F. (2001) Reducing feather pecking when raising laying hen chicks in aviary systems. *Applied Animal Behaviour Science* 73, 59–68.

Hofrogge, W. (2000) Der europäische Eierproduzent im Spannungsfeld zwischen Tierschutz und weltweitem Wettbewerb. *Lohmann Information* 3/2000, 7–12.

Hughes, B.O. (1973) Animal welfare and the intensive housing of domestic fowls. *The Veterinary Record* 93, 658–662.

Johnsen, P.F., Vestergaard, K.S. and Nørgaard-Nielsen, G. (1998) Influence of early rearing conditions on the development of feather pecking and cannibalism in domestic fowl. *Applied Animal Behaviour Science* 60, 25–41.

Käppeli, S., Gebhardt-Henrich, S.G., Fröhlich, E., Pfulg, A. and Stoffel, M.H. (2011) Prevalence of keel bone deformities in Swiss laying hens. *British Poultry Science* 52 (5), 531–536.

KAT (2009) Was steht auf dem EI? http://www.was-steht-auf-dem-ei.de/nc/home/was-steht-auf-dem-ei/ (accessed 30 October 2011).

Knowles, T.G. and Broom, D.M. (1990) Limb bone strength and movements in laying hens from different housing systems. *The Veterinary Record* 126, 354–356.

Knowles, T.G., Kestin, S.C., Haslam, S.M., Steven, N., Brown, S.N., Laura E., Green, L.E., Butterworth, A., Stuart, J., Pope, S.J., Pfeiffer, D. and Nicol, C.J. (2008) Leg Disorders in Broiler Chickens: Prevalence, Risk Factors and Prevention. http://www.plosone.org/article/info:doi/10.1371/journal.pone.0001545 (accessed 28 January 2011).

Lindberg, A.C. and Nicol, C.J. (1997) Dustbathing in modified battery cages: is sham dustbathing an adequate substitute? *Applied Animal Behaviour Science* 55, 113–128.

LAYWEL (2006a) Welfare implications of changes on production systems for laying hens. Deliverable 7.1. Overall strengths and weaknesses of each defined housing system for laying hens, and detailing the overall welfare impact of each housing system. http://www.laywel.eu/web/pdf/deliverable%2071%20welfare%20assessment.pdf (accessed 28 January 2011).

LAYWEL (2006b) Description of housing systems for laying hens. Deliverable 2.3. Description of housing systems for Laying hens. http://www.laywel.eu/web/pdf/deliverable%2023-2.pdf (accessed 28 January 2011).

Maslow, A.H. (1943) A theory of human motivation. *Psychological Review* 50, 370–396.

Nicol, C.J. (1990) Behaviour requirements within a cage environment. *World's Poultry Science Journal* 46, 31–33.

Nicol, C.J., Brown, S.N., Haslam, S.M., Hothersall, B., Melotti, L., Richards, G.J. and Sherwin, C.M. (2009) The welfare of layer hens in four different housing systems in the UK. In: *Proceedings of the 8th European Symposium on Poultry Welfare*, Cervia, Italy, 18–22 May 2009. World's Poultry Science Association, Zollikofen, Switzerland, abstract 12.

Niebuhr, K., Zaludik, K., Gruber, B., Thenmaier, I., Lugmair, A., Baumung, R. and Troxler, J. (2006) Epidemiologische Untersuchungen zum Auftreten von Kannibalismus und Federpicken in alternativen Legehennenhaltungen in Österreich. Endbericht Forschungsprojekt 1313, BMLFUW Wien. https://www.dafne.at/prod/dafne_plus_common/attachment_download/4d4e8f0e88d74438fb1f281ab597de03/Endbericht%20 Forschungsprojekt%20Nr%201313%20ITT%202006N.pdf (accessed 28 January 2011).

Niebuhr, K., Arhant, C., Smaijlhodzic, F., Wimmer, A. and Zaludik, K. (2009) Evaluierung neuer Haltungssysteme am Beispiel von Volieren für Legehennen. EndberichtForschungsprojekt 100184, BMLFUW Wien und BMG Wien. https://www.dafne.at/prod/dafne_plus_common/attachment_download/80373dadc6f5cf343cd1 ec24d2610c3a/Endbericht%20Volierenprojekt_Proj_100184_ITT_2009.pdf (accessed 28 January 2011).

Oester, H. (1985) Die Beurteilung der Tiergerechtheit des Get-Away-Haltungssystems der Schweizerische Geflügelzuchtschule, Zollikofen für Legehennen. PhD thesis, University of Bern, Switzerland.

Oester, H. and Troxler, J. (1999) Die praktische Prüfung auf Tiergerechtheit im Rahmen des Genehmigungsverfahrens in der Schweiz. In: *Beurteilung der Tiergerechtheit von Haltungssystemen*. KTBL-Schrift 377. Kuratorium für Technik und Bauwesen in der Landwirtschaft eV (KTBL), Darmstadt, Germany, pp. 71–80.

Oester, H. and Wiedmer, H. (2005) Evaluation of elevated surfaces and perches for broilers. In: *Proceedings of the 7th European Symposium on Poultry Welfare*, Lublin, Poland, 15–19 June 2005. World's Poultry Science Association, Zollikofen, Switzerland, pp. 231–240.

Pickett, H. (2007) *Alternatives to the barren battery cage for the housing of laying hens in the European Union*. Compassion in World Farming, Godalming, UK.

Prip, M. (1976) Hysteria in laying hens. In: *Proceedings of the 5th European Poultry Conference*, Malta, 5–11 September 1976. World's Poultry Science Association (Malta Branch) vol. 2, pp. 1062–1075.

Rönchen, S. (2007) Evaluation of foot pad health, plumage condition, fat status and behavioural traits in laying hens kept in different housing systems. PhD thesis, Tierärztliche Hochschule Hannover, Germany.

Sandilands, V., Moinard, C. and Sparks, N.H. (2009) Providing laying hens with perches: fulfilling behavioural needs but causing injury? *British Poultry Science* 50, 395–406.

Scholtyssek, N. (1987) *Geflügel*. Eugen Ulmer Verlag GmbH, Stuttgart, Germany.

Schrader, L. (2008) Verhalten und Haltung. In: Legehuhnzucht und Eiererzeugung-Empfehlungen für die Praxis. In: Brade, W., Flachowsky, G. and Schrader, L. (eds) Special Issue. *VTI Agriculture and Forestry Research* 322, 93–117.

Schrader, L. (2010) Entwicklung der 'Celler Kleinvoliere' für Legehennen. In: *Aktuelle Arbeiten zur artgemäßen Tierhaltung 2010*. KTBL-Schrift 482. Kuratorium für Technik und Bauwesen in der Landwirtschaft eV (KTBL), Darmstadt, Germany, pp. 196–206.

Schweizerische Bundesverfassung (1999) Article 120. http://www.admin.ch/ch/d/sr/101/a120.html (accessed 28 January 2011).

Sewerin, K. (2002) Beurteilung des angereicherten Käfigtyps 'Aviplus' unter besonderer berücksichtigung ethologischer und gesundheitlicher Aspekte bei Lohmann Silver Legehennen. Dissertation, Tierärztliche Hochschule Hannover, Germany.

Tauson, R. (1984) Effects of a perch in conventional cages for laying hens. *Acta Agriculturae Scandinavica Section A: Animal Science* 34, 193–209.

Tauson, R. (1998) Health and production in improved cage designs. *Poultry Science* 77, 1820–1827.

Tauson, R. (2002) Furnished cages and aviaries: production and health. *World's Poultry Science Journal* 58, 49–63.

Van de Poel, I. (1998) Why are chickens in battery cages? In: Disco, C. and van der Meulen, (eds) *Getting New Technologies Together*. De Gruyter GmbH and Co., Berlin, pp. 143–178.

Wechsler, B., Fröhlich, E., Oester, H., Oswald, Th., Troxler, J., Weber, R. and Schmid, H., (1997) The contribution of applied ethology in judging animal welfare in farm animal housing Systems. *Applied Animal Behaviour Science* 53, 33–43.

Wegner, R.M. (1981) Choice of production systems for egg layers. In: Sorensen, L.Y. (ed.) *Proceedings of the 1st European Symposium on Poultry Welfare*. World's Poultry Science Association, Copenhagen, pp. 141–148.

CHAPTER 2
The Impact of Legislation and Assurance Schemes on Alternative Systems for Poultry Welfare

D.G. Pritchard

ABSTRACT

Whereas a wide range of policy instruments – legal, economic, education and publicity – are available to effect changes in production systems, legislation and market-led assurance schemes have been the main forces used to protect and enhance the welfare of farm animals. The Conventions of the Council of Europe (COE) and their Recommendations focused on the provisions of resources and duty of care to meet the needs of animals. They are part of the acquis of the European Union (EU) and are incorporated into some national legislations as well as being used as the basis for private standards. The European single market harmonized welfare and health rules, methods of production and labelling for hen's eggs and chicken meat, so allowing consumers to buy foods with known provenance. The EU responded with a legal framework from 'farm to fork' to secure food safety, animal health and welfare. Farmers, food processors and large retailers sought to provide evidence of the quality of their products by private assurance schemes not only to meet new obligations for due diligence but also to gain commercial advantage. These schemes often merely reflected minimum legal standards but some also focused on improved health, welfare or environmental provisions. Membership of assurance schemes is associated with better compliance with legal standards in Great Britain. The EU WELFAREQUALITY project concluded that assurance schemes usually had impact only at a niche market level. The EU ban on conventional cages for hens in 2012 has had the greatest political and economic impact on poultry in Europe. Some countries banned enriched cages ahead of the deadline, despite scientific evidence that supported continued use of enriched cages. New EU legislation to improve the quality of care of intensively kept meat chickens was novel in limiting stocking densities by measuring welfare outcomes at abattoirs. The COE Recommendations on geese, turkeys, ducks and Muscovies have been implemented by national rules and private codes of practice which are often used to support assurance schemes. Major impacts have been on traditional systems through the ban on live plucking of feathers from geese and on gavage and individual cages for the production of foie gras.

INTRODUCTION

Governments can use a wide range of interventions to effect change in the behaviour of animal keepers, consumers and citizens to meet the needs of society and promote political, economic and social goals. The British Farm Animal Welfare Council (FAWC, 2008) categorized the main forms of policy instruments and Table 2.1 gives a general example of each and their application to animal health and welfare.

The potential strengths and weaknesses of each policy instrument are also described in Table 2.1 but in practice governments tend to use them in combination, and often as a 'cascade', for example: primary legislation; secondary legislation; codes of practice and/or guidance; enforcement mechanisms; and publicity campaigns. Recently, several European governments have sought to use a wider range of instruments to effect their policies by incorporating developments in behavioural economics. For example, the Austrian animal welfare law has introduced compulsory public education for animal protection and welfare (http://www.tierschutzmachtschule.at/home.html). The recent Swiss legislation has introduced a requirement that all livestock and pets must only be sold to trained persons. Figure 2.1 highlights policy instruments that appear to have most promise in both improving the care of animals by education, appropriate training and incentives for keepers, and creating informed consumers who may seek to purchase high-welfare products.

Some countries have sought to incorporate animal health and welfare into broader goals of improving food quality, food security and the meeting of environmental and climate change goals. For example, the UK's Health and Welfare Strategy (Defra, 2004) promoted the development of partnerships, the understanding of costs and benefits, and sharing the roles and responsibilities of improving animal health and welfare. The Defra Sustainable Development Strategy 'Securing the future' (Defra, 2002a) included animal health and welfare as an integral part of the sustainability of food production and land use.

The European Union (EU) action plan for animal welfare 2006–2010 (DG SANCO, 2005) stated its main goals as:

- upgrading existing minimum standards for animal protection and welfare (including animal welfare in cross compliance within the reformed Common Agricultural Policy (CAP));
- giving a high priority to promoting policy-oriented future research on animal protection and welfare and application of the 3Rs principle (replacement, reduction and refinement);
- introducing standardized animal welfare indicators;
- ensuring that animal keepers and handlers as well as the general public are more involved and informed on current standards of animal protection and welfare and fully appreciate their role in promoting animal protection and welfare; and

Table 2.1. Types of government intervention and their relative strengths and weaknesses (FAWC, 2008).

Type of policy instrument	General example	Applied to animal welfare and health	Strengths	Weaknesses
1. Legal rights and liabilities	Rules of tort law[a]	Animal Welfare Act 2006 (England and Wales); Animal Health and Welfare (Scotland) Act 2006	Self-help	May not prevent events resulting from accidents or irrational behaviour
2. Command and control	Secondary legislation; Health and safety at work	Minimum space rules for poultry	Force of law; Forceful; Minimum standards set; Immediate; Transparent	Intervention in management; Incentive to meet, not exceed standard; Costly; Inflexible
3. Direct action (by government)	Armed forces	Welfare inspections by state veterinarians and local authorities; Border controls	Can separate infrastructure from operation	Danger of being perceived as 'heavy handed'
4. Public compensation/social insurance	Unemployment benefit	Compensation for animals slaughtered for welfare reasons during 2001 FMD outbreak; Cross compliance; Pillar II monies for farm animal welfare improvements	Insurance provides economic incentives	May provide adverse incentives; Can be costly to tax payers
5. Incentives and taxes	Car fuel tax	Cross compliance; Pillar II monies for farm animal welfare improvements	Low regulator discretion; Low-cost application; Economic pressure to behave acceptably	Rules required; Predicting outcomes from incentives difficult; Can be inflexible
6. Institutional arrangements	Departmental agencies, levy boards, local government	Animal Health, Meat Hygiene Service, Veterinary Laboratories Agency, Local Authorities	Specialist function; Accountability	Potential for narrow focus of responsibility
7. Disclosure of information	Mandatory disclosure in food/drink sector	Reporting of notifiable diseases; Labelling	Low intervention	Information users may make mistakes
8. Education and training	National curriculum	Animal welfare in veterinary education, national school curriculum	Ensures education and skills required by society	Can be too prescriptive and inflexible
9. Research	Research councils	Funding for animal welfare research through BBSRC, Defra, charities, etc.	Provide information to policy	May duplicate or displace private-sector activities

Continued

Table 2.1. Continued.

Type of policy instrument	General example	Applied to animal welfare and health	Strengths	Weaknesses
10. Promoting private markets	Office of Fair Trading	Market power of companies in the food supply chain and prices to farmers to meet production costs	Economies of scale through use of general rules Low level of intervention	No expert agency to solve technical/ commercial problems in the industry Impact of global commodity costs Uncertainties and transaction costs
a) Competition laws	Airline industry Telecommunications			
b) Franchising and licensing	Rail, television, radio.	Veterinary drugs/treatments. Animal husbandry equipment.	Low cost (to public) of enforcement	May create monopoly power.
c) Contracting	Local authority refuse services.	Hire of private vets to provide public services.	Combines control with service provision.	Confusion of regulatory and service roles.
d) Tradable permits	Environmental emissions. Milk quotas.	Permits for intensive livestock production systems (e.g. the Netherlands).	Permits allocated to greatest wealth creators.	Require administration and monitoring.
11. Self regulation			High commitment.	
(a) private	(a) Insurance industry.	(a) Farm assurance schemes, veterinary profession, industry codes of practice.	Low cost to government.	(a) Self-servicing. Monitoring and enforcement may be weak.
(b) enforced.	(b) Income tax.	(b) Defra 'welfare codes'.	Flexible.	

FMD, foot-and-mouth disease; BBSRC, Biotechnology and Biological Sciences Research Council; Defra, Department for Environment, Food and Rural Affairs.
[a]Tort is, for example, a duty of care to an animal; tort law is not part of the criminal law but is part of recognized civil duty.

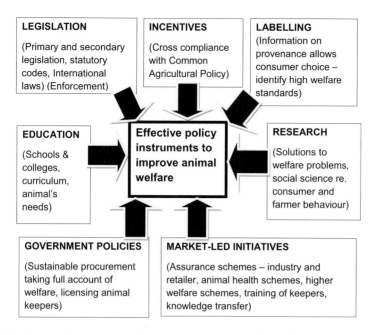

Fig. 2.1. Effective policy instruments to improve animal welfare (FAWC, 2008).

- continuing to support and initiate further international initiatives to raise awareness and create a consensus on animal welfare.

In Europe there is a large body of common health and welfare legislation based on the Council of Europe (COE) and the EU provisions. This is shared by some of its trading partners such as New Zealand and Canada with which the EU has equivalence agreements that include animal welfare. Otherwise, globally, animal welfare legislation tends to be limited to anti-cruelty provisions and to lack equivalent welfare provisions for poultry during the food chain or when used as experimental animals.

Internationally, public standards are promulgated by the Codex Alimentarius Commission (CAC) for food, the International Plant Protection Convention (IPPC) for plant health, and the World Organisation for Animal Health (OIE) for animal health. These are known as 'three sisters' and are officially recognized under the World Trade Organization (WTO) Agreement on the Application of Sanitary and Phytosanitary Standards (SPS). The SPS agreement of the WTO lays down legally binding standards for both public health and animal health for trade in both poultry and poultry products. The Terrestrial Zoo Sanitary Codes of the World Animal Health Organisation provide the standards of animal health for animal diseases to be used in trade disputes. Animal welfare is not included in the SPS and therefore there are no globally binding standards for animal welfare for trade in animal products. The exception is the requirement of the EU for equivalent standards for welfare at slaughter for poultry meat imported from third countries as welfare standards at slaughter affect the quality of the meat and are necessary to protect public health. This measure has had a major impact on improving welfare in poultry processing plants exporting to the EU.

Farmers, food processors and large retailers have sought to provide evidence of the quality of their products by private assurance schemes not only to meet new obligations for due diligence but also to gain commercial advantage by differentiation of their products. These schemes often merely reflect the minimum legal standards but some do focus on improved health, welfare or environmental provisions. Their impact has been hard to demonstrate due to the difficulty of measuring welfare outcomes but in Great Britain membership of assurance schemes is associated with better compliance with legal standards. The lack of implementation of public welfare standards internationally and the perceived need to have private standards which go beyond public standards have led to international private standards which are contentious within the SPS (Wolff and Scannell, 2008). At worst these can be used to place unjustified barriers to commercial trade and improved welfare but at best they can quickly reflect consumer welfare concerns and act as a driver to improve welfare standards of poultry over and above those required by public laws and standards. The following private bodies set standards for poultry: the Global Food Safety Initiative (GFSI), GlobalGAP, the International Poultry Council (IPC), the International Egg Commission (IEC), the International Meat Secretariat (IMS) and the International Federation of Agriculture Producers (IFAP) (Robach, 2010). As governments themselves procure significant quantities of poultry products, they can take a lead in promoting higher-welfare products. For example, the UK government has sought only to procure eggs, poultry meat and pig meat that meet standards of UK assurance schemes which include provision for animal welfare (Defra, 2011). The Food and Agriculture Organization of the United Nations (FAO) has recently promoted animal welfare as having an important role in capacity building by providing benefits for both humans and their animals and that poultry keeping has a key role in developing countries (Fraser, 2008).

Animal welfare legislation sets the legal criteria in which poultry can be kept in conventional and alternative production systems. The legislative framework and assurance schemes for the protection of animal welfare in Europe with particular reference to laying hens and meat chickens will be summarized, followed by their implementation for the minor poultry species and the production of organic poultry. Specific legislative documents are listed in Box 2.1 and Tables 2.2 and 2.3: further details can easily be found on the appropriate web site if required.

LEGISLATION

Early attempts to protect animals from cruelty and abuse (e.g. cock fighting) were driven by moral concerns to prevent animal suffering and used purely legal instruments (see Radford, 2001). Major improvements in the welfare of poultry have resulted from legalisation to control epidemic diseases, licence medicines and set standards for feeds and hygiene, as well as anti-cruelty legislation. More recently, legislation has been adopted to promote welfare by the obligations of a duty of care by defining resources and management. The economic impacts of such legislation occur mainly from the costs of specific provisions, such as a minimum area of floor space and the provision of particular resources e.g. backup generators for mechanically ventilated buildings. Such costs have rarely been reimbursed by the market. However the welfare legislation has also had more general impacts including changes in attitudes to animals. The COE Conventions of the 1970s were strongly based on the ethical duty to care for animals (Pritchard, 2006) and provided a major driver for improvement of poultry welfare. Advances in neuroscience and other scientific evidence of the similarities between animals and humans led to an acceptance of the sentience of animals including poultry. This was formally recognized by the EU Treaty of Amsterdam in 1997. Increased consumer and political pressure to improve animal welfare led to the restatement of these principles more forcefully in the 2009 Treaty for the European Union (TFEU) Lisbon and Article 13 of the treaty extends the concept of sentience and puts animal welfare on an equal footing with other issues in developing EU policies. New Zealand and many European countries have also responded to the demand for more effective welfare legislation by introducing wide-ranging welfare laws. For example, the British Animal Welfare Act (Defra, 2008) now requires animal owners and carers to have a duty of care to go beyond preventing animal cruelty and to promote positive welfare in their animals.

Council of Europe

The COE was established in 1947 and is an intergovernmental organization with currently 47 State Members and five Observers. Its aims are to protect and promote human rights, the rule of law and pluralist democracy. It first considered animal welfare in 1961, noting that 'humane treatment of animals is one of the hallmarks of western civilisation'. The Conventions on animal protection arose from political pressures from both governments and non-governmental organizations in response to concerns related to intensive farming, mainly poultry and pigs, and the use of experimental animals. The COE recognized that animal welfare is important for the contributions animals make to human health and quality of life, and that respect for animals counts among the ideals and principles that are the common heritage of State Members as one of the obligations upon which human dignity is based (Broom, 2006). Ethical principles used for all conventions on animal use and protection are based on the premise that:

> for his own well-being man may, and sometimes must, make use of animals but that he has a moral obligation to ensure, within reasonable limits, that the animal's health and welfare is in each case not unnecessarily put at risk. (COE, 2006.)

Each convention has addressed particular areas of ethical concern which in relation to poultry include slaughter (ETS 102, 1979), international transport (ETS 65, 1968), keeping on the farm (ETS 87, 1976) and their use for research (ETS 123, 1986). The COE has the most comprehensive legislation on poultry welfare internationally.

Box 2.1. European Animal Welfare Legislation

General legalisation

European Convention for the protection of animals kept for farming purposes. *Official Journal* L 323, 17/11/1978, pp. 0014–0022.

78/923/EEC: Council Decision of 19 June 1978 concerning the conclusion of the European Convention for the protection of animals kept for farming purposes. *Official Journal* L 323, 17/11/1978, pp. 0012–0013.

Council Directive 98/58/EC of 20 July 1998 concerning the protection of animals kept for farming purposes. *Official Journal* L 221, 08/08/1998, pp. 0023–0027.

2000/50/EC: Commission Decision of 17 December 1999 concerning minimum requirements for the inspection of holdings on which animals are kept for farming purposes (notified under document number C(1999) 4534) (Text with EEA relevance). *Official Journal* L 019, 25/01/2000, pp. 0051–0053.

Laying hens

Council Directive 88/166/EEC of 7 March 1988 laying down minimum standards for the protection of laying hens kept in battery cages. *Official Journal* L 074, 19/03/1988, pp. 0083–0087.

Council Directive 1999/74/EC of 19 July 1999 laying down minimum standards for the protection of laying hens. *Official Journal* L 203, 03/08/1999, pp. 0053–0057.

Commission Directive 2002/4/EC of 30 January 2002 on the registration of establishments keeping laying hens, covered by Council Directive 1999/74/EC. *Official Journal* L 30, 31/01/2002, pp. 0044–0046.

Protection at the time of slaughter and killing

European Convention for the protection of animals for slaughter. *Official Journal* L 137, 02/06/1988, pp. 0027–0038.

88/306/EEC: Council Decision of 16 May 1988 on the conclusion of the European Convention for the Protection of Animals for Slaughter. *Official Journal* L 137, 02/06/1988, pp. 0025–0026.

Council Directive 93/119/EC of 22 December 1993 on the protection of animals at the time of slaughter or killing. *Official Journal* L 340, 31/12/1993, pp. 0021–0034.

Protection during transport

Council Regulations (EC) No 1/2005 of 22 December 2004, on the protection of animals during transport and related operations and amending Directives 64/432/EEC and 93/119/EC and Regulation (EC) No 1255/97. *Official Journal* L 3, 05/01/2005, pp. 0001–0044.

Council Decision of 21 June 2004 on the signing of the European Convention for the protection of animals during international transport. *Official Journal* L 241, 13/07/2004, p. 0021.

European Convention for the Protection of Animals during International Transport (revised). *Official Journal of the European Communities* L241, 13/07/2004, pp. 0022–0043.

Council Directive 91/628/EEC of 19 November 1991 on the protection of animals during transport and amending Directives 90/425/EEC and 91/496/EEC. *Official Journal* L 340, 11/12/1991, pp. 0017–0027.

Council Directive 95/29/EC of 29 June 1995 amending Directive 91/628/EEC concerning the protection of animals during transport. *Official Journal* L 148, 30/06/1995, pp. 0052–0063.

Commission Decision 2001/298/EEC of 30 March 2001 amending the Annexes to Council Directives 64/432/EEC, 90/426/EEC, 91/68/EEC and 92/65/EEC and to Commission Decision 94/273/EC as regards the protection of animals during transport (Text with EEA relevance). *Official Journal* L 102, 12/04/2001, pp. 0063–0068.

Council Resolution of 19 June 2001 on the protection of animals during transport. *Official Journal* C 273, 28/09/2001, p. 0001.

Protection of animals used for experimental and scientific purposes

Council Directive 86/609/EEC of 24 November 1986 on the approximation of laws, regulations and administrative provisions of the Member States regarding the protection of animals used for experimental and other scientific purposes. *Official Journal* L 358, 18/12/1986, pp. 0001–0028.

Inspections

2006/778/EC: Commission Decision of 14 November 2006 concerning minimum requirements for the collection of information during the inspections of production sites on which certain animals are kept for farming purposes (notified under document number C(2006) 5384) (Text with EEA relevance). *Official Journal* L 314, 15/11/2006, pp. 0039–0047.

COUNCIL REGULATION (EC) No 1099/2009 of 24 September 2009 on the protection of animals at the time of killing (Text with EEA relevance). *Official Journal of the European Union* L303, 18/11/2009, pp. 0001–0030.

Anonymous (2001) Convention on the Protection of Farm Animals – Recommendation concerning turkeys (*Meleagris gallopavo* spp.) (TAP. 95/16). Council of Europe, Strasbourg.

Council Directive 1999/74/EC laying down minimum standards for the protection of laying hens. *Official Journal* L 203, 03/08/1999, pp. 0053–0057.

Anonymous (1996) Recommendation concerning domestic fowl (*Gallus gallus*). European convention on the protection of animals kept for farming purposes. European Treaty Series, No 87. Council of Europe, Strasbourg.

Anonymous (1999) T-AP (95) 5 adopted version. Standing Committee of the European Convention for the Protection of Animals kept for Farming Purposes. Recommendation concerning domestic geese (*Anser anser f. domesticus, Anser cygnoides f. domesticus*) and their crossbreeds. Council of Europe, Strasbourg.

In developing both Conventions and Recommendations, which can have considerable economic and social impact, there is need to resolve conflict between science, technical advice and practice. These may be real or sometimes perceived between the use of animals by man for economic, social, cultural and religious reasons and practices which are not ideal for their protection. The legal texts are firmly based on science and practical experience to determine the essential animal needs. These are carefully defined by the consideration of the biological characteristics of the species, its origin and domestication,

Table 2.2. Council of Europe (COE) Conventions on protection of poultry during transport and at slaughter and relevant European Union (EU) legislation.

COE legislation[a]	EU legislation[b]	Comment
European Convention for the protection of animals during international transport (ETS No. 65) (1968)	Council Directive 77/489/EEC Council Directive 91/628/EEC of 19 November 1991 on the protection of animals during transport and amending Directives 90/425/EEC and 91/496/EEC	ETS 65 required for journeys to/from EU to third countries but superseded by ETS 103 Provided legal basis of welfare in transport for the 'single market'
Council of Europe Convention for the protection of animals during international transport (revised) (ETS No. 193) (2003)	Council Decision 2004/544/EC on the signing of the European Convention for the protection of animals during international transport Council Directive 91/628/EEC Council Regulation 1/2005/EC on the protection of animals during transport and related operations	EU party to treaty Treaty still required for journeys to/from EU to third countries Clarified responsibilities and improved training and enforcement and vehicle standards
Recommendation No. R (90) 6 of the Committee of Ministers to Member States on the transport of poultry (1990)[c]	Council Directive 91/628/EEC Council Regulation 1/2005/EC	R90 provided Code of Conduct for the international transport of poultry with more detailed provisions than EU rules
European Convention for the Protection of Animals for Slaughter (ETS No. 102) (1979) Recommendation No. R (91) 7 on the slaughter of animals (1991)b	Council Directive 74/577/EEC Council Decision 88/306/EEC on the European Convention for the Protection of Animals for Slaughter Council Directive 93/119/EC on the protection of animals at the time of slaughter or killing Council Regulation 1099/2009/ EC on the protection of animals at the time of killing	EU legislation preceded COE. EU party to treaty New EU regulation 1099/2009 in force 2012

[a]Available at: http://www.conventions.coe.int/Treaty/Commun/ListeTraites.asp?CM=8&CL=ENG
[b]Can be found by searching at: http://eur-lex.europa.eu/RECH_naturel.do?ihmlang=en
[c]Available at: http://www.coe.int/document-library/

behaviours in nature including social behaviour and communication, and any special cognitive and physiological aspects including breeding and salient features of farming, transport and slaughter systems.

The EU is party to all these Conventions and has closely cooperated with the COE in the development and application of the Conventions and their secondary legal instruments, i.e. Recommendations, Technical Protocols, Resolutions and Codes of Conduct. Following adoption the Recommendation becomes binding unless the state party has made a declaration to the COE.

Otherwise, each and every party to the Convention shall implement the Convention by legal provisions and/or by administrative provisions (e.g. codes of practice for hens; Defra, 2002b) as is appropriate to meet the objectives of the Conventions. The EU has recently updated its rules on laboratory animals to closely reflect the advances made in revising the ETS 123.

Transport and slaughter

The COE Conventions relating to slaughter and transport and their relations to EU legalisation are summarized in Table 2.2. The EU has brought forward legislation on slaughter and killing, 1099/2009/EU, which go much further than ETS 102.

The revised Convention for International Transport (ETS 165, 2003) covers preparation for journey; loading to unloading; detailed standards for road, sea, air and rail; fitness to travel; and handling and veterinary controls. The COE 1999 Recommendation for the Transport of Poultry has been widely used as the basis of standard operating procedures by the industry and governments. Although the EU revised its transport legislation in 2005 through Regulation 2005/1 EC, it has not fully implemented the Conventions and Recommendations despite introducing additional measures in other areas such as better definition of responsibilities, vehicle standards and a training syllabus. Both the COE and EU Regulation 2005/1 EC have led to improved national legislation but the vertical integration of the industry has also improved the organization, training and equipment for transport and lairage. However FAWC (2009) noted that there were still areas in need of improvement, such as bone breakages and bruising due to injuries during catching and transport and deaths due to thermal stress. Of particular concern are the transport of end-of-lay hens and the assessment of their fitness to travel. In most of Europe, meat chickens, turkeys, ducks and geese usually travel short distances to processing plants but end-of-lay hens need specific processing equipment and tend to travel longer distances. Some countries such as Denmark have addressed this issue by killing such hens on-farm using portable gas killing equipment whereas others have introduced clear guidance for poultry farmers, catching teams and transporters on the fitness to travel of poultry. It is clear that birds which cannot stand or walk should be culled on the farm, but any birds that are severely lame or are showing signs of pain should not be loaded. Furthermore it is recognized that the performance of catchers is dependent on their motivation and management.

The COE legislation on slaughter is limited to animals which are used for human consumption although the principles remain relevant and have been incorporated into EU law. These principles, which must be observed if slaughter or killing of poultry is to be humane, were revised and clarified by FAWC (2008) as follows:

- all personnel involved with slaughter or killing must be trained, competent and caring;
- only those animals that are fit should be caught, loaded and transported to the slaughterhouse;

- any handling of animals prior to slaughter must be done with consideration for the animal's welfare;
- in the slaughterhouse, only equipment that is fit for the purpose must be used;
- prior to slaughter or killing an animal, it must either be rendered unconscious and insensible to pain instantaneously or unconsciousness must be induced without pain or distress; and
- animals must not recover consciousness until death ensues.

Since ETS 102 was adopted there have been considerable technical advances such as controlled atmosphere stunning and these have been included in EU legislation 1099/2009/EU along with rules for killing poultry for disease control. Similarly OIE in 2006 issued guidance on both slaughter and killing of poultry.

Welfare on-farm

The COE Convention ETS 87 (1976) for the protection of farmed animals has provided the basis of legislation for poultry both in the EU and wider Europe (Table 2.3). This Convention, known as the Treaty for Animal Protection (TAP), requires keepers of intensively kept farm animals to have a 'duty of care' to their animals. It is based on the principles that the environment and management have to meet animal needs rather than trying to adapt the animals by procedures such as mutilations like beak trimming. The Protocol of Amendment to the Convention for Protection for Farmed Animals (ETS 145, 1992) extended the scope to extensively kept animals not dependent on automation and introduced new requirements on biotechnology, breeding procedures and genetic selection. Currently the revised Convention lays down 12 welfare criteria for feed, water,

Table 2.3. Council of Europe (COE) Conventions on protection of poultry kept for farming purposes and the relevant European Union (EU) legislation.

COE Conventions[a] and Recommendations[b]	Comments	EU legislation[c]	Comments
European Convention for the Protection of Animals kept for Farming Purposes (TAP) ETS No. 087 (1976)	Treaty requires Standing Committee to monitor the Convention and to produce new and revised Recommendations	Council Directive 78/823/EC	EU party to treaty Convention part of EU aquis[d]
Protocol of Amendment to the European Convention for the Protection of Animals kept for Farming Purposes ETS No. 145 (1992)	Extended scope of TAP to include extensive systems and biotechnology	Council Directive 98/58/EC Regulation (EC) No 882/2004 Commission Decision 2006/778/EC	General welfare requirements for all farmed animals Official Controls on verification of compliance with feed and food law, animal health and animal welfare rules Requires reporting of welfare inspections

COE Conventions[a] and Recommendations[b]	Comments	EU legislation[c]	Comments
Recommendations on fowls 1995/1986	Binding on parties but parties may notify TAP and opt out of some provisions	Council Directive 1999/74/EC Council Directive 2007/43/EC	Scope of EU legislation on poultry is confined to large intensive establishments but provides more detailed prescriptive provisions EU rules have more detailed provisions for hens kept in cage, enriched cage, barn and free range systems EU rules on meat chickens are limited to intensive production but have more detailed provision on stocking density, resources and management
Recommendations on fowls 1995/1986	Requires parties to consider registration of holdings, training and licensing of persons caring for fowls	Council Directive 2002/4/EC Council Regulation (EC) 1234/2007 Commission Regulation 589/2008/EC Council Regulation 2092/91/EEC	Registration of establishments with hens Marketing and marking of hatching eggs and poultry meat Detailed rules for production methods and labelling of eggs and poultry meat Organic production standards, etc.
Recommendations on Ratites, 1997		Council Directive 98/58/EC	98/58/EC provides legal basis for Community measures to follow up Recommendations made under the IAP it necessary for their uniform application within the Community Member States are expected to give effect to the Recommendations
Recommendations on ducks, 1999		Council Directive 98/58/EC	ditto
Recommendation on domestic geese, 1999		Council Directive 98/58/EC	ditto
Recommendation on Muscovy ducks and hybrids of Muscovy and domestic ducks, 1999		Council Directive 98/58/EC	ditto
Recommendations on turkeys, 2001		Council Directive 98/58/EC	ditto

[a]Available at: http://www.conventions.coe.int/Treaty/Commun/ListeTraites.asp?CM=8&CL=ENG
[b]Available at: http://www.coe.int/document-library/
[c]Can be found by searching at: http://eur-lex.europa.eu/RECH_naturel.do?ihmlang=en
[d]The accumulated legislation, legal acts and court decisions that constitute law.

freedom of movement, staffing, inspection, disease treatment, records, housing, environment, equipment, mutilations and breeding procedures.

Article 3 of ETS 87 states that:

> Animals shall be housed and provided with food, water and care in a manner which, having regard to their species and their degree of development, adaptation and domestication, is appropriate to their physiological and ethological needs in accordance with established experience and scientific knowledge.

With respect to farm animals, the Standing Committee of the TAP adopted an extensive set of specific Recommendations for 17 different species or types of animal including poultry (see Table 2.3). Each Recommendation defines the biological characteristics of the species including its origin and domestication, behaviours in nature including social behaviour, communication, any special cognitive and physiological aspects including breeding, and summarizes salient features of farming systems. Each Recommendation details how the essential needs of animals can be meet under commercial farming systems and identifies areas where further research is required.

The EU has ratified and put the Convention on farmed animals into EU law as Directive 98/1998/EC but has developed specific rules only for calves, pigs, domestic fowl and meat chickens. Therefore the aquis of the EU is heavily dependent on the Recommendations of the TAP for welfare rules on turkeys, geese, ducks and Muscovies. There have been considerable advances in knowledge and technology in this area since the Convention and its poultry Recommendations were agreed and they now require revision.

Table 2.3 provides a summary of the legal texts of the COE and their relationship to those of the EU in relation to the protection of poultry kept for farming purposes.

European Union legislation

The European Community regulations for the development of a single market adopted in 1993 for poultry and poultry products provide protection of human health and food supply, animal health and the environment, and have increasingly been driven by the needs of the consumer. Several sections of the Commission of the European Union impact on the poultry industry: the Directorate General (DG) for Health and Consumer Affairs (DG SANCO) is responsible for animal health and the welfare of farmed animals, during transport and at killing and trade aspects; DG Environment for research animals and DG Science for research on animal welfare, DG RELEX (External Relations) for international relations and its Technical Assistance and Information Exchange (TAIEX) office for external training, and DG Enterprise for competitive businesses. The European Food Safety Authority (EFSA; http://www.efsa.europa.eu/) provides advice in the form of risk assessment of welfare to the Commission and Member States. For example the EU scientific committee report on laying hens (Anonymous, 1996) informed the development of legalisation on this sector (Council Directive 1999/74/EEC). Risk management of welfare issues is done by the Council of Ministers and the Standing Committee on the Food Chain and Animal Health where representatives of the Member States discuss current issues in relation to veterinary matters (animal health, animal welfare, public health) and approve urgent measures when necessary.

The Treaty of Amsterdam (Anonymous, 1997) included a 'Protocol on the protection and welfare of animals' which recognized vertebrate animals including poultry as sentient. This was reinforced by the TFEU of Lisbon made in 2009 which created a new political and administrative framework for developing animal welfare law. Under the new arrangements of the TFEU, legislation on animal protection is now developed by the Commission, European Parliament and Council of Ministers. The TFEU lists some key principles the Union should respect and Article 13 has been introduced which states:

> In formulating and implementing the Union's agriculture, fisheries, transport, internal market, research and technological development and space policies, the Union and the Member States shall, since animals are sentient beings, pay full regard to the welfare requirements of animals, while respecting the legislative or administrative provisions and customs of the Member States relating in particular to religious rites, cultural traditions and regional heritage.

This puts animal welfare on an equal footing with other key principles mentioned in the same title, i.e. promote gender equality, guarantee social protection, protect human health, combat discrimination, promote sustainable development, ensure consumer protection and protect personal data. However, the EU operates under the principles of conferred competences and subsidiarity. So competences not conferred upon the Union in the Treaties remain with the Member States and under the principle of subsidiary, in areas that do not fall within its exclusive competence the Union shall act only if and in so far as the objectives cannot be sufficiently achieved by the Member States. As a consequence, certain topics of animal protection remain under the responsibility of the Member States (e.g. the use of animals in competitions, shows, cultural or sporting events). EU legislation affecting poultry welfare sets minimum standards and Member States may introduce higher standards. For example, Sweden, Germany and Austria all introduced bans on conventional battery cages prior to the EU-wide ban in 2012.

The EU has extensive arrangements for ensuring that Member States implement EU rules. Regulation (EC) No 882/2004 lays down the official controls performed to ensure the verification of compliance with feed and food law, animal health and animal welfare rules. Commission Decision of 29th September 2006 sets out the guidelines laying down criteria for the conduct of audits under this Regulation to ensure the verification of compliance with feed and food law, animal health and animal welfare rules.

There remain several drivers for the EU developing action plans for both animal health and welfare, e.g. major animal disease outbreaks such as highly pathogenic avian influenza, food-borne contaminations like the recent dioxin contamination of food, and widespread consumer concern about welfare during transport and in intensive systems such as barren cages for hens. There is also evidence that good poultry welfare reduces food safety risks (Humphrey, 2006) and the EU Commission White Paper on food safety firmly linked animal welfare to food safety. The EU action plan 2006–2010 (Anonymous, 2006) detailed the welfare programme of the Commission and was accompanied by an impact assessment that included the following areas of actions.

- Action 1: upgrading existing minimum standards for animal protection and welfare.
- Action 2: giving a high priority to promoting policy-orientated future research on animal protection and welfare and application of the 3Rs principle.
- Action 3: introducing standardized animal welfare indicators.
- Action 4: ensuring that animal keepers/handlers as well as the general public are more involved and informed on current standards of animal protection and welfare and fully appreciate their role in promoting animal protection and welfare.
- Action 5: continue to support and initiate further international initiatives to raise awareness and create a consensus on animal welfare.

Standards for welfare in farmed poultry are laid down in Directive 98/58/EC which sets out minimum rules for their protection. More detailed rules are laid down for the farming of just a few species, including laying hens and intensively reared meat chickens. Only general requirements of 98/58/EC are in place for ducks, geese, turkeys and Muscovy ducks. The lack of specific EU standards to protect these farmed species is at present difficult to justify given the scientific evidence currently available and the Recommendations made by the COE to which the EU is a signatory.

Animal welfare is one of the strategic priorities related to the development of more sustainable food production policies. It is now accepted as an integral part of the Community's 'farm to fork' policies and has a central place within the reformed CAP. The principles of cross-compliance for the beneficiaries of direct payments from 2007 with various standards include animal welfare requirements. Thus if a breach of EU welfare standards are detected on a holding it will trigger a sanction. The CAP can also be used to promote animal protection by the provision to support farmers who apply animal husbandry practices which go beyond the baseline of good animal husbandry practices, for financial help with farmers' operating costs to adapt to demanding standards based on Community legislation in the fields of environment, public, animal and plant health and animal welfare, the use of farm advisory services, participation in food quality schemes (including schemes based on high animal welfare standards), and for producer groups which undertake information, promotion and advertising activities on the quality schemes supported, including those based on improved animal welfare provisions. Such inspections rely on the assessment of compliance with EU laws as well an assessment of their welfare. The EU WELFAREQUALITY (2007) project involved research teams in most EU countries and made progress in developing detailed animal welfare indicators for laying hens and meat chickens. These have not yet been taken up in the formulation of EU legislation and further advice is being sought from EFSA.

There has been considerable technical and political debate but little progress in developing a new EU label for animal welfare which would classify production systems in relation to the welfare requirements applied. Welfare labelling is still subject to considerable controversy related to welfare issues, standards and the wider issue of food labelling. However existing EU labelling rules for both eggs and poultry meat have had considerable impact. Statutory labelling played an important role in the increase in free range egg production which now approaches 50% of UK retail trade and certain outdoor systems of broiler production such as Label Rouge which has risen to more than 30% in France.

The EU action plan envisaged better education of both animal carers and more involvement of the public in developing and promoting animal welfare standards. The EU Euro-barometer surveys of public opinion have reported that the welfare of both laying hens and broilers needs particular attention and identified a need for clear food labelling (see Toma *et al.*, 2010). The WELFAREQUALITY project provided a greater understanding of the social science of citizens' attitude to animal welfare which often does not match their purchasing decisions. Recent EU legislation has required specific education and certification for transport drivers, slaughtermen and those caring for meat chickens. Some Member States have included an obligation for the provision of education in animal welfare in their legislation (e.g. Austria in 2005). This has resulted in initiatives such as Tierschutz macht Schule (2006) ('Animal Welfare Goes to School') to improve the education of children from preschool to university. DG SANCO (2010) introduced a children's website on animal welfare, 'Farmland', to raise awareness among children about the importance of treating farmed animals in a respectful and humane way.

The EU action plan provided support for further international initiatives to raise awareness and create a consensus on animal welfare. The EU committed to continue to support the work of the COE and has also supported the OIE which, in its animal welfare strategy, has recognized that 'animal welfare is a complex, multi-faceted public policy issue that includes important scientific, ethical, economic and political dimensions'. OIE has produced welfare guidelines applicable to its 197 members on transport by air, sea and road and at slaughter and killing of poultry.

The EU also supported the FAO initiatives on capacity building to implement its Good Animal Welfare Practices recognizing that poultry play an important role in poor communities, that animal welfare is an essential component of animal health, and both contribute significantly to the quality of human life. Within its food chain crisis management framework FAO has recently developed a One Health Programme entitled 'A Comprehensive Approach to Health – People, Animals and the Environment' which provides a strategic framework. Scott and Balogh (2010) summed it up as:

> While more science is necessary to understand the complex relationships among disease emergence, transmission and ecological systems, science alone is not the solution. It is also essential to address the social and cultural dimensions of societies where issues concerning livestock, wildlife, humans and entire ecosystems intersect. Changes in thinking and behaviour must be encouraged, and future decision-making must be cognisant of the repercussions of poor natural resource management and their implications for civilisation.

The EC is using the provision of Regulation (EC) 882/2004 to actively pursue the training of officials of Member States and third countries by a variety of poultry welfare courses under its 'Better Training for Safer Food' programme to supplement their national training. It also uses the TAIEX to assist accession countries and near neighbours which is aimed at improving the competency of authorities and officials. Regulation 882/2004/EU requires Member States to have programmes of surveillance and enforcement of welfare standards and reporting results to the EU Commission. It also provides powers to the Food and Veterinary Office of the EU Commission to audit by on-the-spot inspection both Member States and certain institutions e.g. abattoirs in third countries.

The EU institutions and Member States not only have to pay full regard to the welfare requirements of animals when formulating and implementing other Community policies in the research area, but also have an extensive programme of Community-funded research projects with important animal welfare components. Examples of recent projects relevant to poultry include:

- Consumer concerns about animal welfare and the impact on food choice (FAIR, 2001).
- Welfare implications of changes in production systems for laying hens (LAYWEL, 2006).
- Code of good practice for farm animal breeding and reproduction (EFABAR, 2009).
- Development of welfare indicators (WELFAREQUALITY, 2007).
- Economic analysis of animal welfare (ECONWELFARE, 2008).

Laying hens

Directive 1999/74/EC lays down detailed requirement for housing, feeding, watering and management for the protection of the welfare of laying hens and is supported by marketing and labelling rules. It has standards for conventional cages which had to be phased out by 2012, enriched cages (also known as furnished cages) with 750 cm^2 per bird with nest box, perch and scratching area, barn systems and free range systems. The Directive gives minimum standards but Sweden banned the use of conventional cages for laying hens in 1990 that were largely replaced with enriched cages (Pritchard, 2003). Surprisingly, Switzerland in 1992 and later Germany and Austria banned enriched cages ahead of the EU deadline despite scientific evidence which was reviewed by the EU LAYWEL (2006) project and FAWC (2007) that supported continued use of enriched cages that largely met the welfare needs of hens without the risks of disease, excess mortality and injury problems seen in free range and other alternative systems.

The increase in space allowances and other housing requirements, and in particular the ban on conventional cages, has had considerable economic impact on the laying hen industry as such cages were the most popular system across Europe. This can be best appreciated by examining the relative percentage costs (estimated by Elson, 1985) compared with the conventional cage at 450 cm^2 per hen. The space allowance required by Council Directive 99/74/EC to 560 cm^2 increased cost to 105%, similar to multi-tier housing at

20 birds m^{-2}. The furnished cage with perch, nest, scratching area and 750 cm^2 per hen costs 115%, the same as a single-tier aviary with 10–12 birds m^{-2}. Deep litter at 7–10 birds m^{-2} costs 118%. Free range costs were significantly greater at 135% for stocking at 1000 birds ha^{-1} and 150% for stocking at 400 birds ha^{-1}.

Experience has shown that, when stringent national legislation increases costs, production may move to other countries with less stringent legislation (which post 2012 would be outside the EU). For example in Denmark since the 1970s, the minimum space allowance for caged hens has been 600 cm^2 per hen and only three tiers of caging have been allowed. This was stricter than the EU legislation, which at that time demanded only 450 cm^2 per hen. As a consequence, eggs from caged hens became at least 50% more expensive in Denmark than in most other European countries. So where Denmark was once a major exporter of eggs, by the beginning of the 1990s Danish egg production was barely able to supply the home market. However, in the UK and Denmark, campaigns for free range eggs, together with the interest of supermarkets in promoting products from alternative farming methods, have had a positive effect and resulted in the market share of eggs from conventional cages in Denmark to fall from 92–95% in the 1980s to 60% in 2005 although the price of non-cage eggs was considerably higher. There is a significant risk that higher egg production costs in the EU from 2012 will further weaken the competitive position of the industry compared with producers in other countries and that this will impact on international trade in future, particularly for egg products. This is a particular concern as international trade agreements do not allow discrimination of imports of eggs on their method of production.

Legislation has been developed in the EU for differing hen systems but there have been few studies which compare the impact of all the relevant factors across all systems to assess their impact on bird welfare (however, see Chapter 12, this volume). There is a welfare conflict in housing hens: extensive systems give hens freedom of movement to select their desired environment and the freedom to perform many natural behaviours such as dust bathing, perching, foraging and (where provided) access to range, but increase the risk of injury, cannibalism, predation, disease and thermal cold stress. There is agreement that conventional cages do not provide for the bird's needs and these have been banned in the EU from 2012 as noted above. The EU LAYWEL (2006) project provided a comprehensive meta-analysis of the available data and this was updated by an expert group of mainly US scientists (Lay et al., 2010). They both noted the difficulties of assessing hen welfare in differing housing systems and agreed that, while no system was perfect, enriched cages could provide adequately for the welfare of hens. Enriched cages with groups of 20, 40 or 60 or more birds (so-called colony cages) provide more choice than smaller group sizes (FAWC, 1991). Both reviews found that free range systems had some significant advantages but also some serious problems. Lay et al. (2010) concluded that enriched cages had lower mortality than conventional cages and much less than non-cage systems. Furnished cages also reduce bone breakages which may cause birds pain and reach epidemic proportions in free range and barn systems. Complexity of the environment in loose-housed systems allows more control of the thermal and social environment and more choice. They are believed to be positive for bird welfare but are also associated with greater bone damage (Sandilands *et al.*, 2008). Such systems also have higher feed conversion rates resulting in higher carbon costs for the eggs and therefore may not be optimal for long-term sustainability. Some strains of bird adapt better to some systems and ensuring optimal welfare relies on the selection of appropriate strain, the choice of rearing and laying system, and most importantly high standards of stockmanship. The COE and EU legal framework has assisted in supporting moves to optimal systems and has been assisted by the use of additional codes of best practice based on research conducted in the field, education and continuing development of stockmen, and robust surveillance and enforcement (e.g. Defra, 2002b).

An example of the impact of legislation is the COE Recommendation to generally prohibit mutilations such as beak trimming and that other management procedures should be used to avoid the necessity of this mutilation. Injurious pecking may be a major problem under commercial conditions in the large sized groups common in free range and barn systems. It may be easier to control injurious pecking in enriched colony cage systems and considerable research has been conducted into feeding, lighting regimes, light quality including ultraviolet light, and the effect of environmental enrichment. Beak trimming that is carried out correctly does result in benefits to birds in terms of a reduction in mortality and injuries due to cannibalism and feather loss which outweigh the minor and short-lived adverse effects; however it can be criticized for not dealing with the root cause of feather pecking and cannibalism. Some countries such as Austria have banned beak trimming and developed husbandry systems using strains of hens that limits injurious pecking. Other countries which typically have larger flock sizes have still found it necessary to use beak trimming but placed the practice under the control of the Competent Authority. The development of an infra treatment method to limit the growth of the tip of the upper beaks of very young chicks is preferable to beak trimming by knife and cauterization. Further studies are needed to find other solutions to prevent or control injurious pecking. Breeding companies are continuing work on breeding programmes to prevent feather pecking and cannibalism.

Meat chickens

The EU has laid down complex production requirements for a variety of outdoor husbandry systems for meat chickens but these have had relatively little impact as most of the broilers produced in the EU are reared under intensive indoor systems. The Swedish government and industry introduced a voluntary scheme for such systems which allowed the stocking density of broilers to vary from a basic 25 kg m^{-2} up to 33 kg m^{-2} on the basis of assessments of the management and buildings, and scoring systems for foot health in the abattoir. Litter quality is affected by stocking density and it is well recognized that maintaining litter quality is a key factor in preventing foot, leg and breast skin lesions in broilers. In Sweden the monitoring of foot pad dermatitis in the abattoir has made an important contribution to improving the welfare of broilers and turkeys when incorporated into a welfare-monitoring programme in which the results are fed back to farmers. They are then allowed to increase their stocking densities if they maintain good welfare standards but have to reduce stocking densities and take other actions if welfare levels are not satisfactory (Berg and Algers, 2004). The scheme was adapted by Denmark where it also reduced mortality and improved foot health. This information is consistent with the findings of Dawkins *et al.* (2004) that chicken welfare in large field trials was influenced more by housing conditions than by stocking density. Directive 2007/43/EC for meat chickens came into force in 2010 to improve the quality of care of meat chickens and introduced new standards for husbandry, management and training. It was novel in using on-farm welfare indicators as well as feedback from post-mortem indicators measured at abattoirs to set, and if necessary limit, stocking densities. Three tiers of stocking limits are specified: (i) 32 kg m^{-2} without special conditions; (ii) 38 kg m^{-2} provided certain housing, management and monitoring standards are met; and (iii) up to 42 kg m^{-2} provided that mortality and welfare indicators have been maintained at challengingly low levels for the previous seven crops. The impact on broiler industries of 2007/43/EC is variable. Countries such as Sweden, Denmark, Germany and the UK already largely comply with these stocking rates and the Mediterranean states typically stock at levels below 32 kg m^{-2} because of their climate. For countries such as the Netherlands and Central and Eastern European member states with industries that commonly had very high stocking rates (e.g. 46 kg m^{-2}) the Directive would have a major impact. The Directive has also added additional surveillance and inspections costs. In England these were estimated at around £10 million: Moran and McVittie (2008) used contingent valuation to assess the potential value which the public placed upon improvement of the welfare of meat chickens as proposed by the Directive and found that it was of the order of £7.53 per household, which aggregated to £158 million. Even allowing for the wide confidence limits of the study this demonstrated that the perceived benefits of the Directive would far exceed the likely costs.

Due to the reliance of EU retailers on Quality Assurance schemes it is likely that a large proportion of imported chicken meat will meet standards equivalent to 2007/43/EC. The impact of the Directive on the competitiveness of the EU industry globally is unlikely to change as lower production costs in developing countries generally owe more to differences in labour, feed, climate and other costs than to different animal welfare standards. For example GHK (2010) reported that chicken meat production costs have been estimated to be more than 40% lower in Brazil and 36% lower in the USA than in the Netherlands. The estimated cost of producing and processing poultry meat in the Netherlands is approximately €1.40 kg^{-1} compared with far lower costs in Thailand (€1.10 kg^{-1}), the USA (€1.00 kg^{-1}) and Brazil (€0.90 kg^{-1}). Differences in stocking rates were not considered to be important as the stocking density of chickens exported to the EU from other countries is typically either similar to or lower than in the EU. When the slaughter Regulation 1099/2009 is implemented, EU slaughter costs for poultry may increase, but equivalent standards will be required of third country exporters.

ASSURANCE SCHEMES

Major concerns for food safety during the 1990s, which included salmonella from eggs, led to increased interest by many European consumers in food provenance. In response the EU created a comprehensive legal framework from 'farm to fork' to secure food safety, animal health and welfare. Farmers, food processors and large retailers responded to these new obligations for due diligence and tried to seek commercial advantage to differentiate their products by providing evidence of their quality through a plethora of assurance schemes (Blokhuis *et al.*, 2003). The standards of the majority of these schemes merely reflected the minimum legal standards of animal health and welfare (FAWC, 2005) but some also focused on improved health provisions (e.g. Lion schemes) or welfare standards (e.g. RSPCA Freedom Food), whereas others included environmental and sustainability issues. Large EU research projects (WELFAREQUALITY, 2007) have developed better methods of welfare assessment and improved the understanding of consumer behaviour and market-led assurance schemes. In the European market, such schemes have had a minor impact on animal welfare and were mainly focused on niche markets. It was not clear if consumer pull could be responsible for major changes in welfare standards, but they noted that there were increasing numbers of products with welfare attributes on the market. Private quality standards set by voluntary assurance schemes or actors in the food chain are often used by retailers for the purchase of animal products. Such schemes may react more quickly than national regulations, COE Recommendations and EU law to change welfare standards. The shift to more market-led approaches has also been accompanied by a greater focus on consumer aspects of the animal welfare debate (e.g. McInerney, 2004)

The evidence to date is mostly equivocal as to the potential for adding value to animal products through higher welfare standards. There has for some time been evidence that the public concern over farm animal welfare is not necessarily matched by a willingness of consumers to pay for welfare (e.g. IGD, 2007; Toma *et al.*, 2010). Pritchard *et al.* (2003) reported that farm assurance schemes did influence the outcome of statutory monitoring of animal welfare on UK farms but FAWC (2005) was not impressed by the evidence of higher welfare standards having been promoted through market-led incentives. One reason often given for the slow progress in promoting animal welfare to consumers is a lack of clear labelling information on animal welfare and this in turn has led to considerable research focus on the development of scientifically valid and practical tools for assessing welfare on farms (e.g. Matthews, 2008). A possible explanation was that consumers preferred to assume that animal welfare was taken care of by farmers, the authorities and retailers, rather than be confronted by choices that may be considered upsetting or off-putting. Duffy and Fearne (2009) noted that only one-third of consumers surveyed associated farm-assured products with improved welfare standards. Some consumers also have multiple alternative concerns, e.g. taste, freshness, purity, food miles, local food supply, carbon costs, naturalness, etc. (Castellini *et al.*, 2008).

In contrast, assurance schemes are widely used internationally, especially for wholesale trade. GlobalGAP, formerly known as EUREPGAP, is a private-sector organization with members in more than 100 countries around the world. The governance system is based on 50% producers and 50% retailers and food service, thus providing a match between primary producers and markets. GlobalGAP sets voluntary standards for Good Agricultural Practices (GAP) and for the certification of agricultural products in international trade, thus providing an entry point for producers around the globe (including those in developing countries) to the more affluent food markets. The GlobalGAP standard 'Control Points and Compliance Criteria Integrated Farm Assurance Poultry' contains specific animal housing, husbandry and welfare requirements. While it has a wide range of provisions these tend to reflect current industry practice and minimum legal standards.

It is likely that farmers who choose to join assurance or organic schemes are the more motivated farmers with awareness and knowledge of animal welfare. In addition, regular farm inspections by external auditors could enhance and encourage higher standards of care of livestock and therefore such farms may have higher animal welfare standards. Research into the impact of farm assurance schemes on the welfare of animals has produced mixed results. Main and Green (2000) found that Assured British Pigs were only justified in their claims of providing assurance on some aspects of animal welfare but not others. Main *et al.* (2003) found similar results with dairy cattle in relation to the Freedom Food scheme. Rauw *et al.* (1998) reviewed the evidence that selection for fast growth and/or maximal production has resulted in reduced well-being of farmed animals. Cooper and Wrathall (2010) reported that assurance schemes which specified slow growing strains and environmental enrichment for meat chickens had real welfare benefits for the birds which could be appreciated by informed consumers.

FAWC (2001) proposed that one way of assessing the effectiveness of farm assurance and organic certification schemes was to investigate whether certified farms were more likely to comply with welfare legislation and code when inspected by the State Veterinary Service than non-certified farms. Kilbride *et al.* (2011) concluded there was evidence that membership of a farm assurance scheme was associated with greater compliance with animal welfare legislation at inspections made by British official veterinarians and that the evidence is sufficiently robust that membership of a scheme could be included in the veterinary service's risk-based selection for farm visits. The odds ratio for compliance with the law was estimated at 0.4 (95% confidence limits of 0.2 to 0.8) for poultry enterprises certified as Farm Assured compared with non-certified enterprises. In conclusion, membership of the British poultry assurance schemes which covered hens, meat chickens, turkeys, geese and ducks was associated with better compliance with legal standards.

Assurance schemes have been widely used in some countries in the absence of national legal standards. In the USA the situation is complex as some States have introduced rules on poultry welfare and nationally there is a plethora of schemes for certification of eggs and meat (e.g. Pew Commission Report, 2008; Humane Farm Animal Care, 2009). Some of these schemes have followed the RSPCA Freedom Food standards, and most have developed in conjunction with panels of independent scientists, but there is a lack of evidence as to how such schemes influence the welfare of animals on assured farms.

As part of the ECONWELFARE project, Kilchsperger *et al.* (2010) provided a technical summary of animal welfare initiatives in Europe covering both regulatory (including organic), farmer-initiated assurance schemes and educational programmes. The project used panels representing actors in the food chain to assess perceived success factors in relation to meeting producer needs and consumer demand. It concluded that regulatory (including organic) measures were scored higher than non-regulatory initiatives. They identified several weaknesses with some existing schemes such as:

- goals that were sometimes too narrow (e.g. more focus on technically stable systems than on animal welfare);
- some instruments were not used sufficiently in combination with each other (e.g. labelling schemes with education in non-organic schemes); and
- some important or potentially interesting actors were neglected or not sufficiently involved (e.g. farmers in campaigns or in the design of research projects).

Kilchsperger *et al.* (2010) suggested a dynamic governance model to facilitate the transition to better animal welfare, stimulating and facilitating private initiatives, supporting public–private partnership and, where market mechanisms fail, setting regulatory, labelling or other framework conditions such as financial incentives for farmers and other actors in the food chain.

OTHER POULTRY SPECIES

Ducks and Muscovy ducks

Legislation on ducks and Muscovies has had major impacts relating to provision of water and in relation to individual pens and forced gavage used for foie gras (domestic ducks, Muscovies, Muscovy–domestic duck hybrids and geese may be used for foie gras production).

Ducks need water to drink and for preening and grooming but provision of water without spreading pathogens is a challenge. The COE Recommendations for ducks and those for Muscovies and their cross-breeds state the birds must be provided with water to dip their heads and spread water over their feathers. Some claim that ducks need to swim in baths but this leads to disease. Jones *et al.* (2008) found that commercial farmers may be able to improve duck welfare as much by providing water in troughs or from overhead showers (both clean and also economical of water) as from actual ponds (baths).

The legal requirements for feeding methods of ducks and Muscovies are laid down in Article 3 of the farmed animal COE Convention 1976:

> Animals shall be provided with food, water and care in a manner which, having regard to their species and their degree of development, adaptation and domestication, is appropriate to their physiological and ethological needs in accordance with established experience and scientific knowledge.

The Scientific Committee on Animal Health and Animal Welfare (SCAHAW, 1998) report of the European Communities condemned the practice of force feeding as it is detrimental to the welfare of the birds as currently practised. It results in excess mortality and morbidity rates and the gavage process causes varying amounts of pain and distress. The husbandry systems that use small individual cages do not allow the animals to engage in normal behaviours. The Recommendations on geese and ducks of the TAP (1999) addressed these issues and banned the use of small individual cages for housing these birds. The State of California and many European countries specifically ban gavage, e.g. Austria, Czech Republic, Denmark, Finland, Germany, Italy, Luxembourg, Norway and Poland, but it remains in use in the main producing countries of France and Hungary. Research on developing alternative methods of foie gras production has been successful but most production in Europe still uses gavage.

Turkeys

FAWC (1995) summarized the scientific data and practical experience of keeping turkeys and identified the main issues affecting turkey welfare. This was used by the COE to produce its Recommendation concerning turkeys (*Meleagris gallopavo* ssp.) (2001). This is the principal source of legal guidance on this species (Pritchard, 2002). It includes domestication and breeding, training and stockmanship, enclosures and stocking densities, lighting and mutilations. Implementation of higher standards of turkey welfare occurs by improved stockmanship, which is dependent upon better information and training, and effective management. The Recommendation recognizes that some methods of husbandry at present in commercial use fail to meet the biological needs of turkeys and hence result in poor welfare. Therefore it encouraged research to develop new husbandry systems and methods of breeding and management in line with the Convention so that the needs of the animals can be met.

Regulation 1538/91/EC and Regulation 2891/92 lay down minimum marketing standards for turkeys such as stocking rates of 25 kg m^{-2} for extensive barn-reared turkeys which must be killed at 70 days or older. Additional requirements are laid down for outdoor stocking rates for turkeys labelled 'free range', 'traditional free range' and 'free range total freedom'. European Regulations include standards for organic turkeys which lay down stocking rates but care needs to taken to ensure that limitations placed on diets and use of medicines and vaccines are appropriately interpreted to ensure that the welfare of the turkeys is protected.

Geese

The COE 1999 Recommendations provide the basis of European legal guidelines for the protection of geese. As with the use of gavage for foie gras production, the harvesting of feathers is controversial. The Recommendation states that 'Feathers, including down, shall not be plucked from live birds', but it does not refer to 'harvesting' of feathers. Harvesting is a term which has been used in countries where plucking feathers from live birds is carried out. EFSA (2010) summarized the science and practical experience and concluded that only ripe feathers near moult should be removed. Either a control system should be in place to ensure this is carried out or feathers should be removed by a person using a brushing or combing procedure to remove only 'ripe' feathers.

Grasping and pulling of feathers should be avoided. EFSA recommended some welfare outcome indicators which should be used to monitor the welfare of birds submitted to feather harvesting under commercial conditions. Since bleeding from feather bases and other kinds of skin damage, e.g. tears and blood or tissue on the feather quill, are related only to feather plucking, as opposed to feather gathering, their presence should be used as a criterion to distinguish between plucking and gathering feathers from live geese.

ORGANIC POULTRY

EU Regulation 2092/91 sets out the framework for organic livestock production and Commission Regulation 889/2008/EC lays down the detailed rules for organic production and labelling of organic products. They were formulated by the International Federation of Organic Agriculture Movements (IFOAM). Organic production may be labelled as 'biological' in several countries; KRAV in Sweden; Bioland, Naturland and Demeter in Germany; and SKAL in the Netherlands. Organic production is characterized by husbandry systems based on core values (see Vaarst *et al.*, 2004). Organic standards specify production methods that provide 'natural' conditions for the animal, freedom of movement and access to the outdoors, the restricted use of medical drugs and the production of a healthy product without residues of pesticides or medical drugs. Kijlstra and Eijck (2006) have pointed out these rules were implemented on a national basis before the impact on the health of the animals was evaluated and lack of clarity in the regulations has led to differing national interpretations. As the EU rules are minimum standards, a wide range of additional requirements have been added by some organic schemes without the impact on animal health and welfare being established. Disease prevention in organic farming is based on the promotion of systems which allow animals to exhibit natural behaviour, do not subject animals to stress, provide optimal (organic) feed, and assume that animals will have a higher ability to cope with infections than those reared in a conventional way. Fewer medical treatments would be needed and any diseased animal should be treated with alternative (homeopathy or phytotherapy) treatments instead of conventional drugs. In practice not many organic farmers use these treatment regimens because of lack of scientific evidence of effectiveness. Important health problems in organic livestock farming are often related to exposure to disease and parasites by the requirements for outdoor access (e.g. coccidiosis and external parasites, feather pecking and cannibalism; see Chapter 4, this volume). Outdoor access may also increase risks to the consumer such as through avian influenza and food-borne *Campylobacter* and *Toxoplasma* infections. However, Swarbrick (1986) concluded that disease on organic farms was not different from that seen on other free range systems. Higher animal welfare standards are reported to be one of the main attractions for consumers of organic food (Zander and Hamm, 2010). Kilbride *et al.* (2011) reported that farm assurance membership and organic certification were associated with greater compliance with animal welfare legislation at inspections made by British official veterinarians. They reported an odds ratio for poor welfare of 0.4 (95% confidence limits of 0.2 to 0.8) for assured poultry enterprises and an odds ratio of 0.8 (95% confidence limits of 0.1 to 1.8) for poultry enterprises certified as organic compared with non-certified farms.

CONCLUSIONS

The major source of legislation governing the welfare of poultry in conventional and alternative production systems has been recommendations from the COE. Implementation of these agreements into EU and national legalisation has become increasingly important. Latterly the OIE has added Animal Welfare to its considerations to impact the rest of the world. Legislation covers the whole of the production cycle from hatch to plate and is supplemented by voluntary assurance schemes that are at least as good as the legislation. There is some evidence that these schemes lead to better animal welfare. Legislation for organic poultry is, however, rather confused and may not result in optimum bird welfare.

REFERENCES

Anonymous (1996) *The Welfare of Laying Hens*. Scientific Committee on Animal Health and Welfare, European Commission, Brussels.

Anonymous (1997) Treaty of Amsterdam amending the Treaty on European Union, the Treaties establishing the European Communities and certain related acts – Protocol annexed to the Treaty on the European Community – Protocol on protection and welfare of animals. *Official Journal* C 340, 10/11/1997, p. 0110.

Anonymous (2006) *Commission Working Document on a Community Action Plan on the Protection and Welfare of Animals 2006–2010. Strategic basis for the proposed ın ʃʰnıɪ s7ıɪʃᴘʰıı ı ıɪʃʃᴘʃʃıııɪʃıı ʃᴘııııııttııı uı tıʃu Euʃopʃaıʃ Cuıuıuııtıes, Bıussels, avallable* at: http://ec.europa.eu/food/animal/welfare/comm_staff_work_doc_protection230106_ en.pdf (accessed 3 June 2011).

Berg, C. and Algers, B. (2004) Using welfare outcomes to control intensification: the Swedish model. In: Weeks, C. and Butterworth, A. (eds) *Measuring and Auditing Broiler Welfare*. CABI Publishing, Wallingford, UK, pp. 223–229.

Blokhuis, H.J., Jones, R.B., Geers, R., Miele, M. and Veissier, I. (2003) Measuring and monitoring animal welfare: transparency in the food product quality chain. *Animal Welfare* 12, 445–455.

Broom, D.M. (2006) Introduction – Concepts of animal protection and welfare including obligations. In: *The Ethical Eye – Animal Welfare*. Council of Europe, Strasbourg, France, p. 182.

Castellini, C., Berri, C., Le Bihan-Duval, E. and Martino, G. (2008) Qualitative attributes and consumer perception of organic and free-range poultry meat. *World's Poultry Science Journal* 64, 500–512.

COE (2006) *Building Europe together on the Rule of Law*. Council of Europe, Directorate General I – Legal Affairs, Strasbourg, France; available at: http://www.coe.int/t/e/legal_ affairs/about_us/publications/Brochure_DGI(2006)E.pdf (accessed 10 October 2011).

Cooper, M.D. and Wrathall, J.H.M. (2010) Assurance schemes as a tool to tackle genetic welfare problems in farm animals: broilers. *Animal Welfare* 19, 51–56.

Dawkins, M.S., Donnelly, C.A. and Jones, T.A. (2004) Chicken welfare is influenced more by housing conditions than by stocking density. *Nature* 427, 342–344.

Defra (2002a) *Foundations for Our Future. Defra's Sustainable Development Strategy*. PB 7175. Department for Environment, Food and Rural Affairs, London; available at: http:// www.defra.gov.uk/corporate/sdstrategy/sdstrategy.pdf (accessed 3 June 2011).

Defra (2002b) *Code of Recommendations for the Welfare of Livestock, Laying Hens.* PB 7274. Department for Environment, Food and Rural Affairs, London; available at: http://www.defra.gov.uk/publications/files/pb7274-laying-hens-020717.pdf (accessed 3rd June 2011).

Defra (2004) *Animal Health and Welfare Strategy.* Department for Environment, Food and Rural Affairs, London; available at: http://www.defra.gov.uk/animalh/ahws/default.htm (accessed 3 June 2011).

Defra (2008) *Animal Welfare Act (2006).* Department for Environment, Food and Rural Affairs, London; available at: http://www.defra.gov.uk/animalh/welfare/act/index.htm (accessed 3 June 2011).

Defra (2011) Guidance on Government Buying Standards for Food and Catering Services. http://sd.defra.gov.uk/documents/GBS-guidance-food.pdf (accessed 10 October 2011).

DG SANCO (2005) *Community Action Plan 2006–2010.* Commission of the European Communities, Brussels; available at: http://ec.europa.eu/food/animal/welfare/work_doc_strategic_basis230106_en.pdf (accessed 3 June 2011).

DG SANCO (2010) Farmland – the Game. http://www.farmland-thegame.eu/ (accessed 3 June 2011).

Duffy, R. and Fearne, A. (2009) Value perceptions of farm assurance in the red meat supply chain. *British Food Journal* 111, 669–685.

ECONWELFARE (2008) Socio Economic Aspects of Farm Animal Welfare. http://www.econwelfare.eu/Default.aspx (accessed 3 June 2011).

EFABAR (2009) European Forum for Farm Animal Breeders. http://effab.org/Home.aspx (accessed 3 June 2011).

EFSA (2010) Scientific Opinion on the welfare aspects of the practice of harvesting feathers from live geese for down production. *EFSA Journal* 8(11), 1886; available at: http://www.efsa.europa.eu/en/efsajournal/pub/1886.htm (accessed 3 June 2011).

Elson, H.A. (1985) The economics of poultry welfare. In: Wegner, R.M. (ed.) *Proceedings of the 2nd European Symposium on Poultry Welfare.* World's Poultry Science Association, Celle, Germany, pp. 244–253.

FAIR (2001) Consumer Concerns about Animal Welfare and the Impact on Food Choice. http://ec.europa.eu/food/animal/welfare/research/fair_project.pdf (accessed 3 June 2011).

FAWC (1991) *Farm Animal Welfare Council's Report on the Welfare of Laying Hens in Colony Systems.* PB 0734. Ministry of Agriculture, Fisheries and Food, Tolworth, UK.

FAWC (1995) *Report on the Welfare of Turkeys.* Farm Animal Welfare Council, Surbiton, UK.

FAWC (2001) *Interim Report on the Animal Welfare Implications of Farm Assurance Schemes.* Farm Animal Welfare Council, London; available at: http://www.fawc.org.uk/pdf/farmassurance.pdf (accessed 3 June 2011).

FAWC (2005) *Report on the Welfare Implications of Farm Assurance Schemes.* Farm Animal Welfare Council, London; available at: http://www.fawc.org.uk/pdf/fas-report05.pdf (accessed 3 June 3rd 2011).

FAWC (2007) *Opinion on Enriched Cages for Laying Hens.* Farm Animal Welfare Council, London; available at http://www.fawc.org.uk/pdf/enriched-cages.pdf (accessed 3 June 2011).

FAWC (2008) *Opinion on Policy Instruments for Protecting and Improving Farm Animal Welfare.* Farm Animal Welfare Council, London; available at: http://www.fawc.org.uk/pdf/policy-instr-081212.pdf (accessed 3 June 2011).

FAWC (2009) *Welfare of Farmed Animals at Slaughter or Killing Part 2: White Meat Animals.* Farm Animal Welfare Council, London; available at: http://www.fawc.org.uk/pdf/report-090528.pdf (accessed 3 June 2011).

Fraser, D. (2008). *Animal Welfare and the Intensification of Animal Production: An Alternative Interpretation.* Food and Agriculture Organization of the United Nations,

Rome; available at: http://www.fao.org/docrep/009/a0158e/a0158e00.HTM (accessed 3 June 2011).

GHK (2010) *Evaluation of the EU Policy on Animal Welfare and Possible Options for the Future*. DG SANCO Final Report 2010. GHK Consulting, in association with ADAS UK (Food Policy Evaluation Consortium), London; available at: http://ec.europa.eu/food/animal/welfare/actionplan/3%20Final%20Report%20-%20EUPAW%20Evaluation.pdf (accessed 3 June 2011).

Humane Farm Animal Care (2009) *Animal Care Standards: Egg Laying Hens*. Humane Farm Animal Care, Herndon, Virginia; available at: http://www.certifiedhumane.org/uploads/pdf/Standards/English/Std09.Layers.2J.pdf (accessed 3 June 2011).

Humphrey, T. (2006) Are happy chickens safer chickens? Poultry welfare and disease susceptibility. *British Poultry Science* 47, 379–391.

IGD (2007) *Consumer Attitudes to Animal Welfare: A Report for Freedom Food*. Institute for Grocery Distribution, Watford, UK; available at: http://www.igd.com/index.asp?id=1&fid=1&sid=8&tid=33&cid=311 (accessed 3rd June 2011).

Jones, T.A, Waitt, C.D. and Dawkins, M.S. (2008) Water off a duck's back: showers and troughs match ponds for improving duck welfare. *Applied Animal Behaviour Science* 116, 52–57.

Kijlstra, A. and Eijck, I.A.J.M. (2006) Animal health in organic livestock production systems: a review. *NJAS – Wageningen Journal of Life Sciences* 54, 77–94.

Kilbride, A.L., Mason, S.A., Honeyman, P.C., Pritchard, D.G., Hepple, S. and Green, L.E. (2011) Membership of a farm assurance scheme is associated with higher compliance with animal welfare legislation when inspected by Animal Health. In: *Proceedings of UFAW International Animal Welfare Symposium*, Portsmouth, UK, 28–29 June 2011; available at: http://www.ufaw.org.uk/documents/UFAW2011posterabstractsamended.pdf (accessed 3 June 2011).

Kilchsperger, R., Schmid, O. and Hecht J. (2010) Animal welfare initiatives in Europe. *Technical report on grouping method for animal welfare standards and initiatives*. http://www.econwelfare.eu/publications/EconWelfareD1 1-final_updateOct2010.pdf (accessed 3 June 2011).

Lay, D.C., Fulton, R.M., Hester, P.Y., Karcher, D.M., Kjaer, J.B., Mench, J.A., Mullens, B.A., Newberry, R.C., Nicol, C.J., O'Sullivan, N.P. and Porter, R.E. (2010) Hen welfare in different housing systems. *Poultry Science* 90, 278–294.

LAYWEL (2006) Welfare implications of changes in production systems for laying hens. http://www.laywel.eu/web/pdf/deliverable%2071%20welfare%20assessment.pdf (accessed 3 June 2011).

Main, D.C.J. and Green, L.E. (2000) Descriptive analysis of the operation of the farm assured British pigs scheme. *The Veterinary Record* 147, 162–163.

Main, D.C.J., Whay, H.R., Green, L.E. and Webster, A.J.F. (2003) Effect of the RSPCA Freedom Food scheme on the welfare of dairy cattle. *The Veterinary Record* 153, 227–231.

Matthews, L.R. (2008) Methodologies by which to study and evaluate welfare issues facing livestock systems of production. *Australian Journal of Experimental Agriculture* 48, 1014–1021.

McInerney, J. (2004) *Animal Welfare, Economics and Policy*. Department for Environment, Food and Rural Affairs, London; available at: https://statistics.defra.gov.uk/esg/reports/animalwelfare.pdf (accessed 3 June 2011).

Moran, D. and McVittie A. (2008) Estimation of the value the public places on regulations to improve broiler welfare. *Animal Welfare* 17, 43–52.

Pew Commission Report (2008) *Putting Meat on the Table: Industrial Farm Animal Production in America*. http://www.ncifap.org/bin/e/j/PCIFAPFin.pdf (accessed 3 June 2011).

Pritchard, D.G. (2002) Improving turkey welfare. *Turkeys* 50, 1–5.

Pritchard, D.G. (2003) Government views on the welfare of laying hen. In: Perry, G.C. (ed.) *Welfare of the Laying Hen*. Poultry Science Symposium Series vol. 27. CABI Publishing, Wallingford, UK, pp. 23–29.

Pritchard, D.G. (2006) Introduction to the development of the Council of Europe Conventions for animal protection – ethics, democratic processes and monitoring. In: *Proceedings of Workshop on 'Animal Welfare in Europe: Achievements and Future Prospects'*, Council of Europe, Strasbourg, 23–24 November 2006. http://www.coe.int/t/e/legal_affairs/legal_co-operation/biological_safety_and_use_of_animals/seminar/Presentation%20D%20PRITCHARD%20%20COE%2022_11_2006.pdf (accessed 3 June 2011).

Pritchard, D.G., Clarke, C.H., Dear, H.L. and Honeyman, P.C. (2003) Statutory monitoring of animal welfare on UK farms and influence of farm assurance schemes. In: Ferrante, V. and Scientific Committee (eds) *Proceedings of the 37th International Congress of the ISAE*. International Society for Applied Ethology, Abano Terme, Italy, p. 103.

Radford, M. (2001) *Animal Welfare Law in Britain: Regulation and Responsibility*. Oxford University Press, Oxford, UK.

Rauw, W.M., Kanis, E., Noordhuizen-Stassen, E.N. and Grommers, F.J. (1998) Undesirable side effects of selection for high production efficiency in farm animals: a review. *Livestock Production Science* 56, 15–33.

Robach, M. (2008) A private sector perspective on private standards – some approaches that could help to reduce current and potential future conflicts between public and private standards. In: *Proceedings of 78th OIE Plenary Sessions*, OIE, Paris. http://www.oie.int/fileadmin/Home/eng/Internationa_Standard_Setting/docs/pdf/A_78SG_9_.pdffgh (accessed 3 June 2011).

Sandilands, V., Nevison, I. and Sparks, N.H.C. (2008) The welfare of laying hens at depopulation. In: *Proceedings XXIII World's Poultry Congress Book of Abstracts*, Suppl. 2. World's Poultry Congress, Brisbane, Australia, pp. 267–268.

SCAHAW (1998) *Welfare Aspects of the Production of Foie Gras in Ducks and Geese. Report of the Scientific Committee on Animal Health and Animal Welfare*. Adopted 16th December 1998. http://ec.europa.eu/food/fs/sc/scah/out17_en.pdf (accessed 3 June 2011).

Scott, N.H. and de Balogh, K. (2010) FAO Embraces the One Health Challenge at the wildlife–livestock–human–ecosystem interfaces. http://www.ewda-2010.nl/Shared%20Documents/EWDA_Abstractbook_2010.pdf (accessed 3 June 2011).

Swarbrick, O. (1986) Clinical problems in 'free range' layers. *The Veterinary Record* 118, 363.

Tierschutz macht Schule (2006) Home page. http://www.tierschutzmachtschule.at/ (accessed 3 June 2011).

Toma, L., Kupiec-Teahan, B., Stott, A.W. and Revoredo-Giha, C. (2010) Animal welfare, information and consumer behaviour. In: *Proceedings of the 9th European IFSA Symposium*, Vienna, Austria, 4–7 July 2010. http://ifsa.boku.ac.at/cms/fileadmin/Proceeding2010/2010_WS4.5_Toma.pdf (accessed 3 June 2011).

Wolff, C. and Scannell, M. (2008) Implication of private standards in international trade of animals and their products. *Proceedings of OIE 76th General Session*. OIE. Paris; available at: http://www.oie.int/fileadmin/Home/eng/Internationa_Standard_Setting/docs/pdf/A_private_20standards.pdf (accessed 3 June 2011).

Vaarst, M., Roderick, N., Lund, V., Lockeretz, W. and Hovi, M. (2004) Organic principles and values the framework for organic husbandry. In: Vaarst, M., Roderick, S., Lund, V. and Lockeretz, W. (eds) *Animal Health and Welfare in Organic Agriculture*. CABI Publishing, Wallingford, UK, pp. 389–403.

WELFAREQUALTY (2007) Science and Society Improving Animal Welfare. http://www.welfarequality.net/everyone (accessed 3 June 2011).

Zander, K. and Hamm, U. (2010) Consumer preferences for additional ethical attributes of organic food. *Food Quality and Preference* 21, 495–503.

CHAPTER 3
Politics and Economics

M.C. Appleby

ABSTRACT

Choice of production systems for poultry is complex given the different attitudes and needs of different stakeholders, including producers, retailers, consumers and governments, with welfare and environmental considerations playing an important role. Political and legal decisions both affect and are affected by the attitudes of people to poultry and their management. Increasingly, legislation in European countries originates from the European Union, notably the 1999 Directive on laying hens and the 2007 Directive on broilers. Legislation and other decisions are strongly influenced by the activities of stakeholder groups, including trade associations, scientific societies and animal welfare organizations. These groups also influence the economic context of poultry production. In egg production, costs are generally higher in systems perceived to have higher welfare, but the demand for eggs is inelastic and sales of eggs from systems such as free range have led the way for welfare improvements in all livestock production. High-welfare poultry meat production has also expanded in recent years, helped by overlap with other criteria such as organic standards, but sales are less reliable than for eggs. Free trade outside Europe threatens welfare-friendly production within Europe, but voluntary agreements and emphasis on local origin may combat such free market pressures. The way in which decisions are made about poultry production systems will alter over the next few years as legislation changes and other stakeholder forums increase their impact.

INTRODUCTION

A common tendency in developed countries in the second half of the 20th century was the drive for efficiency in agriculture, for cutting the cost of producing each egg or kilogram of meat. This was initiated by public policies – before, during and after World War II – in favour of more abundant, cheaper food. It subsequently became market driven, with competition between producers and between retailers to sell food as cheaply as possible, and thereby acquired its own momentum. Nevertheless, variation in poultry production systems persisted, partly because some producers used other criteria in addition to efficiency in their decision making. And towards the end of that century, the drive for efficiency began to be limited by those other criteria, including the effects of production systems on the environment and on animal welfare, and these limitations tended to favour alternatives to the most efficient, intensive systems. This process has involved all stakeholders, including producers, retailers, consumers and governments, so any consideration of the politics and economics of alternative systems is an attempt to describe complex developments in comprehensible terms.

POLITICS

Legislation

Political and legal decisions concerning how poultry are treated both affect and are affected by the attitudes of people to poultry and their management. That is demonstrated by variation in concern about animal welfare between European countries. Concern has historically been stronger in the north of Europe – particularly the UK, the Netherlands, Germany and Scandinavia – and weaker in the south. The most persuasive explanation is that concern has largely developed in urban people whose involvement with animals differed from that in rural areas. The UK and the Netherlands, for example, were more industrialized than many other countries, and pressure for animal protection mostly came from city dwellers rather than from those involved in farming. Correspondingly, legislation affecting poultry welfare in individual European countries also shows a dichotomy. Northern countries have detailed laws, with codified lists of actions that are prohibited. Southern countries tend simply to state that animals must not be ill-treated. Legislation is also enforced more strictly in some countries than in others.

Increasingly, however, legislation in European countries originates from the European Union (EU). This includes legislation on employment, on the environment and on animal welfare itself. It is influenced by the broader grouping of the Council of Europe, in which 47 countries and the EU are now members. In 1976 the Council produced the Convention on the Protection of Animals kept for Farming Purposes, which was concerned with the care, husbandry and housing of farm animals, especially those in intensive systems. A Standing Committee elaborates specific requirements, and one of the first areas in which it became active was that of poultry welfare. This placed responsibility for action on the EU. To take laying hens as an example, in 1985 the EU produced Trading Standards Regulations for labelling of eggs, and in 1986 a Directive laying down minimum standards for the protection of hens in battery cages. This was superseded by a new Directive in 1999, under which barren battery cages had to be phased out by 2012. All cages must then be furnished, and requirements for non-cage alternatives are also specified. Some countries such as Germany have considered banning cages altogether. Regarding broiler production, the EU passed a Directive in 2007.

In no other country has legislation advanced as far as in Europe. That is partly because of different attitudes to animal welfare, and partly because of different legal systems. For example, in the USA there are only three federal laws that apply to animal welfare; two (on slaughter and general welfare) specifically exclude poultry and the other (on transport) has never been applied to poultry. In the country as a whole the industry and the retail sector have achieved more in improving how poultry are kept than has any legislation to date (Mench, 2004). However, some states are acting on this independently. California passed a ballot initiative in 2008 that will effectively ban battery cages.

Stakeholder activities

Politics does not just involve the actions of professional politicians but all developments in policy and public affairs. In agribusiness this includes the activities of trade associations, which recruit a high proportion of producers as members. In the UK the main players are the National Farmers Union, the British Egg Industry Council (BEIC) and the British Poultry Council (representing meat producers). While there has in the past sometimes been resistance to pressure for change from those organizations, they have become more active on animal welfare in recent years to reflect increased concern for this issue. As one example, BEIC (1999) issues guidelines on safeguarding welfare at depopulation of laying hens and breeders.

In the USA the poultry trade associations are even more influential: the United Egg Producers (UEP), the National Chicken Council (which deals with meat producers) and the National Turkey Federation. They have sometimes tended to criticize calls for greater consideration of welfare. However, in about 1999 UEP started the process of drawing up detailed Guidelines for their members on husbandry and welfare (UEP, 2010). The other associations have since followed suit, partly because from 2000, retailers started putting pressure on them to require humane treatment of animals.

International trade-related associations have also tended to be conservative in this area. Following the 1999 European Directive, the International Egg Commission, representing 33 countries including all of the major producing countries, resolved to fight the ban on conventional laying cages. One reason must have been solidarity in the face of what was perceived as an attack on their European members, and in addition 'a domino effect is feared by the US, Canada and Australia' (Farrant, 1999: 1).

There are also trade associations for other groups who contribute to discussion and negotiation. Poultry scientists, for example, form societies such as the Poultry Science Association (in North America) and the World's Poultry Science Association. The latter has organized European Symposia on Poultry Welfare every four years from 1981 (following a predecessor in Denmark in 1977). The eighth was in Italy in May 2009. There has also been one North American Poultry Welfare Symposium (Mench and Duncan, 1998).

Many members of the public write to politicians about animal treatment, and this highlights the extent to which involvement in this subject is not confined to professionals. Among farm animals, a considerable proportion of this attention has been paid to poultry, including by the large numbers of societies and groups that have been set up in most countries. The core staff of these organisations is generally professional, but they need to retain the support of their amateur supporters for their actions. In the UK, for example, the Royal Society for the Prevention of Cruelty to Animals (RSPCA) is a mainstream animal protection society that is active on behalf of all animals including livestock, and among many other activities lobbies for improved housing and conditions for farm animals. Other societies tend to have special interests and concentrate their efforts on more specific issues. They include Compassion in World Farming, which campaigns for a ban on the export of live animals and is also very active on housing conditions. It gives an annual Good Egg Award to European food companies whose policies on egg sourcing it approves.

There are also, of course, animal protection societies in other European countries, and international groups. One organization playing a strong role in the EU is Eurogroup for Animals, which lobbies the EU's politicians on behalf of member societies from each of the EU states. In addition, there are some societies that are active on the wider, world scene, such as the World Society for Protection of Animals and the International Fund for Animal Welfare.

In the USA, as well, there are many societies that seek dialogue with the industry to negotiate change, while also lobbying for change through other routes. This approach started in the 19th century, as in Europe. The American Society for the Prevention of Cruelty to Animals was formed in 1866. Today the largest US animal protection society is the Humane Society of the United States (HSUS), with around 10 million supporters. Other groups such as the American Humane Association are also active on poultry welfare and are mentioned again below.

In 2011, UEP and the HSUS announced that they would work jointly for legislation in the USA to phase out barren battery cages and to make other provisions for laying hen welfare (Smith, 2011).

ECONOMICS

The attitudes of the public, members of the poultry industry and others involved are primarily expressed, of course, within the economic context of agricultural production and sales. How commercial poultry are housed and treated will always be affected by monetary considerations, although other factors are also important. Economics is not just about money: 'Economics is concerned with how we in society make decisions about using resources to achieve the things that we want' (Bennett, 1997: 235).

We consider first the 'supply side economics' of interactions between supply and demand, between producers, retailers and customers, before returning to the broader mechanisms by which society makes decisions relevant to production systems for poultry.

Finances of egg production

The finances of table egg production by laying hens have received considerable attention, perhaps because such a variety of systems is available and the choice between them is controversial. Costs are generally higher in systems perceived to have higher welfare: greater space allowances in cages, as well as production in different systems, increase costs. This is because many of the factors that influence cost are less favourable in furnished cages and non-cage systems than in conventional cages: housing, labour, feed intake, hygiene, mortality and predictability of performance (Fisher and Bowles, 2002; Appleby *et al.*, 2004). Outdoor systems also incur costs for land, although these are generally treated separately because land can often be shared with other uses and usually appreciates in value rather than depreciating like housing and equipment. Overall, alternative production systems are estimated to cost anything up to 70% more than conventional cages (Appleby *et al.*, 2004). However, such comparisons can never be definitive and will always be affected both by circumstances at the time of the study (such as the cost of inputs such as feed) and by the particular parameters studied.

On the other side of the equation, welfare considerations have played a greater role in egg sales than in any other sector of poultry production. Indeed, sales of eggs from systems such as free range have led the way for welfare improvements in all livestock production. This mostly applies, though, to eggs sold whole. Few ready-made meals or other products containing eggs indicate how the hens were kept (an exception may be organic products), and few customers think to ask.

The demand for eggs is inelastic: eggs are not readily interchangeable with other items in the diet and people tend to buy a set number whatever the price. This background may explain why, uniquely among animal production sectors, the system in which eggs were produced became a selling point. A niche market developed, particularly in northern Europe, for eggs that did not come from cages: free range eggs in some countries, deep litter or 'scratching' eggs in others. Some people bought them – at a higher price – because they perceived them to be more nutritious, tastier or healthier. Some were also concerned about the welfare of the hens, and this concern led to the development of other non-cage systems. A similar trend began in North America in the 1990s and increased after 2000, influenced by a campaign by groups including the Humane Society of the United States.

Free range eggs are sold in the shops at up to twice the price of cage eggs, or even more, and other categories such as barn eggs are also generally priced much higher than cage eggs. It is important to recognize that this is not just a reflection of higher production costs, as shop prices include grading, packing, transport and so on. Some of these processes vary in cost between systems, but not proportionally, while others are similar for all producers. So packers or retailers or both make a higher profit on eggs from alternative systems. It is not clear how this will change once conventional cages are banned from 2012. Nevertheless, the fact that so many people were prepared to pay more for eggs that they believed were associated with higher welfare was probably one of the most important factors that led European governments and the EU to legislate for improved hen welfare, leading up to the 1999 Directive. Indeed, it led the way for improved welfare of all farm animals, as social, economic and legislative pressure for changes in the treatment of farm mammals tended to follow those for the treatment of poultry.

Finances of meat production

Most broilers, turkeys and ducks are loose housed on litter and reared as rapidly as possible to obtain maximum growth rate and feed conversion. However, an alternative approach developed, initially in France and the UK (particularly under the trade name Label Rouge) but also to some extent in the USA, for slower growing, free range broilers and turkeys. In recent years, free range poultry meat production expanded rather like free range egg production did in Europe, including under other trade names. Sales were boosted in the UK by television programmes featuring celebrity chefs. However, persistence of this change in purchasing has been uneven, especially during the latest financial crises.

Housing and production costs of 'alternative' poultry meat are higher, because of lower stocking density within the building, the fact that in many cases the birds have access to the outside for at least part of their lives, and the fact that birds are kept for longer. Feed consumption is higher, both on a daily basis and because it takes longer for birds to reach selling weight, and feed conversion is therefore less economic. However, selling prices are higher, either through supermarkets or small-scale outlets.

Two other approaches that overlap with such speciality production are organic production and welfare-labelled products. Organic standards set by the EU are applied in the UK by bodies such as the Organic Food Federation, Organic Farmers & Growers and the Soil Association, while the USA has a National Organic Standards Board. Birds to be sold as organic should be raised on organic feed and without synthetic drugs. However, these organizations also recognize that both consumers and most producers expect organic livestock to have outdoor access and to have reasonable standards of welfare. The costs of organic production are therefore higher than normal commercial costs and organically produced goods need to be identified as such by suitable labelling.

Welfare labels

It has been emphasized that the part played in all this by people particularly concerned about poultry welfare includes the buying of products perceived as beneficial for welfare, such as free range eggs and meat. It also includes the activities of the animal protection societies in launching schemes that directly address welfare concerns. The leader in this field is the RSPCA, which launched its Freedom Food programme in 1994. There are detailed criteria that must be met by producers who want to join the programme. They can then use the Freedom Food label. This includes the name 'RSPCA', which has widespread recognition and confidence from the British public. The RSPCA also helps with marketing. The programme has grown steadily, helped by the overlap in criteria between this and other schemes. Thus if egg producers are already certified organic or are producing free range eggs, they do not usually have to make many additional changes to be able to use the Freedom Food label, which is therefore well worthwhile. Similar programmes have begun in North America. The American Humane Association started its Free Farmed scheme in 2000, Certified Humane was launched by Humane Farm Animal Care in 2003, and in Canada SPCA Certified food was launched in British Columbia in 2002. The criteria of such programmes are not identical, but all require alternative production systems for both laying hens and meat birds rather than conventional production methods. Furthermore, they have tended to lead the way to wider change in poultry production and marketing.

DECISION MAKING

Change in poultry production and marketing will not cease in the foreseeable future, not even in the EU with the implementation of the 1999 Laying Hen Directive and the 2007 Broiler Directive. This is partly because of globalization, which increasingly means that decisions taken within a country cannot be wholly independent of those in other countries. They are affected by international trade and other trans-national issues such as disease control and environmental sustainability, as well as global communication.

Regarding trade, there have been some concerns that the 1999 Directive will weaken EU competitiveness so that a significant proportion of domestic consumption will be substituted by imported eggs (Wolffram et al., 2002). However, others suggest that such fears are overstated. There is a danger that imports of processed eggs, which make up 25% of European egg production, will rise in the absence of protection. But current analysis suggests that most egg trade of European countries will continue to be intra-EU (Windhorst, 2009). In any case, these pressures emphasize that decisions will continue to be needed about poultry production systems, including choices between whatever alternatives continue to be available, and that those decisions will continue to involve political and economic processes.

The best known agricultural economist writing on animal welfare, Professor John McInerney (1998: 124), considers that: 'If animal welfare is a public good, regulation (not market forces) has to determine standards'. Regulation does not necessarily have to be by legislation, as self-regulation by industries is also possible, but self-regulation is variable in its effectiveness and in its responsiveness to public opinion. Control by regulation, taking public opinion into account, avoids the limitations inherent in 'purchasing power', for example the tendency of the latter to apply to whole eggs but not to egg products. This is important because an increasing proportion of food is sold in processed form. In fact the shift towards sale of pre-processed food in developed countries increases the de-linkage between production costs and shop prices discussed above. McInerney concludes that it should be possible for the farmers to maintain their profits, offsetting the increased costs of alternative systems to improve welfare with increased selling prices. His calculations illustrate two general propositions (McInerney, 1998: 127, 130):

> The economic costs of reasonable improvements in animal welfare are likely to be relatively small;
> Higher animal welfare standards are not an economic imposition on farmers.

However, major questions remain about how decisions on poultry production systems can be made better in future, taking into account the needs of all the stakeholders involved, when these decisions affect important outcomes including animal welfare and environmental sustainability.

ACKNOWLEDGEMENT

Some of the text in this chapter is adapted from Appleby et al. (2004).

REFERENCES

Appleby, M.C., Mench, J.A. and Hughes, B.O. (2004) Poultry Behaviour and Welfare. CABI Publishing, Wallingford, UK.

BEIC (1999) Joint Industry Welfare Guide to the Handling of End of Lay Hens and Breeders. British Egg Industry Council, London.

Bennett, R.M. (1997) Economics. In: Appleby, M.C. and Hughes, B.O. (eds) Animal Welfare. CABI Publishing, Wallingford, UK, pp. 235–248.

Farrant, J. (1999) IEC's world action to keep cages. Poultry World November, 1–4.

Fisher, C. and Bowles, D. (2002) Hard-Boiled Reality: Animal Welfare-Friendly Egg Production in a Global Market. Royal Society for the Protection of Animals, Horsham, UK.

McInerney, J.P. (1998) The economics of welfare. In: Michell, A.R. and Ewbank, R. (eds) Ethics, Welfare, Law and Market Forces: The Veterinary Interface. Universities Federation for Animal Welfare, Wheathampstead, UK, pp. 115–132.

Mench, J.A. (2004) Assessing animal welfare at the farm and group level: a US perspective. Animal Welfare 12, 493–503.

Mench, J.A. and Duncan, I.J.H. (1998) Poultry welfare in North America: opportunities and challenges. Poultry Science 77, 1763–1765.

Smith, R. (2011) HSUS, UEP reach agreement to transition to colonies. Feedstuffs 7 July; available at: http://www.feedstuffs.com/ME2/dirmod.asp?sid=F4D1A9DFCD974EAD8C D5205E15C1CB42&nm=Breaking+News&type=news&mod=News&mid=A3D60400B4 204079A76C4B1B129CB433&tier=3&nid=924A050BC1E846CDBD0637361C9B 42C0 (accessed 11 October 2011).

UEP (2010) *United Egg Producers Animal Husbandry Guidelines for U.S. Egg Laying Flocks*. United Egg Producers, Alpharetta, Georgia; available at: http://www.unitedegg. org/information/pdf/UEP_2010_Animal_Welfare_Guidelines.pdf (accessed 25 January 2011).

Windhorst, H.W. (2009) Recent patterns of egg production and trade: a status report on a regional basis. *World's Poultry Science Journal* 65, 685–708.

Wolffram, R., Simons, J., Giebel, A. and Bongaerts, R. (2002) Impacts of stricter legal standards in the EU for keeping laying hens in battery cages. *World's Poultry Science Journal* 58, 365–370.

CHAPTER 4

The Effects of Alternative Systems on Disease and Health of Poultry

S. Lister and B. van Nijhuis

ABSTRACT

Poultry health management is a pivotal component of successful poultry production. Disease and its effects on poultry health can damage productive performance and have an adverse effect on bird welfare and food safety. A whole host of factors can affect disease incidence and its impact on poultry health. These include the prevalence and interaction of many pathogens, availability and use of vaccines and medicines, standards of husbandry and management and levels of stockmanship. One area with potential to have the most dramatic influence is the birds' environment and how the birds respond to it. This impact has been well known throughout the development of the global poultry industry as it adapted to varying climates and market requirements. This involved considerable advances in technology and husbandry techniques. The first major changes tended to intensify poultry production. As such systems became the norm, they have often been described as 'conventional'. Key drivers in poultry production have changed in recent years including a re-evaluation of the welfare impact of such production systems for both egg laying and meat birds. Part of this has been some move away from conventional systems and a re-introduction of more traditional systems or the development of novel alternative systems. The list of diseases that can affect poultry is the same regardless of the system of production. However, the clinical effects of those disease challenges and impacts on health, performance and welfare can be specific to a particular system. In order to ensure health and welfare is maintained, the interaction of the bird with the environment and the effect this can have on poultry health and the bird's response to disease challenges must be understood.

INTRODUCTION

Disease and its impact on bird health, welfare and food safety is multifactorial. It involves a complex interaction between the birds, the production system and the environment in which the birds are kept and thus which range of pathogens

© CAB International 2012. *Alternative Systems for Poultry –*
Health, Welfare and Productivity (eds V. Sandilands and P.M. Hocking)

and challenges the birds are exposed to. A key factor in all production systems is the level of management and stockmanship such that this is likely to have a bigger influence on bird health and welfare than the production system itself. This clash of 'nature versus nurture' means that the plusses and minuses of different production systems must not only consider perceived welfare benefits of alternative systems in providing 'a good life' (FAWC, 2009), but must also consider any adverse repercussions on poultry health and food safety that may be a consequence of such systems.

Maintaining bird health requires an understanding of the pathogens and challenges, the birds and how they respond or adapt to each other. In many situations this is an inherent reaction between pathogen and bird and therefore general disease control and biosecurity requirements are fundamentally the same whatever production system the birds find themselves in, from backyard flocks through to large production systems. However, environmental factors do play a major part in how birds may be exposed to pathogens, their ability to respond and resist challenges and other insults to the birds' immune system.

In addition to the direct effects on the birds' environment in terms of space allowance, air quality, temperature variation and thermal comfort, the introduction of alternative systems is often intimately tied into particular farm assurance systems or specifications, including what are known in the European Union (EU) as 'special marketing terms'. The attributes or requirements of such schemes can also include the provision of particular diets, requirements on the composition of feed (e.g. percentage of whole grains or maize in the diet), reduced reliance or restrictions on the use of medicines and a requirement for outdoor access, all of which can have additive influences, impacts or adverse effects on bird health. For example outdoor access can result in an increase in contact with wild birds and their faeces.

While there may be differences of opinion as to whether a move from 'conventional' housing to more extensive or free range 'alternative' systems represents progress, it is clear that there must be an awareness and appreciation of the potential impacts on poultry health and disease. There is nothing to prevent the development of satisfactory alternative systems in terms of maintaining disease control, but it requires a holistic approach considering all aspects of such systems that may influence bird health.

The Environment and Its Impact on Disease and Bird Health

There is a range of alternative systems available for meat birds (broilers and turkeys) and layers. Other species such as ducks, geese or game birds have always tended to remain as more traditional extensive outdoor systems. Even in the case of game birds there has been the introduction of more confined cage systems, predominantly for breeding pheasant and partridge, bringing with it issues concerning bird welfare as well as disease control (see Pennycott et al. Chapter 9, this volume). However, the present chapter focuses on commercial meat birds and laying stock.

The term 'alternative' can encompass a wide range of husbandry systems that usually involve a cage-free environment and/or access to outdoor free range paddocks or pens. In addition, in line with EU special marketing terms or the requirements for specific retail or other farm assurance schemes (e.g. RSPCA Freedom Food, organic, etc.), this often also involves specification of maximum flock size, stocking density or other aspects of environmental enrichment. A common driver of these standards and schemes is to define systems that are hoped to offer animal welfare benefits. Such standards can have an effect on disease control, bird health and food safety. There is a whole range of interrelated factors to consider in these different systems in terms of their effects on bird health and welfare. For example, the EU LAYWEL project (http://www.laywel.eu) has attempted to assess the health and welfare implications of different categories of housing systems for laying hens (Blokhuis et al., 2007). This project assessed various key indicators that demonstrated that more extensive alternative systems had a beneficial impact on behaviour such as foraging, dust bathing and access to nesting areas and perches. However, non-cage and outdoor systems faired less well in relation to mortality, feather pecking, cannibalism, incidence of foot problems and, in non-cage indoor systems, aspects of air quality. This work demonstrated the clear trade-offs in relation to different indicators of health and welfare. It also importantly stressed the pivotal role of management and stockmanship in all systems.

Some of the more significant influences of production systems affecting bird health that require consideration in the planning and implementation of any production system include some or all of the following.

Exposure to pathogens

In the case of layers alternative systems mean a move from cage to indoor floor systems or into free range systems. For meat birds, alternative systems most often mean a move from housed systems to free range. In the move from cage to floor or free range systems birds will have increased and constant exposure to litter and other substrates contaminated with the birds' faeces and, in the case of free range flocks, faeces of wild birds and other animals. De Reu et al. (2005) demonstrated that accommodation with litter had ten times more air-borne bacteria in the environment and 20 to 30 times more bacteria on egg shells as compared with cage housing systems. This exposure to pathogens may increase the challenge 'load' on birds in terms of bacterial, viral, parasitic or fungal organisms.

Temperature and other external environmental fluctuations

Access to outdoor conditions can lead to thermal stress due to extremes of temperature or significant diurnal fluctuations, although this needs to be balanced against the potential for significant heat stress issues in housed flocks. These fluctuations and extremes can cause physiological stress that can affect the birds' immune response to pathogens. Low environmental can increase feed maintenance requirements that can have nutritic implications. Excessive rain can result in deterioration of the range and contribute to heavy pathogen challenge.

Air quality

Alternative systems, which are often naturally ventilated, can present challenges in maintaining a stable environment within houses or night-time accommodation. Lack of automation, the use of multiple small houses or arks can make it difficult for stockmen to respond to fluctuations in climatic conditions, although many of these can be addressed with good management of suitable baffles and vents. Any shortcomings in ventilation can cause exposure to high levels of ammonia, carbon dioxide and dust that can have adverse health effects (see Hartung, 1994; Wathes, 1994). These noxious insults to the upper respiratory tract can cause potent damage to the birds' immune system, acting as a trigger to allow infectious agents a way of circumventing the birds' normal defence mechanisms. For example, the delicate lining of the nose and trachea is covered in microscopic cilia, which are fine hairs that help to waft inhaled dust and pathogens back up the respiratory tract out of the trachea such that they can be harmlessly swallowed or eliminated through sneezing or coughing. Proper functioning of this defence mechanism depends on the integrity of this tracheal lining. Ammonia levels as low as 20 ppm and relative humidity above 75% (or indeed too dry, e.g. below 50%) in any system can either alone or in combination rapidly induce ciliostasis (i.e. a loss of movement in the cilia) and their eventual destruction if the insult persists. Therefore inhaled dust and pathogens become trapped and fall deeper into the respiratory tract, causing direct damage to the trachea and/or lungs and increasing the likelihood of respiratory disease (e.g. Anderson et al., 1964). Research has indicated that different housing systems can markedly affect the numbers of air-borne bacteria, fungi and dust. For example, Vucemilo et al. (2010) demonstrated that levels of air-borne bacteria and fungi determined in an aviary system far exceeded those of a conventional cage system.

Increased bird activity

It is well known that systems that allow the provision of additional space, increased light intensity and environmental enrichment (e.g. provision of perches, access to range, etc.) can increase bird or flock activity (SCAHAW, 2000; Dawkins et al., 2004). Indeed this is frequently a rationale for these provisions and there are many perceived welfare benefits of this increased activity. In addition to the ability of birds to express the fullest range of natural behaviours, e.g. dust bathing, wing flapping, use of perches and ranging, there is evidence that bone strength, notably in layers, can be improved with a reduction in fractures associated with osteoporosis. Gregory et al. (1990)

demonstrated that at slaughter the incidence of recent fractures in end-of-lay hens was reduced in percheries (10%) and for free range systems (14%) compared with cage systems (31%). Sandilands *et al.* (2005) also demonstrated that wing fractures can be less common in free range and barn systems. Even within cage systems, new wing fractures in birds in enriched cages (2%) were far less than those from conventional cages (17%). These effects are likely to relate to a combination of improved bone strength through activity (perching and wing flapping) and aspects of cage design, including generally wider cage openings in enriched cages (Sandilands *et al.*, 2005). However this increased activity can have adverse effects on skeletal health mainly associated with collisions and injuries. Gregory *et al.* (1990) also showed that although new fractures were reduced in more extensive systems, old fractures were more common (5% in cages, 12% on free range and 25% in percheries).

In the case of broilers, increased activity is one of the factors that can increase the incidence of skin damage through scratches and skin necrosis (Randall *et al.*, 1984; Norton *et al.*, 1997). This breach in the birds' defence system can lead to systemic bacterial disease, mortality and increased rejections at processing, frequently known as infectious process (IP) in the USA and Canada and skin necrosis in the UK (Norton, 1997).

Stocking density and flock size

Reductions in stocking density in housed meat chicken flocks can be associated with increased activity and reduction in incidence of leg problems (Dawkins *et al.*, 2004). In broilers and turkeys this may also be related to the ability to control litter conditions at lower stocking density and reduce the impact of dysbacteriosis and other conditions associated with wet litter (Lister, 2006). In free range flocks, overall flock size can have an influence on ranging behaviour. For example, Bubier and Bradshaw (1998) demonstrated a significant effect of flock size on the mean percentage of birds outside on range during the day (12% for flock sizes ranging from 1432 to 2450 versus 42% for a flock of 490 birds). In such research, it was difficult to apportion this wholly to flock size as other aspects of range quality and management probably had a significant influence (e.g. feeding regimes, type of vegetation on range, etc.). Ranging behaviour can be expected to have a beneficial effect on bird health if range use results in dilution of faecal contamination across the range, reducing bacterial or parasitic load. Hegelund *et al.* (2005) examined 37 organic flocks in Denmark and while there was a tendency for the percentage of hens outside to decrease as flock size increased, flock size did not significantly influence use of the range in terms of distance travelled from the house. It is possible therefore that providing a theoretically large range area may only result in overuse of land close to the house. This can inadvertently lead to significant faecal contamination, parasite and microorganism build-up or heavy poaching of land, all of which can have an adverse effect on bird health and welfare. This clearly demonstrates the interplay of what is expected to be achieved by environmental enrichment in alternative systems and how birds actually react, or are managed, in relation to effects on bird health. Therefore those stockmen who are actively encouraging birds to range should be encouraging them to range over the whole area. All the factors influencing range use, such as rearing system, early access to range, flock size, shelters on range, etc., are aimed at getting birds outside and not overusing space adjacent to the house.

Pasture access in free range systems

One of the major perceived benefits of free range systems is the ability for birds to express the full range of natural behaviours on range. As previously discussed, the successful fulfilling of these aims depends on how well birds actually range, together with the quality and management of vegetation on the range. Practical experience shows that much of this is related to early range access, original siting of the range relative to the poultry house and the aspect and drainage of the land. One of the biggest risks to health is in exposure to pathogens on the range or significant viral diseases such as avian influenza, Newcastle disease or mycoplasma associated with the presence of wild birds or possible local wind-borne spread. These are discussed in more detail later but the presence of poorly drained, poached areas resulting in standing water would be attractive to birds to drink and can increase exposure to bacterial organisms such as *Escherichia coli* and *Brachyspira*. This is one element of land becoming 'fowl sick' due to excessive or repeated use of range areas by flocks. This can be countered by effective pasture rotation, better drainage, temporary fencing off of heavily poached areas and keeping grass on pasture short. The latter allows more access of ultraviolet rays from sunlight known to reduce the persistence of pathogens on the pasture. In addition, short herbage length will reduce the likelihood of gizzard and intestinal impactions.

Cover on range

Cover provided as small shelters or vegetation, especially trees, is known to promote ranging behaviour presumably through birds feeling innately safer from the dangers of predation (Jones et al., 2007). Shelter also provides shade from the sun, opportunities to roost and, depending on design, possibly also windbreaks (Dawkins et al., 2003). As a result, such additions to the range are frequently a component of farm assurance scheme requirements for free range birds. The provision of shelter can have beneficial effects on growth rate and mortality rates. Organic broilers in France given access to range with herbaceous cover as oak trees demonstrated improved body weight gain and lowered mortality than broilers having access to a simple meadow range (Germain et al., 2010). The reduction in mortality was attributed to less predation.

ᵣsus multi-age sites

ᵢgle age, all in/all out sites are the preferred option for poultry production as they permit effective cleansing and disinfection between flocks to eliminate any residual or persistent pathogen challenge. It can also avoid younger flocks being exposed to pathogens before the completion of any vaccination or other control programme. This is established good practice in conventional poultry production. While there is nothing to prevent similar implementation for alternative or more extensive systems, such systems frequently require small flock or unit size to comply with other aspects of the system or farm assurance standard. This requires considerably greater manpower input if flocks are to be kept effectively isolated. Personnel, vehicle and equipment movements between sites are known to be a significant potential biosecurity risk and hence a strict movement and hygiene policy, with the use of effective personal protective equipment dedicated to specific sites, is essential if such a policy is to be implemented effectively (Lister, 2008).

BIOSECURITY AND DISEASE CONTROL

Biosecurity can be defined as:

> a set of management practices which, when followed, collectively reduce the potential for the introduction and spread of disease causing organisms on and between, sites. (Lister, 2008.)

What is clear from this definition is that biosecurity describes a whole range of management interventions including the use of therapeutic agents and vaccines, effective terminal cleansing and disinfection and ongoing pathogen control throughout the life of the flock, effective vermin control, environmental control and general levels of stockmanship. The aim of an effective biosecurity programme is to prevent the introduction of disease-causing organisms at various levels, be they national, regional, company, farm or house, and to prevent or reduce spread between these compartments.

The biosecurity programme can be directed at a range of scenarios, including:

- keeping lethal highly contagious diseases out of the premises, e.g. avian influenza, Newcastle disease, Gumboro disease;
- reducing challenge by endemic disease causing mortality and reduced productivity, e.g. *E. coli*, coccidia, worms;
- reducing or eliminating background immunosuppressive agents that leave birds more susceptible to other diseases, e.g. Gumboro disease, Marek's disease, chick anaemia virus; and
- reducing contamination with agents of public health significance, e.g. *Salmonella* and *Campylobacter*.

The most effective biosecurity programmes are based on sound Hazard Analysis and Critical Control Point principles assessing critical control points at different levels. Cooperation between producers and veterinarians can help to establish the most effective intervention strategies through practical veterinary health planning. Some of these interventions are considered in more detail under the following separate disease headings. However, general points to consider for all systems include (after Lister, 2008):

- staff movements;
- source of poultry;
- vehicle movement and disinfection;
- equipment movement and disinfection;
- feed source;
- litter source;
- water quality;
- vermin control;
- wild bird exclusion; and
- site decontamination and terminal cleansing and disinfection.

The EU LAYWEL project, while assessing the relative risk to bird welfare of different categories of housing system for laying hens, highlighted differences in mortality rates between different systems. Part of this mortality was attributed to cannibalism or foot abnormalities (e.g. bumble foot). Other studies, e.g. Fossum *et al.* (2009), have shown the potential for higher occurrence of disease (predominantly bacterial and parasitic disease) in litter-based housing systems on free range systems than in hens kept in cages. They also demonstrated that the occurrence of viral diseases was significantly higher in indoor litter-based housing systems than in cages. These findings and those of work in other countries have shown the potential for increased disease challenges in alternative systems. As already discussed, part of this may relate to intrinsic factors such as pathogen exposure in more extensive systems but it is likely that attention to management, stockmanship and the effective biosecurity procedures can reduce the impact of these disease challenges (Lister, 2008).

The cornerstone of successful veterinary health planning involves a working knowledge of the disease, risks and challenges active in a particular region, company or farm. This can be established by prompt veterinary diagnosis of specific outbreaks, underpinned by strategic disease monitoring to act as an early warning system of disease challenge or an assessment of the success of intervention strategies.

SPECIFIC DISEASE CONSIDERATIONS

Diseases of public health significance

These aspects are well reviewed under Chapter 5.

Bacterial infections and soil-borne pathogens

In extensive alternative systems, the presence and persistence of pathogens in soil on range can be of potential significance especially relative to enclosed housed or cage systems. Fossum *et al.* (2009) demonstrated greater incidence of disease when birds had contact with litter and faeces and that bacterial and parasitic diseases such as colibacillosis, erysipelas, pasteurellosis, coccidiosis and red mite infestation were the most common diagnoses. Christensen *et al.* (1998) demonstrated that approximately 80% of cases of pasteurellosis in Danish flocks occurred in free range flocks that had contact with wild birds.

Erysipelas

Erysipelas is a bacterial infection causing sudden death or morbidity that can infect a number of different animal species. Vermin are a significant risk factor for introducing infection into a flock. Among farm animals, pigs and sheep are most commonly affected. Turkeys are the most susceptible bird species but most avian species, even wild birds, can act as carriers or vectors of infection. It is suspected that the organism can survive for up to 6 months outside the host especially in soil contaminated by previous livestock. It is known for erysipelas to recur on a site that has not been used for livestock for several years. Mortality can be acute and dramatic or more insidious and long lasting. As this is a bacterial infection antibiotics can be used to treat infection. In terms of prevention, effective vermin control and vaccination of known risk sites are the most important control measures. There is some evidence that red mite (see later) can be a significant trigger factor in some flocks due to the general stress and anaemia they cause or as mechanical vectors of the bacteria. Effective red mite control is therefore highly significant.

Brachyspira

Brachyspira infection, the cause of avian intestinal spirochaetosis, has become more prevalent in recent years. The incidence of the known pathogenic strains of *B. intermedia*, *B. pilosicoli* and *B. alvinipulli* is reportedly higher and occurs earlier in the life of free range flocks and barn systems than in caged flocks (Burch *et al.* 2009). Clinical signs of *Brachyspira* infection include the presence of watery pasty brown faeces, and pale caecal droppings have increased, sometimes in fowl sick premises but also in more newly established sites (Lister, personal observation). These have been associated with reduced egg production in breeders and layers especially on free range and are sometimes associated with poor body weights, increased mortality and reduced appetite (Griffiths *et al.* 1987; Burch *et al.*, 2006). Birds appear to respond to specific therapy aimed at eliminating *Brachyspira* (e.g. with tiamulin, in countries where this product is licensed for use in laying hens) and re-establishing gut function. This must be coupled with environmental control of poached areas on range and often dietary control to address weight loss and unevenness while maintaining appropriate egg size. This is a really good example of where there must be an integrated approach to control through veterinary intervention and measures aimed at correcting environmental deficiencies.

Histomoniasis

Histomoniasis (blackhead) is caused by a protozoal parasite, *Histomonas meleagridis*. It affects the liver and caeca and can cause either sudden severe mortality in turkeys or more chronic infections in chickens and game birds. The *Heterakis* worm can be important as a source of infection and range areas can become severely contaminated with both the *Heterakis* and *Histomonas* parasites. Disease is often precipitated by disturbance of soil on range either through human excavation work or by the birds' activities themselves when foraging. Currently there are no therapeutic or preventative medications licensed for food producing animals in the EU, so control is aimed at exclusion of the parasites and maintaining the drainage and quality of range areas.

Viral diseases

The most significant perceived risk of viral infection in alternative systems, most specifically free range, must relate to the notifiable avian diseases of avian influenza (AI) and Newcastle disease and the risk of introduction and spread by wild birds, although failures in biosecurity can also result in such infections gaining access to housed flocks. Indeed, the UK Government appreciates the risks involved with this, with the provision to enforce movement restrictions and even housing orders in evaluated wild bird risk areas, when netting and other controls to exclude contact with wild birds are not feasible. Control strategies for avian influenza in the UK centre on the exclusion of infection from the country, while for Newcastle disease this is reinforced by a voluntary vaccination policy. General biosecurity measures highlighted earlier are important but specific additional control aspects include:

* exclusion of wild birds to prevent contact with commercial stock;
* clearing up of all feed spillages to discourage wild birds; and
* avoidance of siting free range flocks near open water that may attract wild birds.

Other viruses such as infectious bronchitis (IB) and avian pneumovirus (APV, also known as turkey rhinotracheitis, TRT) are capable of limited spread between sites depending on the proximity of other susceptible flocks or spreading of poultry litter adjacent to free range sites. Vaccination policies for both viruses are commonly practised in housed and free range flocks with live and inactivated programmes for birds in rear, often supplemented by live 'top up' vaccination in lay. Certain viruses are resistant to many disinfectants and can persist in litter, soil or on surfaces in houses or on equipment. These can be a potential source of infection for meat birds and laying stock.

Enteric viruses such as rotavirus, astroviruses and coronaviruses can lead to significant stunting and unevenness in turkeys and in broilers (Saif *et al.*, 1985).

Marek's disease is a herpes virus which can result in tumour formation in broilers and layers. Effective vaccination programmes have been in place for many years in commercial layers and breeders but overwhelming challenge in rear can put pressure on these vaccines and over the years new, more virulent pathotypes have emerged as the virus adapts to these pressures. In laying stock, disease is usually manifested as significant mortality from internal tumours. In housed broilers in the UK, Marek's disease is rare and vaccination is seldom practised. In the USA, where it is more commonplace to reuse litter between crops, there is heavier challenge pressure leading to skin Marek's tumours or internal lesions resulting in poorer performance, mortality and significant downgrading at processing. As a result, it is common for housed broilers in the USA to be vaccinated at day old in the hatchery. In the case of free range broilers in the EU, slower growing breeds are usually used, with killing at much older ages (84 days). This, along with the likely increased challenge pressure through an inability to decontaminate pasture, means that disease is more likely to be seen or for there to be subclinical effects on flock performance. As a result, day-old vaccination of free range broilers against Marek's disease is more common in UK and Europe. In turkeys, infection, although rare, can be seen with significant mortality where turkeys are ranged on land previously populated by broilers. In such situations, Marek's disease vaccination can be used in turkeys but, more usually, measures to prevent turkeys having access to houses or pasture previously used for chickens is advised.

Parasitic diseases

In alternative systems where birds have more access to their own faeces, internal intestinal parasites are always going to be an issue (Permin *et al.*, 1999). Indeed, control of coccidiosis and intestinal worm infestations was one of the main disease drivers for the move from traditional free range systems into cages for layers and, to a lesser extent, broilers and turkeys. In the case of game birds, both breeders and rearing birds may be raised on wire specifically to exclude contact with faeces. In these birds, this is aimed at controlling helminth worms and coccidia but also the protozoal agents of *Hexamita* spp. and *Trichomonas* spp. that can cause significant disease in floor pens on earth.

Coccidiosis

Coccidial oocysts (eggs) are sticky and persistent organisms often present in massive numbers in faeces. The strains involved tend to be species specific to chickens, turkeys or game birds. Explosive outbreaks of disease can occur in pens especially during warm, wet conditions. Control is aimed at reducing challenge pressure and litter for free range systems by maintaining dry, friable litter, reducing stocking density and good pasture management with rotation and resting periods. This can be coupled with either the use of in-feed coccidiostats or, in the case of chickens, a live vaccine. Vaccination is gaining in popularity as a result of its efficacy but also the ability to avoid the use of therapeutic medication. As with all vaccines, they must be properly applied with care and awareness that overwhelming environmental challenge can still overcome vaccination responses.

Hexamita *and* Trichomonas

In the case of the other protozoal infections in game birds and turkeys, *Hexamita* and *Trichomonas*, control is centred on good management, good litter control, optimal stocking rates and the strategic use of therapeutic medication. Climatic conditions, especially rain, can predispose to problems, so siting of pens can be highly significant.

Roundworms

Whenever birds have contact with their own droppings for any length of time, there is a likelihood of exposure to intestinal and other worms (Permin *et al.*, 1999). A survey in 2000 organized by SAC Auchincruive and Janssen Animal Health identified worm infestations in 96% of samples from free range flocks, which would require treatment (D. Cunnah, personal communication, 2000). The infection cycle starts with eggs in droppings that are then picked up by the birds eating or foraging on anything contaminated with such droppings. These eggs hatch in the birds' intestine, maturing into adults ready to pass out in the droppings again. The likelihood and severity of disease problems depends on the balance between challenge and control. Damage associated with worm infestations can include;

- loss of shell colour, strength, yolk colour and egg size;
- poor body weight leading to unevenness;
- poor feed conversion;
- increased cannibalism due to vent pecking; and
- increased risk of egg peritonitis.

The main worms involved are the following.

- Roundworms (*Ascaridia* spp. – the biggest and most common). They are white, up to 5 cm long and may be visible in droppings in heavy infestations.
- Hairworms (*Capillaria* spp.). Much smaller, hair-like worms, rarely visible with the naked eye but can cause significant damage even in only moderate infestations.
- Caecal worms (*Hetarakis* spp.). These populate the lower intestine and caecae, frequently causing little direct damage or harm. However they can carry the protozoal parasite, *Histomonas* spp., which is the cause of histomoniasis (Blackhead).
- Gapeworm (*Syngamus trachea*). These are the cause of 'gapes' and respiratory signs in chickens and game birds.

Worm control is a combination of effective pasture management and rotation with monitoring to identify the incidence and severity of burdens. This may be achieved by:

- targeted and regular worming in the water or in the feed on the basis of previous experience and detection of worm burdens;
- effective paddock rotation to reduce worm build-up and prevent land becoming fowl sick;
- use of well-drained land;
- avoiding access to poached muddy areas;
- use of stones close to pop holes to help clean feet and allow droppings passed there to dry, be broken up and be exposed to ultraviolet in sunlight which is lethal to worms; and
- keeping pasture short, especially close to the house, again to allow ultraviolet in sunlight access to droppings.

Red mite

Red mites are voracious blood-sucking parasites that affect a variety of avian species probably having their origins as a nest mite of wild birds. They are nocturnal in their activities and spend most of their time in dark recesses within the house, only coming out to feed during the night. In heavy infestations there can be over half a million mites for every bird in the house. Infestations can cause physical irritation, reduced in egg production, blood staining of eggs and even mortality through anaemia and secondary diseases. The mites can also harbour a variety of pathogens, including Salmonella, Erysipelas and Newcastle disease virus (Moro *et al.*, 2009). House design providing safe harbourage for the parasites is a significant issue with respect to incidence and control and surveys indicate higher incidence in some alternative systems (Hoglund *et al.*, 1995). Control is through the use of chemical treatments and possibly future vaccines, but a fundamental aspect is in the design of alternative systems to avoid providing places for the mites to hide under perches and in other equipment where they can persist and multiply.

CONCLUSION

The maintenance of poultry health and prevention of disease in commercial poultry flocks is a constant challenge in all production systems. Control of disease in alternative systems can present additional or unique challenges, either through inherent aspects of the specific production system or through the bird's response to such systems. This can be manifest in a number of ways ranging from additional or novel management and husbandry aspects requiring a reappraisal of stockmanship requirements and techniques, or additional biosecurity measures aimed at controlling access to faecal contamination, wild birds, vermin, extremes of weather and a reassessment of cleansing and disinfection procedures. None of these problems are insurmountable in alternative systems if there is good knowledge of the risks presented and the intervention policies likely to avoid the adverse effects of such systems on health and response of the birds to disease. The keys to success include good house design, effective management of the birds' total environment, levels of stockmanship and access to effective veterinary advice through well-planned communication and veterinary health and welfare planning.

REFERENCES

Anderson, D.P., Beard, C.W. and Hansen, R.P. (1964) The adverse effects of ammonia on chickens including resistance to infection with Newcastle disease virus. *Avian Diseases* 8, 369–379.

Blokhuis, H.J., Fiks van Niekerk, T., Bessei, W., Elson, A., Guemene, D., Kjaer, J.B., Maria Levrino, G.A., Nicol, C.J., Tauson, R., Weeks, C.A. and van de Weerd, H.A. (2007) The LayWel project: welfare implications of changes in production systems for laying hens. *World's Poultry Science Journal* 63, 101–113.

Bubier, N.E. and Bradshaw, R.H. (1998) Movement of flocks of laying hens in and out of the hen house in four free range systems. *British Poultry Science* 39, S5–S18.

Burch, D.G.S., Harding, C., Alvarez, R. and Valks, M. (2006) Treatment of a field case of avian intestinal spirochaetosis caused by *Brachyspira pilosicoli* with tiamulin. *Avian Pathology* 35, 211–216.

Burch, D.G.S., Strugnell, B.W., Steventon, A., Watson, E.N. and Harding, C. (2009) Survey of 222 flocks in Great Britain for the presence of *Brachyspira* species and their effects on production. Presented at *5th International Conference on Colonic Spirochaetal Infections in Animals and Humans*, Leon, Spain, 8–10 June 2009, abstract 21, pp. 55–56; available at: http://www.octagon-services.co.uk/articles/poultry/spiroconference.pdf (accessed 10 October 2011).

Christensen, J.P., Dietz, H.H. and Bisgaard, M. (1998) Phenotypic and genotypic characteristics of isolates of *Pasteurella multocida* obtained from backyard poultry and from outbreaks of avian cholera in avifauna in Denmark. *Avian Pathology* 27, 373–381.

Dawkins, M.S., Cook, P.A., Whittingham, M.J., Mansell, K.A. and Harper, A.E. (2003) What makes free range broiler chickens range? *In situ* measurement of habitat preference. *Animal Behaviour* 66, 151–160.

Dawkins, M.S., Donnelly, C.A. and Jones, T.A. (2004) Chicken welfare is influenced more by housing conditions than by stocking density. *Nature* 427, 342–344.

De Reu, K., Grijspeerdt, K., Heyndrickx, M., Zoons, J., de Baere, K., Uyttendaele, M., Debevere, J. and Herman, L. (2005) Bacterial egg shell contamination in conventional cages, furnished cages and aviary housing systems for laying hens. *British Poultry Science* 46, 149–155.

FAWC (2009) *Farm Animal Welfare in Great Britain: Past, Present and Future.* Farm Animal Welfare Council, London; available at: http://www.fawc.org.uk/pdf/ppf-report091012.pdf (accessed 9 September 2010).

Fossum, O., Jansson, D.S., Etterlin, P.E. and Vagsholm, I. (2009) Causes of mortality in laying hens in different housing systems in 2001 to 2004. *Acta Veterinaria Scandinavica* 51, 3; doi:10.1186/1751-0147-51-3.

Germain, K., Juin, H. and Lessire, M. (2010) Effect of the outdoor run characteristics on growth performance in broiler organic production. In: Duclos, M. and Nys, Y. (eds) *Proceedings of the 13th European Poultry Conference*, Tours, France, 23–27 August 2010. World's Poultry Science Association, Nouzilly, France, p. 318.

Gregory, N.G., Wilkins, L.J., Eleperuma, S.D., Ballantyne, A.J. and Overfield, N.D. (1990) Broken bones in domestic fowls: effect of husbandry system and stunning method in end of lay hens. *British Poultry Science* 31, 59–69.

Griffiths, I.B., Hunt, B.W., Lister, S.A. and Lamont, M.H. (1987) Retarded growth rate and delayed onset of egg production associated with spirochaete infection in pullets. *The Veterinary Record* 121, 35–37.

Hartung, J. (1994) The effect of airborne particulates on livestock health and production. In: Dewi, I., Axford, R.F.E., Fayez, I.M.M. and Omed, H.M. (eds) *Pollution in Livestock Production Systems*. CABI Publishing, Wallingford, UK, pp. 55–69.

Hegelund, L., Sorensen, J.T., Kjaer, J.B. and Kristensen, I.S. (2005) Use of the range area in organic egg production systems: effect of climatic factors, flock size, age and artificial cover. *British Poultry Science* 46, 1–8.

Hoglund, J., Nordenfors, H. and Uggla, A. (1995) Prevalence of poultry red mite in different types of production systems for egg layers Sweden. *Poultry Science* 74, 1793–1798.

Jones, T.A., Feber, R., Hemery, G., James, K., Cook, P., Lambeth, C. and Dawkins, M.S. (2007) Welfare and environmental benefits of integrating commercially viable free-range broiler chickens into newly planted woodland: a UK case study. *Agricultural Systems* 94, 177–188.

Lister, S.A. (2006) Gut problems: the field experience and what it means to the poultry farmer. In: Perry, G.C. (ed.) *Avian Gut Function in Health and Disease*. CABI Publishing, Wallingford, UK, pp. 350–360.

Lister, S.A. (2008) Biosecurity in poultry management. In: Pattison, M., McMullin, P.F., Bradbury, J.M. and Alexander, D.J. (eds) *Poultry Diseases*, 6th edn. Saunders/Elsevier Ltd, Philadelphia, Pennsylvania, pp. 48–65.

Moro, C.V., De Luna, C.J., Tod, A., Guy, J.H., Sparagano, O.A.E. and Zenner, L. (2009) The poultry red mite (*Dermanyssus gallinae*): a potential vector of pathogenic agents. *Experimental and Applied Acarology* 48, 93–104.

Norton, R.A. (1997) Avian cellulitis. *World's Poultry Science Journal* 53, 337–349.

Norton, R.A., Bilgili, S.E. and McMurtrey, B.C. (1997) A reproducible model for the induction of avian cellulitis in broiler chickens. *Avian Diseases* 41, 422–428.

Permin, A., Bisgaard, M., Frandsen, F., Pearman, M., Kold, J., and Nansen, P. (1999) Prevalence of gastrointestinal helminths in different poultry production systems. *British Poultry Science* 40, 439–443.

Randall, C.J., Meakins, P.A., Harris, M.P. and Watt, D.J. (1984) A new skin disease of broilers? *The Veterinary Record* 114, 246.

Saif, L.J., Saif, Y.M. and Theil, K.W. (1985) Enteric viruses in diarrheic turkey poults. *Avian Diseases* 29, 798–811.

Sandilands, V., Sparks, N., Wilson, S. and Nevison, I. (2005) Laying hens at depopulation: the impact of the production system on bird welfare. *British Poultry Abstracts* 1, 23–24.

SCAHAW (2000) *The Welfare of Chickens Kept for Meat Production (Broilers)*. Report of the Scientific Committee on Animal Health and Welfare. Adopted 21 March 2000. European Commission, Brussels.

Vucemilo, M., Vinkovic, B., Matkovic, K., Stokovic, I., Jaksic, S., Radovic, S., Granic, K. and Stubian, D. (2010) The influence of housing systems on the air quality and bacterial eggshell contamination of table eggs. *Czech Journal of Animal Science* 55, 243–249.

Wathes, C.M. (1994) Air and surface hygiene. In: Wathes, C.M. and Charles, D.R. (eds) *Livestock Housing*. CABI Publishing, Wallingford, UK, pp. 123–148.

CHAPTER 5

Production Systems for Laying Hens and Broilers and Risk of Human Pathogens

S. Van Hoorebeke, J. Dewulf, F. Van Immerseel and F. Jorgensen

ABSTRACT

There is evidence that the type of production systems used for laying hens and broilers can affect the likelihood of the chickens being colonized by human pathogens. The most significant public health risk associated with layers is transmission of *Salmonella* to humans via eggs. Based on experimental and epidemiological data, however, it seems unlikely that the move from conventional cages to enriched cages and non-cage systems will result in an increase in the prevalence and/or shedding of *Salmonella* in laying hen flocks. Studies on broiler chickens suggest that free range and organic flocks are significantly more likely to be positive for *Campylobacter* at slaughter in comparison with first depopulated batches of conventionally reared broilers. Data in relation to broiler rearing system and the likelihood of birds being infected by other pathogens including *Salmonella* are scarce but there is no significant evidence to suggest that organic and free range broilers are more likely to be infected with *Salmonella* than are conventionally reared ones. There is some evidence for a higher level of antibiotic-resistant *Campylobacter* strains in conventional broilers compared with those found in organic ones and other antibiotic-resistant human pathogens are also more commonly isolated from conventionally reared broilers.

INTRODUCTION

This chapter focuses on the transmission of human pathogens from chickens in relation to production systems for laying hens and broilers. While there is an assumption that chickens kept with access to range have a higher risk of being infected with human pathogens, some consumers may perceive that organic and free range chickens are less likely to shed pathogens in comparison with chickens held in conventional production systems. This perception may relate to beliefs e.g. that higher stocking densities are associated with unsanitary conditions or substandard diets (Bailey and Cosby, 2005).

Transmission of *Salmonella* to humans via eggs is recognized as a significant public health risk associated with layers and this is discussed in the following section. Salmonellosis in humans is usually characterized by diarrhoea, nausea, stomach cramps, fever, headache and sometimes vomiting. Most infections are self-limiting, lasting on average from 4 to 7 days, but they can be prolonged and in some patients the infection may be more serious; moreover, long-term sequelae such as reactive arthritis are known. While laying hens can be colonized with human pathogens other than *Salmonella* such as *Campylobacter* (Schwaiger *et al.*, 2008), there are very few reports concerning these, reflecting the lower human health risk thought to be associated with their transmission from this reservoir. Campylobacteriosis in humans is usually characterized by fever and headache followed by severe abdominal pain and diarrhoea which may be bloody. Most cases settle within one week but infrequently complications such as reactive arthritis and neurological disorders occur. Much recent emphasis has been placed on the human health burden caused by *Campylobacter* infections; these are thought to be the main human health risk associated with broilers and this is discussed in the latter part of this chapter.

THE PREVALENCE OF *SALMONELLA* IN DIFFERENT LAYING HEN HOUSING SYSTEMS

Inspired by a growing body of consumer aversion to eggs produced by laying hens housed in conventional cages and evidence concerning animal welfare, the European Union (EU) adopted Council Directive 1999/74/EC, stating that from 1 January 2012 onwards the housing of laying hens in all EU member states must be restricted to enriched cages and non-cage systems (Appleby, 2003). The housing in enriched cages implies that the hens must have at least 750 cm^2 of floor space per hen, a nest, perches and litter. The non-cage systems consist of a single- or multi-level indoor area that may be combined with covered ('winter garden') or uncovered ('free range') outdoor facilities (EFSA, 2005; LAYWEL, 2006) (see also chapters 11 and 12, this volume). In non-caged single-level systems such as floor raised or 'barn' systems, the maximum stocking density must not exceed 9 hens m^{-2} usable area. The influence of these alternatives for conventional cages on the prevalence of zoonotic pathogens has been the topic of debate. One of the best studied pathogens is *Salmonella*, a bacterium that is worldwide still a very important cause of human disease (EFSA, 2009). Eggs and egg-related products are the main sources of infection of humans with *Salmonella* Enteritidis (Crespo *et al.*, 2005; De Jong and Ekdahl, 2006; Delmas *et al.*, 2006).

The aim of this section is to present an overview of the currently available information on the effects of the housing system on the occurrence and epidemiology of *Salmonella* in laying hen flocks.

Effect of the housing system on *Salmonella* prevalence

The prevalence of *Salmonella* in different laying hen housing systems has been described in a number of observational and experimental studies. Because of large differences in sample size and methodology used, the conclusions of these studies differ greatly, ranging from a preventive effect of the conventional cage system over no influence up to a higher risk of *Salmonella* in conventional cages in comparison with non-cage systems (Table 5.1). The estimated odds ratios were either available or calculated from data presented in the papers. The majority of the studies indicate that housing of laying hens in conventional cages significantly increases the risk of detecting *Salmonella* compared with the housing in non-cage systems. However, the observed influence of the housing type does not necessarily mean that there is a causal relationship between the housing system and the level of *Salmonella* infection and excretion.

One study found a protective effect of conventional cage systems but this was only observed on farms with hens of different ages (Mollenhorst *et al.*, 2005). When all the hens were of the same age, the protective effect was restricted to conventional cage systems with wet manure, whereas flocks housed in conventional cage systems with dry manure had a higher chance of infection with *S.* Enteritidis compared with deep litter systems. Second, the

Table 5.1. Overview of observational studies evaluating the effect of layer housing system on the prevalence of *Salmonella*.

Comparison	No. of flocks	OR[a]	95% CI	Comment	Reference
Cage versus Deep litter	1642	0.48	N/A[b]	Serology	Mollenhorst *et al.* (2005)
Cage versus Free range	34	0.61	0.2–2.3	Bacteriology	Schaar *et al.* (1997)
Cage versus Aviary	8	1.28	0.5–3.2	Bacteriology	Pieskus *et al.* (2008b)
Cage versus Non-cage	292	4.69	1.9–11.9	Bacteriology	Van Hoorebeke *et al.* (2010c)
Cage versus Non-cage	329	2.34	1.4–3.9	Bacteriology (EFSA)	Methner *et al.* (2006)
Cage versus Non-cage	3768	5.12	4.1–6.5	Bacteriology (EFSA)	EFSA (2007)
Cage versus Free range	148	10.27	2.1–49.6	Bacteriology (EFSA)	Namata *et al.* (2008)
Cage versus Floor raised	148	20.11	2.5–160.5	Bacteriology (EFSA)	Namata *et al.* (2008)
Cage versus Floor raised	519	35.1	12.2–101.1	Bacteriology (EFSA)	Huneau-Salaün *et al.* (2009)

OR, odds ratio; CI, confidence interval; N/A, not available; EFSA, European Food Safety Authority.
[a]Estimated ORs presented or calculated from data available.
[b]Could not be calculated due to lack of data.

categorization of the sampled flock into S. Enteritidis-positive or -negative was based on serology rather than on the bacteriological isolation of the pathogen. Prior to this study, Garber et al. (2003) found that flocks that had been primarily floor reared as pullets were more likely to be positive for S. Enteritidis during their productive lifespan than were flocks that had been cage reared. In other studies, no significant influence of the housing system on the prevalence of Salmonella could be demonstrated based on either bacteriology (Schaar et al., 1997; Pieskus et al., 2008b) or serology (Hald et al., 2002). Mølbak and Neimann (2002) found that eggs from conventional cages yielded a higher risk for infection of humans with S. Enteritidis than eggs from non-cage housing systems, a theory which was later confirmed by several other studies (Methner et al., 2006; EFSA, 2007; Namata et al., 2008; Huneau-Salaün et al., 2009; Van Hoorebeke et al., 2010c). The sampling for the publications by Methner et al. (2006), Namata et al. (2008) and Huneau-Salaün et al. (2009) were all performed in 2004–2005 in the framework of the European baseline study on the prevalence of Salmonella in laying hen flocks (EFSA, 2007). The aim of that study was to determine the prevalence of Salmonella spp. in laying hen flocks in all European member states and to determine risk factors for the presence of Salmonella on laying hen farms. Both on the EU level and on the level of the individual member states the housing in conventional cages turned out to be a risk factor. The study of Van Hoorebeke et al. (2010c) was specifically designed to investigate the influence of the housing type on the prevalence of Salmonella on laying hen farms. For this purpose five main housing types, i.e. conventional cages, aviaries, floor raised systems, free range systems and organic systems, were sampled in equal proportions. In total, 292 laying hen flocks from as many different laying hen farms in Belgium, Germany, Greece, Italy and Switzerland were sampled. Through such studies it appears that a number of laying hen husbandry characteristics may be related to both the housing system and the probability of a Salmonella infection, and these are discussed in the following section.

Factors related to housing system and *Salmonella* prevalence

Farm and flock size

The number of flocks on the farm and the number of hens in a flock have been identified as risk factors for Salmonella infections in laying hens (Heuvelink et al., 1999; Mollenhorst et al., 2005; EFSA, 2007; Snow et al., 2007; Carrique-Mas et al., 2008; Huneau-Salaün et al., 2009) and several studies have shown that conventional cage farms are in general larger farms, not only with more hens per flock but also with more flocks on the farm (EFSA, 2007; Carrique-Mas et al., 2008; Van Hoorebeke et al., 2010c). This could be one of the factors explaining why conventional cage farms are more frequently positive for Salmonella than non-cage housing systems. The presence of multiple flocks on one farm may enhance cross-contamination from one flock to another, especially when the different flocks and laying hen houses on the farm are linked through egg conveyor belts, feed pipes, passageways, etc. (Carrique-

Mas *et al.*, 2008). Furthermore, as is often the case on farms flocks, not all the hens are of the same age. Multistage produ been identified as a risk factor for *Salmonella* in laying hens *al.*, 2005; Wales *et al.*, 2007; Carrique-Mas *et al.*, 2008; Huneau *al.*, 2009).

Stocking density

Stocking density is often related to both the housing type and the flock size. For many infectious diseases in production animals it has been demonstrated that a higher stocking density increases the prevalence of disease and the ease of spread (Dewulf *et al.*, 2007). Possibly the high density of laying hens in conventional cages and in connection with this the large volume of faeces and dust increases the incidence of *Salmonella* infections in this type of housing (Davies and Breslin, 2004). High stocking densities could also indirectly interact with *Salmonella* infections due to stress in birds as discussed below.

Stress

The immunosuppressive effect of stress in laying hens (El-Lethey *et al.*, 2003; Humphrey, 2006) can have negative consequences with respect to *Salmonella* infection and shedding. The move from the rearing site to the egg producing plant (Hughes *et al.*, 1989), the onset of lay (Jones and Ambali, 1987; Humphrey, 2006), the final stages of the production period, thermal extremes (Thaxton *et al.*, 1974; Marshally *et al.*, 2004) and transportation to the slaughterhouse (Beuving and Vonder, 1978) are all moments in the laying hen's life where the bird is subjected to stress. In some European countries induced moulting is practised. The effects on *S.* Enteritidis infections during moult have been extensively studied: moulted hens shed more *S.* Enteritidis in their eggs and faeces (Holt, 2003; Golden *et al.*, 2008) and have higher levels of internal organ colonization (Holt *et al.*, 1995). Moulting causes the recurrence of previous *S.* Enteritidis infections (Holt and Porter, 1993). There are some contradictory data on the influence of the housing type on the stress levels in laying hens. Some studies suggest that laying hens have less stress in conventional cages (Craig *et al.*, 1986; Koelkebeck *et al.*, 1987) whereas other authors state that hens housed in non-cage systems experience less stress (Hansen *et al.*, 1993; Colson *et al.*, 2008). With regard to the housing system, the age of the hens (Singh *et al.*, 2009) and the breed of the hens could also play a role: certain hen breeds exhibit significantly higher stress responses when raised in deep litter versus free range systems, compared with other breeds (Campo *et al.*, 2008).

Carry-over of infections and age of the infrastructure

An adequate cleaning and disinfection policy is essential in modern laying hen husbandry since it has been stated that the major part of *Salmonella* infections on laying hen farms is not newly introduced on the farm but is the result of re-introduction of the pathogen from the farm's environment (van de Giessen *et al.*, 1994; Gradel *et al.*, 2004; Carrique-Mas *et al.*, 2009a). However, in particular conventional cage systems are thought to be extremely hard to clean and disinfect sufficiently because of the restricted access to cage interiors, feeders, egg belts, and so forth (Davies and Breslin, 2003; Carrique-Mas *et al.*, 2009b).

Besides the specific difficulties in cleaning and disinfecting different housing systems, the age of the current infrastructure might also play a role. Due to the wear of the materials and the inherent difficulties to thoroughly clean and disinfect them, older equipment increases the risk for *Salmonella*. At the present time, most conventional cages are older than floor raised, free range and organic installations (Van Hoorebeke *et al.*, 2010c). This finding could also contribute to the fact that farms with conventional cages are more frequently found positive for *Salmonella*.

Pests

The role of rodents, flies and beetles as vectors in the transfer of *Salmonella* has been extensively studied (Guard-Petter, 2001; Davies and Breslin, 2003; Kinde *et al.*, 2005; Carrique-Mas *et al.*, 2009a). It has been suggested that non-cage housing systems present a less attractive environment to these pests because laying hens can interfere more with their movements since the birds are not restrained to cages (Carrique-Mas *et al.*, 2009a). Another important pest in laying hens' houses is the poultry red mite (*Dermanyssus gallinae*). It has been shown under experimental conditions that mites could play a role in the persistence of *Salmonella* in laying hens, either by transferring the bacterium from hen to hen or by hens consuming contaminated mites leading to a persisting infection (Valiente Moro *et al.*, 2007, 2009). Yet under field conditions no *Salmonella* were detected on or in red mites sampled from herds in the Netherlands (Van Hoorebeke *et al.*, 2010b). Nevertheless mass red mite infestations can lead to immunosuppression (Kowalski and Sokol, 2009), increasing the susceptibility for infections. This could also be the case with gastrointestinal helminths: the prevalence of helminth infections in free range and deep litter systems can be higher than in conventional cage systems (Permin *et al.*, 1999; Marcos-Atxutegi *et al.*, 2009) and this may make birds more susceptible to *Salmonella* infections. However, in well managed non-cage systems birds are de-wormed regularly and under such circumstances poor body condition of the birds is much less likely.

Finally, a recent review concluded that the prevalence of *Salmonella* in laying hen flocks is unlikely to increase when moving from conventional cage systems to non-cage housing systems (Van Hoorebeke *et al.*, 2010b).

Presence of *Salmonella* serotypes other than *S.* Enteritidis in outdoor production systems

Since *S.* Enteritidis and *Salmonella* Typhimurium are responsible for the lion's share of human salmonellosis cases in Europe and North America (CDC, 2007;

EFSA, 2009), so far the focus of *Salmonella* control programmes has been mainly on these two serovars. Nevertheless some differences in epidemiology are reported between these two serotypes. Because S. Typhimurium is much more common in wildlife, pigs and cattle it has been stated that free range layer flocks will be at greater risk of becoming infected with S. Typhimurium than flocks housed in systems without an outdoor run (Carrique-Mas *et al.*, 2008). However, this could not be confirmed in the EFSA baseline study (EFSA, 2007) or in another large-scale study in Belgium, Germany, Greece, Italy and Switzerland (Van Hoorebeke *et al.*, 2010c).

BROILER SYSTEMS AND HUMAN PATHOGENS

Campylobacter, *Salmonella*, *Listeria* and verotoxin-producing *Escherichia coli* (VTEC) can be present in the gastrointestinal tract of poultry and, as such, this reservoir constitutes a risk for human infection (Esteban *et al.*, 2008). Transmission from the broiler reservoir via meat is thought to be the most important route for human infection but other routes may also be significant e.g. occupational exposure or contact with contaminated recreational waters.

Isolation of VTEC from chickens is very rare, however, and the broiler reservoir is not thought to contribute significantly to human VTEC infections (EFSA, 2010c). Isolation of *Listeria monocytogenes* from broilers is less common than isolation from sheep, goats and cattle and broiler meat is not thought to represent a major reservoir for human *Listeria* infections (Chinivasagam *et al.*, 2010; EFSA, 2010c). Very little has been published on the prevalence of either of these human pathogens in relation to different broiler production systems. In one study of 60 free range flocks from Spain, *L. monocytogenes* was detected in nine and *E. coli* O157 in none (Esteban *et al.*, 2008).

The presence of *Campylobacter* spp. and *Salmonella* spp. in poultry meat is of significant public health concern. Contaminated poultry meat is recognized as an important risk factor for human *Campylobacter* infection in industrialized countries and many studies have investigated factors affecting the likelihood of flocks being colonized with *Campylobacter*. Poultry meat also remains an important source for human *Salmonella* infection in some countries despite recent intervention (EFSA, 2009).

While the large majority of fresh poultry meat at retail sale in the UK originates from birds reared inside houses with a controlled environment, there has been an increasing demand for free range and organic chicken meat. It is difficult to obtain accurate data for the number of broiler chickens that are reared with access to pasture but they may represent 2–3% of the chickens slaughtered annually in the UK with <1% representing organic birds. Free range and organic chickens in the UK are sourced from slower growing breeds than those used for conventional production and are also fed different diets using less energy-dense feed. There are strict standards governing the use of antimicrobials and feed for birds reared to certified organic standards. Free range and organic birds are also usually slaughtered later than conventionally reared broilers at around 56 and 73 days compared with 38 days of age, respectively. Initially, however, free range and organic birds are reared inside houses until around 25 days of age after which they have access to pasture (Allen *et al.*, 2011).

The increase in consumption of free range and organic chicken meat has prompted studies to investigate whether the type of rearing systems used can influence the likelihood of birds being colonized with *Campylobacter* spp. and *Salmonella* spp. The extent to which birds are exposed to these pathogens via the environment (e.g. by access to pasture), stocking density and susceptibility to infection (e.g. affected by differences in immunity, breed, gut microflora and stress levels) has been suggested to influence the chances of broiler chickens being positive for these pathogens at slaughter (Rivoal *et al.*, 2005; Humphrey, 2006). In the following sections studies concerned with this are presented and discussed.

Campylobacter spp. in relation to broiler systems

Colonization of conventionally reared chickens with *Campylobacter* spp. has been widely investigated but it is also well recognized that organic and other free range flocks reared throughout the world can be colonized with *Campylobacter* spp. (Kazwala *et al.* 1993; Rivoal *et al.* 1999; Uyttendaele *et al.*, 1999; Heuer *et al.*, 2001; Van Overbeke *et al.*, 2006). A recent systematic review concluded that the prevalence of *Campylobacter* spp. was higher in organic broiler chickens at slaughter compared with conventionally reared ones, but no difference was found between types of chicken meat at the retail level (Young *et al.*, 2009). In UK retail studies, *Campylobacter* spp. have been isolated from conventional, free range and organic chicken meat but valid comparisons of prevalences were hampered by small sample sizes (FSA, 2003, 2009; CLASSP, 2007). Nevertheless, *Campylobacter* spp. were isolated from 43, 51 and 60% of conventional, free range and organic chicken meat samples, respectively (FSA, 2009).

In a recent EU-wide survey from 2008 the presence of *Campylobacter* spp. was investigated in broiler batches and carcasses from conventional and free range flock types (EFSA, 2010b). While no statistically significant difference was found using a multivariate approach, a higher prevalence was generally found in flock types with access to the outside than in conventionally housed flocks (Fig. 5.1). The report questioned whether too few data in the non-conventional flock type categories may have affected the outcome of the statistical results. A number of other investigations also found that conventionally reared flocks tended to be less likely to be colonized than flocks reared with access to pasture (Table 5.2).

In a study from the UK a lower prevalence of *Campylobacter* spp. colonization was found in conventional than in organic and free range flocks reared within one company in 2004 (Allen *et al.*, 2011). A study carried out on flocks reared in Denmark also showed that conventionally reared flocks were significantly less likely to be colonized at slaughter but there was no difference in

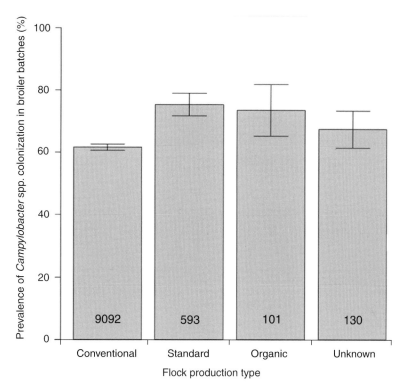

Fig. 5.1. Prevalence of *Campylobacter*-colonized broiler batches by flock production type in the European Union in 2008. (Data taken from EFSA, 2010b; the 'standard' flock production type is standard free range.)

Table 5.2. Prevalence of *Campylobacter* in relation to broiler rearing system, showing (where available) the number of flocks testing positive for *Campylobacter*/number of flocks examined.

Conventionally housed flocks	Free range flocks	Organic flocks	Location of flocks	Reference
22/40 (55%)	19/20 (95%)	21/21 (100%)	UK	Allen *et al.* (2011)
29/79 (37%)	–[a]	22/22 (100%)	Denmark	Heuer *et al.* (2001)
3/11 (27%)	–	7/9 (78%)	Belgium	Van Overbeke *et al.* (2006)
–	12/18 (67%)	–	Belgium	Vandeplas *et al.* (2010)
230/403 (57%)	50/62[b] (80%)	–	France	Avrain *et al.* (2001)
–	52/73 (71%)	–	France	Huneau-Salaün *et al.* (2007)
–	(77%)	–	France	Rivoal *et al.* (1999)
–	46/60 (77%)	–	Spain	Esteban *et al.* (2008)
51/125 (40%)	12/19[c] (63%)		Germany	Näther *et al.* (2009)

[a]No data presented.
[b]These flocks were the Label Rouge type.
[c]Combined data for flocks from seven free range and three organic farms.

the proportion of birds colonized within conventional (68%) or organic flocks (65%) (Heuer et al., 2001). The number of Campylobacter spp. cells in caecal contents collected at slaughter from different flock types was not significantly different between conventionally reared birds ($\log_{10}$ 6.2) and birds reared with access to pasture ($\log_{10}$ 6.7 and 6.5 for free range and organic birds, respectively) (Allen et al., 2011). Sources of Campylobacter for organic and free range flocks include the environment, e.g. ante areas and ground between houses (Allen et al., 2011), and cows (Zweifel et al., 2008). Campylobacter strains may persist for some time in the environment and in one study they were isolated up to 24 days after flocks had been removed from pasture (Morris et al., 2009). One Campylobacter strain identified as one of the colonizing strains in an extensively reared flock was isolated from pasture land 9 days after the flock had been removed for slaughter (Morris et al., 2009). In this study the longer rearing time was not thought to be the most important factor in causing the differences in colonization rate as flocks were colonized early in the rearing period (Allen et al., 2011). Twice-weekly examination of the Campylobacter status showed that 14 flocks reared according to organic standards were colonized on average at 14.2 days of age whereas 14 free range flocks were colonized on average at 31.6 days of age (Allen et al., 2011). Thus, organic birds were often colonized while still in brooding sheds but the free range flocks were mostly colonized only when put out on pasture. The study concluded that husbandry practices and farm conditions were the most likely factors explaining the more frequent Campylobacter colonization of organic and other free range flocks. Further investigation of organic flocks concluded that the earlier colonization of organic flocks was likely to relate to husbandry factors and house conditions, e.g. in houses in good conditions with hygiene barriers or with anterooms with foot dips, birds were more likely to remain negative until the end of the brooding period.

Both Campylobacter jejuni and Campylobacter coli can be isolated from conventional as well as extensively reared flocks. A large majority (88%) of 64 free range flocks reared on two farms near Oxford in 2003 were colonized with both C. jejuni and C. coli although in a few flocks only C. jejuni or C. coli strains were detected and overall ~50% of strains isolated were C. coli (Colles et al., 2008). Similar proportions of C. jejuni and C. coli were also recovered from 28 organic and free range flocks reared within another UK company in 2004 (Allen et al., 2011). These studies also found that within one flock a number of different Campylobacter genotypes could usually be identified. Studies investigating Campylobacter genotypes in conventionally reared broilers in the UK could suggest that these flocks are colonized with a more limited number of genotypes (Bull et al., 2006). In a recent EU survey where the majority of flocks were conventionally reared, C. jejuni and C. coli were found in 74 and 26% of UK flocks, respectively (Table 5.3). Taken together these data could suggest that C. jejuni may be more common in conventionally reared broiler chickens than in organic or free range chickens in the UK. Studies from the UK have shown that while C. jejuni is more commonly causing human infection, C. coli is responsible for a not insubstantial proportion of human Campylobacter cases (Sheppard et al., 2009).

Table 5.3. Proportions of *Campylobacter jejuni* and *Campylobacter coli* identified in broiler chickens in the European Union[a].

Campylobacter species isolated	Area	% of samples with species (no. of samples with species/total no. of samples with *C. jejuni* or *C. coli*)	
		In broiler batches	In carcasses
C. jejuni	UK	74 (225/304)	72 (266/369)
	France	43 (180/422)	72 (304/422)
	Denmark	17 (69/396)	28 (112/396)
	EU (average)	59 (3060/5184)	63 (3830/6040)
C. coli	UK	26 (79/304)	28 (103/366)
	France	39 (163/422)	54 (226/422)
	Denmark	1.8 (7/396)	2.8 (11/396)
	EU (average)	41 (2124/5184)	37 (2210/6040)

[a]Data from EFSA (2010b).

In 26 free range flocks studied in 2003–2006 in France, *C. jejuni* was detected in nine flocks and *C. coli* in four flocks, while the remaining 13 were colonized in almost equal measure by *C. jejuni* and *C. coli* strains (Denis *et al.*, 2008). In an earlier study from France, the large majority of free range flocks were colonized by both *C. jejuni* and *C. coli* but on one of the seven farms studied birds were exclusively colonized with *C. coli* (Rivoal *et al.*, 2005). In a study of five organic flocks reared in the USA, four were colonized with both *C. jejuni* and *C. coli* and one exclusively with *C. jejuni* (Luangtongkum *et al.*, 2006). On the contrary, only *C. jejuni* isolates were detected in the large majority of flocks from a Danish study regardless of rearing system (86, 86 and 91% of standard, free range and organic flocks, respectively, while only *C. coli* was detected in 10, 14 and 4.5%, respectively) (Heuer *et al.*, 2001). It is possible that this could reflect differences in the proportion of the flock types reared in different countries; for instance, in France a much larger proportion of birds are reared as free range compared with e.g. Denmark, and this could affect the relative extent to which birds are exposed to these *Campylobacter* species.

Salmonella spp. in relation to broiler systems

Broiler meat and associated products were reported to be the fifth most frequent cause of food-borne *Salmonella* outbreaks in the EU in 2007 (EFSA, 2009). EU data from 2008 indicated that 2.8% of broiler flocks were positive for *Salmonella* (EFSA, 2010c). EU Community targets for the reduction of *Salmonella* in broiler flocks have been laid down in Community legislation (EC, 2003, 2007) and the first year of implementation of mandatory control programmes by member states was in 2009. A reduction in flock prevalence has been achieved in some member states already (EFSA, 2010c). Despite this, at the EU level 15.7% of carcasses were positive for *Salmonella* in the EU baseline survey from 2008 (EFSA, 2010a).

It is recognized that both conventionally reared and free range broilers can be colonized with *Salmonella* (McCrea *et al.*, 2006; EFSA, 2009). Data from UK retail surveys of chicken have shown that *Salmonella* can be isolated from standard, free range and organic chicken meat (CLASSP, 2007; FSA, 2009). Various risk factors have been identified for *Salmonella* infection on broiler farms including farm management factors, biosecurity, flock size, age of chickens and season (Namata *et al.*, 2009; Le Bouquin *et al.*, 2010). In Spain, 41% of 388 broiler batches most likely reared in conventional broiler production systems were positive for *Salmonella* in 2006 (EFSA, 2009) while 14% of carcasses from Spain were positive in the 2008 EU survey (EFSA, 2010a). In a report from northern Spain however, *Salmonella* was detected in only one of 60 free range broiler flocks (originating from 34 farms) examined between 2004 and 2006 (Esteban *et al.*, 2008). Data from the Netherlands and Italy did not find any evidence for higher *Salmonella* infection rates in organic flocks, and four of 108 (3.7%) organic flocks studied from the Netherlands were infected with *Salmonella* (Pieskus *et al.*, 2008a).

From 2006 to 2008, between 0.8 and 1.6% of broiler batches from the Netherlands were reported to be positive for *Salmonella* (EFSA, 2010c), but 10% (43/429) of carcasses were positive for *Salmonella* in survey data from 2008 (EFSA, 2010a). There was no significant difference in the prevalence of *Salmonella* in a small study of organic and conventional broilers reared in Belgium in 2004 (Van Overbeke *et al.*, 2006). A study of broiler flocks reared in Germany indicated that the incidence of *Salmonella* was lower on organic than on conventional broiler farms (Woll Reuter *et al.*, 2002). In the USA, a recent study suggested that there was a higher prevalence of *Salmonella* in samples collected from four farms rearing conventional broilers than in samples collected from three farms rearing organic birds (Alali *et al.*, 2010). However, older US studies found that the *Salmonella* prevalence in organic chickens was higher than in conventionally raised ones at the farm level (Bailey and Cosby, 2005) and in chicken meat at retail (Cui *et al.*, 2005).

In summary, there are few current studies with sufficient data to make firm conclusions regarding the role of broiler production system for flock infection with *Salmonella*. Nevertheless, the studies which have been done imply there is little evidence to suggest that there is any significant difference in *Salmonella* infection rates between conventionally reared broilers and broilers reared with access to range.

ANTIBIOTIC RESISTANCE AND REARING SYSTEM

There are concerns related to the level of antibiotic-resistant bacteria in the food chain and human infections with antibiotic-resistant strains are likely to present a more serious problem as there may be a reduced efficacy of antimicrobial drugs. It is generally accepted that antimicrobial usage is a risk factor for the development of antibiotic resistance and it could be hypothesized that there would be a lower level of resistance in rearing systems with a stricter usage policy. However, this is complicated by other factors including disease prevalence, husbandry, breed, and varying exposure to multidrug-resistant bacterial strains (Harada and Asai, 2010).

In a recent study the potential association between laying hen housing type and multidrug resistance (MDR) in *E. coli* and *Enterococcus faecalis* isolates was studied (Van Hoorebeke *et al.*, 2010a). There was no difference in the level of MDR *E. coli* strains in free range compared with conventional cage systems, but MDR *E. faecalis* levels were significantly lower in free range than in conventional cage systems. On laying farms in Switzerland results indicated that resistance rates of bacteria isolated from organic systems were lower than in those isolated from conventional ones, particularly for strains of *E. coli* (Schwaiger *et al.*, 2008, 2010). Differing outcomes of such comparisons may be explained by factors not captured in such studies such as antibiotic usage and whether farms recently switched from a conventional to a non-conventional system. Increased exposure to a wider range of organisms in free range systems has been suggested to decrease the likelihood of birds being colonized by antibiotic-resistant strains (Van Hoorebeke *et al.*, 2010a). On the other hand if there is a higher incidence of disease in laying hens in alternative housing systems, as some studies have indicated, this could affect usage and hence levels of resistance to antimicrobials (Van Hoorebeke *et al.*, 2010a).

In a recent review, bacterial isolates from conventional broilers exhibited more antimicrobial resistance and MDR than isolates from organic production. *Campylobacter* isolates from conventional retail chicken, in particular, were more likely to be resistant to ciprofloxacin (odds ratio=9.6, 95% confidence interval—6 16) (Young *et al.*, 2009). Conventionally reared chickens harboured more antimicrobial-resistant *Campylobacter* strains than chickens reared to organic standards in a US study of isolates from five organic and ten conventional broiler farms. This was particularly true for fluoroquinolone resistance but also significant for erythromycin resistance (Luangtongkum *et al.*, 2006). In another study from the USA investigating chickens at retail sale *Campylobacter* strains tended to be more resistant to antibiotics if isolated from conventional chickens compared with organic chickens (Cui *et al.* 2005). In a study of French free range and standard chickens *C. jejuni* isolates from standard chickens were more often resistant to tetracycline and ampicillin while *C. coli* strains from standard chickens were more often resistant to erythromycin and tetracycline (Desmonts *et al.*, 2004). Two US studies found that the prevalence of *Salmonella* with resistance to antibiotics was higher in conventionally reared broiler flocks than in organic broiler flocks (Siemon *et al.* 2007; Alali *et al.*, 2010). However, another US study showed no evidence for this (Lestari *et al.*, 2009).

The presence of MDR *E. coli* strains was significantly higher ($P<0.0001$) in conventional (61 samples) compared with organic (55 samples) poultry meat samples in a study from Spain (Miranda *et al.*, 2008). Only resistance to doxycycline in *Staphylococcus aureus* and *L. monocytogenes* isolates was, however, significantly higher in strains from conventional compared with organic poultry meat samples.

CONCLUSIONS

Based on experimental and epidemiological data, it seems highly unlikely that the move from conventional cages to enriched cages and non-cage systems will result in an increase in *Salmonella* infections and shedding in laying hen flocks. The underlying reasons causing the prevalence of *Salmonella* to be generally lower in alternative housing systems are not known but may relate to factors such as the age of the infrastructure, flock size and sanitary conditions.

The data from studies on broiler chickens suggest that free range and organic flocks are significantly more likely to be positive for *Campylobacter* at slaughter, certainly in comparison with first depopulated batches of conventionally reared broilers. The reason for this difference may relate to increased exposure of extensively reared birds to sources associated with the longer outdoor rearing period compared with birds reared indoors throughout their life. Data in relation to rearing system and the likelihood of birds being colonized by *Salmonella* and other human pathogens are scarce, but nevertheless there is little evidence that organic and free range broilers are substantially more infected with *Salmonella* than are conventionally reared broiler birds.

Recent studies suggest that bacteria isolated from conventional animal production in general seem to exhibit a higher prevalence of resistance to antimicrobials but some resistant strains can also be identified in organic animal production. The level of antibiotic-resistant *Campylobacter* strains isolated from conventionally reared broilers, in particular, seems to be higher compared with that found in organic broilers and there is also some evidence that other antibiotic-resistant human pathogens are more commonly isolated from conventionally reared broilers.

REFERENCES

Alali, W.Q., Thakur, S., Berghaus, R.D., Martin, M.P. and Gebreyes, W.A. (2010) Prevalence and distribution of *Salmonella* in organic and conventional broiler poultry farms. *Foodborne Pathogen Disease* 7, 1363–1371.

Allen, V.M., Ridley, A.M., Harris, J.A., Newell, D.G. and Powell, L. (2011) Influence of production system on the rate of onset of *Campylobacter* colonisation in chicken flocks reared extensively in the United Kingdom. *British Poultry Science* 52, 30–39.

Appleby, M.C. (2003) The European Union ban for conventional cages for laying hens: history and prospects. *Journal of Applied Animal Welfare* 6, 103–121.

Avrain, L., Humbert, F., L'Hospitalier, R., Sanders, P. and Kempf, I. (2001) Etude de l'antibiorésistance des campylobacters de la filière avicole. In: *Compte-rendu des 4è Journées de la Recherche Avicole*, Nantes, France, 27–29 mars 2001. Institute de Recherche et de Developpement en Agroenvironment, Québec, Canada, pp. 281–284.

Bailey, J.S. and Cosby, D.E. (2005) *Salmonella* prevalence in free-range and certified organic chickens. *Journal of Food Protection* 68, 2451–2453.

Beuving, G. and Vonder, G.M.A. (1978) Effects of stressing factors on corticosterone levels in the plasma of laying hens. *General Compendium of Endocrinology* 35, 153–159.

Bull, S.A., Allen, V.M., Domingue, G., Jorgensen, F., Frost, J.A., Ure, R., Whyte, R., Tinker,

D., Corry, J.E.L., Gillard-King, J. and Humphrey, T.J. (2006) Sources of *Campylobacter* spp. colonising housed broiler flocks during rearing. *Applied and Environmental Microbiology* 72, 645–652.

Campo, J.L., Prieto, M.T. and Davila, S.G. (2008) Effects of housing system and cold stress on heterophil-to-lymphocyte ratio, fluctuating asymmetry, and tonic immobility duration of chickens. *Poultry Science* 87, 621–626.

Carrique-Mas, J.J., Breslin, M., Snow, L., Arnold, M.E., Wales, A., Laren, I. and Davies, R.H. (2008) Observations related to the *Salmonella* EU layer baseline survey in the United Kingdom: follow-up of positive flocks and sensitivity issues. *Epidemiology and Infection* 136, 1537–1546.

Carrique-Mas, J.J., Breslin, M., Snow, L., McLaren, I., Sayers, A. and Davies, R.H. (2009a) Persistence and clearance of different *Salmonella* serovars in buildings housing laying hens. *Epidemiology and Infection* 137, 837–846.

Carrique-Mas, J.J., Marin, C., Breslin, M., McLaren, I. and Davies, R.H. (2009b) A comparison of the efficacy of cleaning and disinfection methods in eliminating *Salmonella* spp. from commercial egg laying houses. *Avian Pathology* 38, 419–424.

CDC (Centers for Disease Control and Prevention) (2007) Preliminary FoodNet data on the incidence of infection with pathogens transmitted commonly through food – 10 states. *MMWR. Morbidity and Mortality Weekly Report* 57, 366–370.

Chinivasagam, H.N., Redding, M., Runge, G. and Blackall, P.J. (2010) Presence and incidence of food-borne pathogens in Australian chicken litter. *British Poultry Science* 51, 311–318.

CLASSP (2007) LACORS/HPA Coordinated Local Authority Sentinel Surveillance of Pathogens (CLASSP) Two Year Surveillance Report 1st November 2004 to 31st October 2006. http://www.lacors.gov.uk/LACORS/ContentDetails.aspx?id=17579 (accessed 1 October 2010).

Colles, F.M., Jones, T.A., McCarthy, N.D., Sheppard, S.K., Cody, A.J., Dingle, K.E., Dawkins, M.S. and Maiden, M.C. (2008) *Campylobacter* infection of broiler chickens in a free-range environment. *Environmental Microbiology* 10, 2042–2050.

Colson, S., Arnould, C. and Michel, V. (2008) Influence of rearing conditions of pullets of space use and performance of hens placed in aviaries at the beginning of the laying period. *Applied Animal Behaviour Science* 111, 286–300.

Craig, J.V., Craig, J.A. and Vargas, J.V. (1986) Corticosteroids and other indicators of hen's well-being in four laying-house environments. *Poultry Science* 65, 856–863.

Crespo, P.S., Hernandeze, G., Echeita, A., Torres, A., Ordonez, P. and Aladuena, A. (2005) Surveillance of foodbourne disease outbreaks associated with consumption of eggs and egg products: Spain, 2002–2003. *Eurosurveillance Weekly* 10(6):pii=2726; available at: http://www.eurosurveillance.org/ViewArticle.aspx?ArticleId=2726 (accessed 1 October 2010).

Cui, S., Ge, B., Zheng, J. and Meng, J. (2005) Prevalence and antimicrobial resistance of *Campylobacter* spp. and *Salmonella* serovars in organic chickens from Maryland retail stores. *Applied and Environmental Microbiology* 71, 4108–4111.

Davies, R. and Breslin, M. (2003) Observations on *Salmonella* contamination of commercial laying farms before and after cleaning and disinfection. *The Veterinary Record* 152, 283–287.

Davies, R. and Breslin, M. (2004) Observations on *Salmonella* contamination of eggs from infected commercial laying flocks where vaccination for *Salmonella enterica* serovar Enteritidis had been used. *Avian Pathology* 33, 133–144.

De Jong, B. and Ekdahl, K. (2006) Human salmonellosis in travelers is highly correlated to the prevalence of *Salmonella* in laying hen flocks. *Eurosurveillance Weekly* 11(27):pii=2993; available at: http://www.eurosurveillance.org/ViewArticle.aspx?ArticleId=2993 (accessed 1 October 2010).

Delmas, G., Gallay, A., Espié, E., Haeghebaert, S., Pihier, N., Weill, F.X., De Valk, H., Vaillant, V. and Désenclos, J.C. (2006) Foodborne-diseases outbreaks in France between 1996 and 2005. *Bulletin Epidemiologique Hebdomadaire* 51/52, 418–422.

Denis, M., Rose, V., Huneau-Salaün, A., Balaine, L. and Salvat, G. (2008) Diversity of pulsed-field gel electrophoresis profiles of *Campylobacter jejuni* and *Campylobacter coli* from broiler chickens in France. *Poultry Science* 87, 1662–1671.

Desmonts, M., Dufour-Gesbert, F., Avrain, L. and Kempf, I. (2004) Antimicrobial resistance in *Campylobacter* strains isolated from French broilers before and after antimicrobial growth promoter bans. *Journal of Antimicrobial Chemotherapy* 54, 1025–1030.

Dewulf, J., Tuyttens, F., Lauwers, L., Van Huylebroeck, G. and Maes, D. (2007) Influence of pen density on pig meat production, health and welfare. *Vlaams Diergeneeskundig Tijdschrift* 76, 410–416.

EC (European Commission) (2003) Regulation (EC) No 2160/2003 of the European Parliament and the Council of 17 November 2003 on the control of salmonella and other specified food-borne zoonotic agents. *Official Journal of the European Union* L 325, 12/12/2003, 1–15.

EC (European Commission) (2007) Regulation (EC) No 646/2007(3) implementing Regulation (EC) No 2160/2003 as regards a Community target for the reduction of the prevalence of *Salmonella enteridis* and *Salmonella typhimurium* in broilers. *Official Journal of the European Union* L 151, 13/06/2007, 21–25.

EFSA (European Food Safety Authority) (2005) Welfare aspects of various systems for keeping laying hens. Scientific Report: p 143. *Annex of the EFSA Journal* 197, 1–23.

EFSA (European Food Safety Authority) (2007) Report of the Task Force on Zoonoses Data Collection on the Analysis of the baseline study on the prevalence of *Salmonella* in holdings of laying hen flocks of *Gallus gallus*. *The EFSA Journal* 97.

EFSA (European Food Safety Authority) (2009) The Community Summary Report on trends and sources of zoonoses and zoonotic agents in the European Union in 2007. *The EFSA Journal* 223, doi:10.2903/j.efsa.2009.223.

EFSA (European Food Safety Authority) (2010a) Analysis of the baseline survey on the prevalence of *Campylobacter* in broiler batches and of *Campylobacter* and *Salmonella* on broiler carcasses in the EU, 2008, Part A: *Campylobacter* and *Salmonella* prevalence estimates. *The EFSA Journal* 8, 1503; doi:10.2903/j.efsa.2010.1503.

EFSA (European Food Safety Authority) (2010b) Analysis of the baseline survey on the prevalence of *Campylobacter* in broiler batches and of *Campylobacter* and *Salmonella* on broiler carcasses, in the EU, 2008; Part B: Analysis of factors associated with *Campylobacter* colonization of broiler batches and with *Campylobacter* contamination of broiler carcasses; and investigation of the culture method diagnostic characteristics used to analyse broiler carcass samples. *The EFSA Journal* 8, 1522; doi:10.2903/j.efsa.2010.1522.

EFSA (European Food Safety Authority) (2010c) The Community Summary Report on trends and sources of zoonoses and zoonotic agents in the European Union in 2008. *The EFSA Journal* 8, 1496; doi:10.2903/j.efsa.2010.1496.

El-Lethey H., Huber-Eicher, B. and Jungi, T.W. (2003) Exploration of stress-induced immunosuppression in chickens reveals both stress-resistant and stress-susceptible antigen responses. *Veterinary Immunology and Immunopathology* 95, 91–101.

Esteban, J.I., Oporto, B., Aduriz, G., Juste, R.A. and Hurtado, A. (2008) A survey of food-borne pathogens in free-range poultry farms. *International Journal of Food Microbiology* 123, 177–182.

FSA (2003) *UK-wide Survey of Salmonella and Campylobacter Contamination of Fresh and Frozen Chicken on Retail Sale.* London: Food Standards Agency; available at: http://www.food.gov.uk/multimedia/pdfs/campsalmsurvey.pdf (accessed 1 October 2010).

FSA (2009) *FSA Report for the UK Survey of* Campylobacter *and* Salmonella *Contamination of Fresh Chicken at Retail Sale*. Foods Standards Agency, London; available at: http://www.foodbase.org.uk/results.php?f_report_id=351 (accessed 1 October 2010).

Garber, L., Smeltzer, M., Fedorka-Cray, P., Ladely, S. and Ferris, K. (2003) *Salmonella enterica* serotype Enteritidis in table egg layer house environments and in mice in US layer houses and associated risk factors. *Avian Diseases* 47, 134–142.

Golden, N.J., Marks, H.M., Coleman, M.E., Schroeder, C.M., Bauer, N.E. Jr, and Schlosser, W.D. (2008) Review of induced molting by feed removal and contamination of eggs with *Salmonella enterica* serovar Enteritidis. *Veterinary Microbiology* 131, 215–228.

Gradel, K.O., Sayers, A.R. and Davies, R.H. (2004) Surface disinfection tests with *Salmonella* and a putative indicator bacterium, mimicking worst-case scenarios in poultry houses. *Poultry Science* 83, 1636–1646.

Guard-Petter, J. (2001) The chicken, the egg and *Salmonella enteritidis*. *Applied and Environmental Microbiology* 3, 421–430.

Hald, T., Kabell, S. and Madsen, M. (2002) The influence of production on the occurrence of *Salmonella* in the Danish table-egg production. In: Smulders, F.J.M. and Collin, J.D. (eds) *Food Safety Assurance in the Pre-Harvest Phase*, vol. 1. Wageningen Academic Publishers, Wageningen, The Netherlands, pp. 276–279.

Hansen, I., Braastad, B.O., Storbraten, J. and Tofastrud, M. (1993) Differences in fearfulness indicated by tonic immobility between laying hens in aviaries and cages. *Animal Welfare* 2, 105–112.

Harada, K. and Asai, T. (2010) Role of antimicrobial selective pressure and secondary factors on antimicrobial resistance prevalence in *Escherichia coli* from food-producing animals in Japan. *Journal of Biomedical Biotechnology* doi:10.1155/2010/180682.

Heuer, O.E., Pedersen, K., Andersen, J.S. and Madsen, M. (2001) Prevalence and antimicrobial susceptibility of the thermophilic *Campylobacter* in organic and conventional broiler flocks. *Letters in Applied Microbiology* 33, 269–274.

Heuvelink, A.E., Tilburg, J.J.H.C., Voogt, N., van Pelt, W., van Leeuwen, J,M,, Sturm ,J M ,J und Van de Ulessen, A.W. (1999) *Surveillance van bacteriele zoonoseverwekkers bij landbouwhuisdieren: Periode april 1997 tot en met maart 1998*. RIVM, Bilthoven, The Netherlands.

Holt, P.S. (2003) Molting and *Salmonella enterica* serovar Enteritidis infection: the problem and some solutions. *Poultry Science* 82, 1008–1010.

Holt, P.S. and Porter, R.E. Jr, (1993) Effect of induced molting on the recurrence of a previous *Salmonella enteritidis* infection. *Poultry Science* 72, 2069–2078.

Holt, P.S., Macri, N.P. and Porter, R.E. Jr, (1995) Microbiological analysis of the early *Salmonella enteritidis* in molted and unmolted hens. *Avian Diseases* 39, 55–63.

Hughes, C.S., Gaskell, R.M., Jones, R.C., Bradbury, J.M. and Jordan, F.T.W. (1989) Effects of certain stress factors on the re-excretion of infectious laryngotracheitis virus from latently infected carrier birds. *Research in Veterinary Science* 46, 274–76.

Humphrey, T. (2006) Are happy chickens safer chickens? Poultry welfare and disease susceptibility. *British Poultry Science* 47, 379–391.

Huneau-Salaün, A., Denis, M., Balaine, L. and Salvat, G. (2007) Risk factors for *Campylobacter* spp. colonization in French free-range broiler-chicken flocks at the end of the indoor rearing period. *Preventive Veterinary Medicine* 80, 34–48.

Huneau-Salaün, A., Chemaly, M., Le Bouquin, S., Lalande, F., Petetin, I., Rouxel, S., Michel, V., Fravallo, P. and Rose, N. (2009) Risk factors for *Salmonella enterica* subsp. *enterica* contamination in 519 French laying hen flocks at the end of the laying period. *Preventive Veterinary Medicine* 89, 51–58.

Jones, R.C. and Ambali, R.G. (1987) Re-excretion of an entero-tropic infectious-bronchitis virus by hens at point of lay after experimental-infection at day old. *The Veterinary Record* 120, 117–118.

Kazwala, R.R., Jiwa, S.F. and Nkya, A.E. (1993) The role of management systems in the epidemiology of thermophilic Campylobacters among poultry in eastern zone of Tanzania. *Epidemiological Infection* 110, 273–278.

Kinde, H., Castellan, D.M., Kerr, D., Campbell, J., Breitmeyer, R. and Ardans, A. (2005) Longitudinal monitoring of two commercial layer flocks and their environments for *Salmonella enterica* serovar Enteritidis and other salmonellae. *Avian Diseases* 49, 189–194.

Koelkebeck, K.W., Amoss, M.S. and Cain, J.R. (1987) Production, physiological and behavioral responses of laying hens in different management environments. *Poultry Science* 63, 2123–2129.

Kowalski, A. and Sokol, R. (2009) Influence of *Dermanyssus gallinae* (poultry red mite) invasion on the plasma levels of corticosterone, cathecholamines and proteins in layer hens. *Polish Journal of Veterinary Science* 12, 231–235.

LAYWEL (2006) Welfare implications of changes in production systems for laying hens. Periodic Final Activity Report. http://www.laywel.eu/web/pdf/final%20activity%20report. pdf (accessed 1 October 2010).

Le Bouquin, S., Allain, V., Rouxel, S., Petetin, I., Picherot, M., Michel, V. and Chemaly, M. (2010) Prevalence and risk factors for *Salmonella* spp. contamination in French broiler-chicken flocks at the end of the rearing period. *Preventive Veterinary Medicine* 97, 245–251.

Lestari, S.I., Han, F., Wang, F. and Ge, B. (2009) Prevalence and antimicrobial resistance of *Salmonella* serovars in conventional and organic chickens from Louisiana retail stores. *Journal of Food Protection* 72, 1165–1172.

Luangtongkum, T., Morishita, T.Y., Ison, A.J., Huang, S., McDermott, P.F. and Zhang, Q. (2006) Effect of conventional and organic production practices on the prevalence and antimicrobial resistance of *Campylobacter* spp. in poultry. *Applied and Environmental Microbiology* 72, 3600–3607.

Marcos Arautogi, C., Gandolfi, B., Aranguena, T., Sepulveda, R., Arévalo, M. and Simon, F. (2009) Antibody and inflammatory responses in laying hens with experimental primary infections of *Ascaridia galli*. *Veterinary Parasitology* 161, 69–75.

Marshally, M.M., Hendricks, G.L., Kalama, M.A., Gehad, A.E., Abbas, A.O. and Patterson, P.H. (2004) Effect of heat stress on production parameters and immune responses of commercial laying hens. *Poultry Science* 83, 889–894.

McCrea, B.A., Tonooka, K.H., VanWorth, C., Boggs, C.L., Atwill, E.R. and Schrader, J.S. (2006) Prevalence of *Campylobacter* and *Salmonella* species on farm, after transport, and at processing in specialty market poultry. *Poultry Science* 85, 136–143.

Methner, U., Diller, R., Reiche, R. and Böhland, K. (2006) Occurrence of Salmonellae in laying hens in different housing systems and conclusion for the control. *Münchner Tierärztliche Wochenschrift* 119, 467–473.

Miranda, J.M., Vázquez, B.I., Fente, C.A., Calo-Mata, P., Cepeda, A. and Franco, C.M. (2008) Comparison of antimicrobial resistance in *Escherichia coli*, *Staphylococcus aureus*, and *Listeria monocytogenes* strains isolated from organic and conventional poultry meat. *Journal of Food Protection* 71, 2537–2542.

Mølbak, K. and Neimann, J. (2002) Risk factors for sporadic infection with *Salmonella* Enteritidis, Denmark 1997–1999. *American Journal of Epidemiology* 156, 654–661.

Mollenhorst, H., van Woudenbergh, C.J., Bokkers, E.G.M. and de Boer, I.J.M. (2005) Risk factors for *Salmonella* Enteritidis infections in laying hens. *Poultry Science* 84, 1308–1313.

Morris, V.K., Allen, V.M., Edge, S. and Canning, P. (2009) Review of current practices and recommendations for *Campylobacter* reduction extensively-reared flocks. Food Standards Agency Final Report. http://www.foodbase.org.uk/admintools/reportdocuments/361-1-624_B15014_Final_Report.pdf (accessed 1 October 2010).

Namata, H., Méroc, E., Aerts, M., Faes, C., Abrahantes, J., Imberechts, H. and Mintiens, K. (2008) *Salmonella* in Belgian laying hens: an identification of risk factors. *Preventive Veterinary Medicine* 83, 323–336.

Namata, H., Welby, S., Aerts, M., Faes, C., Abrahantes, J.C., Imberechts, H., Vermeersch, K., Hooyberghs, J., Méroc, E. and Mintiens, K. (2009) Identification of risk factors for the prevalence and persistence of *Salmonella* in Belgian broiler chicken flocks. *Preventive Veterinary Medicine* 90, 211–222.

Näther, G., Alter, T., Martin, A. and Ellerbroek, L. (2009) Analysis of risk factors for *Campylobacter* species infection in broiler flocks. *Poultry Science* 88, 1299–1305.

Permin, A., Bisgaard, M., Frandsen, F., Pearman, M., Kold, J. and Nansen, P. (1999) Prevalence of gastrointestinal helminths in different poultry production systems. *British Poultry Science* 40, 439–443.

Pieskus, J., Franciosini, M.P., Proietti, P., Reich, F., Kazeniauskas, E., Butrimaite-Ambrozeviciene, C., Mauricas, M. and Bolder, N. (2008a) Preliminary investigations on *Salmonella* spp. incidence in meat chicken farms in Italy, Germany, Lithuania and the Netherlands. *International Journal of Poultry Science* 7, 813–817.

Pieskus, J., Kazeniauskas, E., Butrimaite-Ambrozeviciene, C., Stanevicius, Z. and Mauricas, M. (2008b) *Salmonella* incidence in broiler and laying hens with the different housing systems. *Journal of Poultry Science* 45, 227–231.

Rivoal, K., Denis, M., Salvat, G., Colin, P. and Ermel, G. (1999) Molecular characterisation of the diversity of *Campylobacter* spp. isolates collected from a poultry slaughter house: analysis of cross-contamination. *Letters in Applied Microbiology* 29, 370–374.

Rivoal, K., Ragimbeau, C., Salvat, G., Colin, P. and Ermel, G. (2005) Genomic diversity of *Campylobacter coli* and *Campylobacter jejuni* isolates recovered from free range broiler farms and comparison with isolates of various origins. *Applied and Environmental Microbiology* 71, 6216–6227.

Schaar, U., Kaleta, E.F. and Baumbach, B. (1997) Comparative studies on the prevalence of *Salmonella enteritidis* and *Salmonella typhimurium* in laying chickens maintained in batteries or on floor using bacteriological isolation techniques and two commercially available ELISA kits for serological monitoring. *Tierärzlichen Praxis* 25, 451–459.

Schwaiger, K., Schmied, E.M. and Bauer, J. (2008) Comparative analysis of antibiotic resistance characteristics of Gram-negative bacteria isolated from laying hens and eggs in conventional and organic keeping systems in Bavaria, Germany. *Zoonoses and Public Health* 55, 331–341.

Schwaiger, K., Schmied, E.M. and Bauer, J. (2010) Comparative analysis on antibiotic resistance characteristics of *Listeria* spp. and *Enterococcus* spp. isolated from laying hens and eggs in conventional and organic keeping systems in Bavaria, Germany. *Zoonoses and Public Health* 57, 171–180.

Sheppard, S.K., Dallas, J.F., MacRae, M., McCarthy, N.D., Sproston, E.L., Gormley, F.J., Strachan, N.J., Ogden, I.D., Maiden, M.C. and Forbes, K.J. (2009) *Campylobacter* genotypes from food animals, environmental sources and clinical disease in Scotland 2005/6. *International Journal of Food Microbiology* 134, 96–103.

Siemon, C.E., Bahnson, P.B. and Gebreyes, W.A. (2007) Comparative investigation of prevalence and antimicrobial resistance of *Salmonella* between pasture and conventionally reared poultry. *Avian Diseases* 51, 112–117.

Singh, R., Cook, N., Cheng, K.M. and Silversides, F.G. (2009) Invasive and noninvasive measurement of stress in laying hens kept in conventional cages and in floor pens. *Poultry Science* 88, 1346–1351.

Snow, L.C., Davies, R.H., Christiansen, K.H., Carrique-Mas, J.J., Wales, D., O'Connor, J.L., Cook, A.J.C. and Evans, S.J. (2007) Survey of the prevalence of *Salmonella* species on commercial laying farms in the United Kingdom. *The Veterinary Record* 161, 471–476.

Thaxton, P., Wyatt, R.D. and Hamilton, P.B. (1974) The effect of environmental temperature on paratyphoid infection in the neonatal chicken. *Poultry Science* 53, 88–94.

Uyttendaele, M., De Troy, P. and Debevere, J. (1999) Incidence of *Salmonella, Campylobacter jejuni, Campylobacter coli,* and *Listeria monocytogenes* in poultry carcasses and different types of poultry products for sale on the Belgian retail market. *Journal of Food Protection* 62, 735–740.

Valiente Moro, C., De Luna, C.J., Tod, A., Guy, J.H., Sparagano, O.A.E. and Zenner, L. (2009) The poultry red mite (*Dermanyssus gallinae*): a potential vector of pathogenic agents. *Experimental and Applied Acarology* 48, 93–104.

Valiente Moro, C., Fravalo, P., Amelot, M., Chauve, C., Zenner, L. and Salvat, G. (2007) Colonization and organ invasion in chicks experimentally infected with *Dermanyssus gallinae* contaminated by *Salmonella* Enteritidis. *Avian Pathology* 36, 307–311.

Van de Giessen, A.W., Ament, A.J.H.A. and Notermans, S.H.W. (1994) Intervention strategies for *Salmonella* enteritidis in poultry flocks: a basic approach. *International Journal of Food Microbiology* 21, 145–154.

Van Hoorebeke, S., Van Immerseel, F., Berge, A.C., Persoons, D., Schulz, J., Hartung, J., Harisberger, M., Regula, G., Barco, L., Ricci, A., De Vylder, J., Ducatelle, R., Haesebrouck, F. and Dewulf, J. (2010a) Antimicrobial resistance of *Escherichia coli* and *Enterococcus faecalis* in housed laying-hen flocks in Europe. *Epidemiology and Infection* 139, 1610–1620.

Van Hoorebeke, S., Van Immerseel, F., Haesebrouck, F., Ducatelle, R. and Dewulf, J. (2010b) The influence of the housing system on *Salmonella* infections in laying hens: a review. *Zoonoses and Public Health* 58, 304–311.

Van Hoorebeke, S., Van Immerseel, F., Schulz, J., Hartung, J., Harisberger, M., Barco, L., Ricci, A., Theodoropoulos, G., Xylouri, E., De Vylder, J., Ducatelle, R., Haesebrouck, F., Pasmans, F., de Kruif, A. and Dewulf, J. (2010c) Determination of the within and between flock prevalence and identification of risk factors for *Salmonella* infections in laying hen flocks housed in conventional and alternative systems. *Preventive Medicine* 94, 94–100.

Van Overbeke, I., Duchateau, L., De Zutter, L., Albers, G. and Ducatelle, R. (2006) A comparison survey of organic and conventional broiler chickens for infectious agents affecting health and food safety. *Avian Diseases* 50, 196–200.

Vandeplas, S., Dubois-Dauphin, R., Palm, R., Beckers, Y., Thonart, P. and Théwis, A. (2010) Prevalence and sources of *Campylobacter* spp. contamination in free-range broiler production in the southern part of Belgium. *Biotechnology, Agronomy, Society and Environment* 14, 279–288.

Wales, A., Breslin, M., Carter, B., Sayers, R. and Davies, R. (2007) A longitudinal study of environmental *Salmonella* contamination in caged and free-range layer flocks. *Avian Pathology* 36, 187–197.

Wolf-Reuter, M., Matthes, S. and Ellendorf, F. (2002) Salmonella prevalence in intensive, free and organic production systems. *Archiv für Geflügelkunde* 66, 158.

Young, I., Rajič, A., Wilhelm, B.J., Waddell, L., Parker, S. and McEwen, S.A. (2009) Comparison of the prevalence of bacterial enteropathogens, potentially zoonotic bacteria and bacterial resistance to antimicrobials in organic and conventional poultry, swine and beef production: a systematic review and meta-analysis. *Epidemiology and Infection* 137, 1217–1232.

Zweifel, C., Scheu, K.D., Keel, M., Renggli, F. and Stephan, R. (2008) Occurrence and genotypes of *Campylobacter* in broiler flocks, other farm animals, and the environment during several rearing periods on selected poultry farms. *International Journal of Food Microbiology* 125, 182–187.

CHAPTER 6
Introduction to Village and Backyard Poultry Production

R.A.E. Pym and R.G. Alders

ABSTRACT

Small-scale family poultry farming involving semi-scavenging flocks of mostly indigenous breed poultry in rural regions of many developing countries contributes in a very meaningful way towards the social and financial needs of rural families. While productivity is relatively low, so too are inputs; which makes the production system reasonably viable, as evidenced by the many millions of such flocks worldwide. The principal constraint to profitability is the high mortality rate in young chicks, due to a combination of disease, predation, malnutrition and climatic exposure, combined with moderate to high mortality rates in grower and adult stock due to the effects of disease, of which Newcastle disease is a common cause. Simple cost-effective interventions, involving vaccination of the flock against Newcastle disease with heat-tolerant vaccines combined with early confinement of the chicks with the hen and creep feeding over the first three to four weeks, have been demonstrated to impact dramatically on survival of the birds and on household food security and profitability. Such improvements are fully compatible with programmes aimed at development of the commercial poultry meat and egg industries in developing countries to meet the needs of the urban and peri-urban populations. Family poultry-raising is experiencing a resurgence in many 'developed' countries. The number of families raising backyard poultry is on the increase due to both a growing enthusiasm for organic poultry products and the economic downturn. Backyard production systems vary in accordance with local government regulations, producer preferences, household residential circumstances and climatic conditions.

INTRODUCTION

Much has been written in recent times about the role and importance of small-scale family poultry production in terms of food security and in contributing towards the social and financial needs of families in many developing countries (Alders and Pym, 2009). The majority of families in the poorer rural regions of many developing countries have small scavenging flocks of indigenous breed birds. Productivity is low, but very importantly, inputs are also very few, with generally no purchased feed. The birds generally survive on household scraps and on what they can scavenge from the environment, which can be quite meagre. The overwhelming number of birds kept in this manner globally demonstrates that the system can be economically viable, albeit because financial inputs are minimal. This production system is still often referred to as 'sector 4' production, which was essentially a biosecurity classification system proposed by the Food and Agriculture Organization of the United Nations (FAO, 2008). This classification, as it relates to production systems, has also recently been referred to as 'safety net' production (McLeod et al., 2009), to distinguish this from small-scale commercial 'asset builder' production and large-scale commercial 'industrialized' production. While the distinction between the latter two systems is somewhat blurred, according to size of operation, 'safety net' production is clearly separated from the others by feeding and confinement (free-ranging scavenging versus confinement with provision of compounded diets) and genotype (indigenous versus genetically selected 'improved' meat or egg strains).

While there are commonalities, there are also significant differences between small-scale backyard 'pure-breed fancier' operations in developed countries and the above 'safety net' production systems in developing countries. These include genotype, climate, nature of housing, level and nature of supplemental feeding, disease prevention and control procedures. It should also be emphasized that large-scale 'free range' egg or meat production systems in both developed and developing countries differ substantially from the above 'safety net' production, principally in terms of scale of operation. They also differ in the genotypes used and the housing and feeding provided, as the supplementary feed, usually provided in the house, typically accounts for at least 90% of the nutrient intake in the former.

The focus of the present chapter is on small-scale 'safety net' production systems in developing countries; however, reference and comparison will be made to small-scale commercial 'asset builder' production, as this is also a production system encountered in villages in developing countries. The final section provides an overview of backyard poultry production systems in 'developed' countries.

SEMI-SCAVENGING PRODUCTION SYSTEMS

In the rural regions of nearly all developing countries worldwide, a significant proportion of families keep a small semi-scavenging flock of between five and 30 indigenous breed poultry (Aini, 1990; Gueye, 1998). Chickens are predominant, but ducks, turkeys and guinea fowl also contribute meaningfully in many areas. Many of the flocks are composed of more than one species (see Plate 1 in colour plate section). In the poorer developing countries, poultry in these small family flocks often constitute more than 80% of the country's total poultry population (Pym et al., 2006).

There are a number of reasons why people keep poultry in these small family flocks: they provide the family with eggs and meat – a chicken provides a single meal for the family without the need for refrigeration; both can be readily sold locally for cash to purchase household items, other food or educational needs for the children; eggs and chickens are widely used in many traditional and religious ceremonies; chickens may be given as gifts to honoured guests; and the birds are active in pest control and produce manure (Alders and Pym, 2009).

In many developing countries, the meat and eggs from indigenous breeds are preferred to those from commercial broilers and layers by a significant proportion of the populace, who will often pay a premium for them (Sonaiya *et al.*, 1999). In relation to poultry meat, this is partly due to the older, firmer and more flavoursome meat from the indigenous birds, which is more suitable for traditional forms of cooking than broiler meat (Aini, 1990). On a per kilogram basis, the premium paid is even greater than per unit chicken or egg, due to the normally low egg weight and body weights at slaughter of the indigenous breeds.

One of the particular incentives for keeping poultry in this manner is that the production is achieved at very little cost. Feed, which is a major cost in commercial production, accounting for between 70 and 75% of production costs, is not a factor here, as, apart from household scraps, the birds source their own feed from the environment. A small amount of grain is also often provided, but this usually makes up only a small fraction of their total daily energy and protein intake (Sonaiya, 2004). The amount of feed available in the environment (the 'scavenging feed resource base', SFRB) (Roberts, 1992), which includes plant seeds and fruits, grain, earthworms, snails, frogs, insects, etc., is influenced by geographic location, local vegetation and microclimate, physical restrictions on the scavenging area, the poultry and other scavenging animal population density, as well as season. In many situations, the SFRB is limited which restricts the overall effective scavenging poultry population.

The composition of the household flock varies considerably from region to region and even from household to household. The typical ratio of chicks to growers to adults in scavenging systems in Africa and Asia has been estimated in a number of studies to range from 2:1:2 (Awuni, 2002), 2:1:1 (Minga *et al.*, 1996), 3:2:2 (NAFRI, 2005) to 1:1:2 (Khalafalla *et al.*, 2002; Njue *et al.*, 2002). The ratio of adult males to females is typically about 1:3 (e.g. Sonaiya *et al.*, 1999; Bamhare, 2001; Chitate and Guta, 2001; Mavale, 2001; Ekue *et al.*, 2002; NAFRI, 2005). From these studies the average proportion of adult hens in the total indigenous chicken flock including young chicks is thus about 25%. Males are kept to adulthood for the purposes of breeding, but also in many regions for cock-fighting, a 'sport' providing entertainment and an opportunity for gambling, indulged in mainly by the male members of the communities in question (Marvin, 1984).

There is often limited provision of shelter for the household flock. This ranges from a dedicated structure to provide overnight shelter, and constructed normally from local materials (see Plate 2 in colour plate section), through an area under or even inside the human dwelling for the same purpose, to no provision at all, where the birds usually roost in the trees (Gueye, 1998; Sonaiya *et al.*, 1999). Sometimes coops made of split bamboo or similar materials are used to house individual birds (e.g. cock-fighting males) or hens with their chicks (see Plate 2 and Plate 5 in colour plate section). Overnight secure housing provides an opportunity for preventing or reducing predation and theft, but can contribute to increased problems with external parasitism if good sanitation is not practised. Where food e.g. household scraps or grain is provided, it is usually placed or scattered on the ground, and water may be provided in pots, but not infrequently the birds are required to source water from the environment or make do with water that has already been used by the household. Where water is scarce and is not provided for the birds, the need to seek it in the environment can impact significantly on performance and also increase the risk of predation. Young chicks are particularly susceptible to predation from rats, snakes, cats, dogs, hawks and other animals, reptiles and birds (Farrell *et al.*, 2000).

Because of the risks from predation, nests constructed of cane, woven banana leaves or some similar material are usually located high on the walls and under the eaves of the human dwelling or other nearby buildings (see Plate 3 in colour plate section). On hatching, the chicks are usually transferred with the hen to small pens or coops on the ground.

RELATIVE PRODUCTIVITY OF INDIGENOUS AND COMMERCIAL BREEDS

Growth rate and egg production of indigenous breeds are usually much lower than from commercial meat or egg genotypes (FAO, 2010), but the typical limitations to the SFRB mean that commercial strains usually perform very poorly under typical village semi-scavenging conditions (Besbes, 2009). Part of this is to do with the limited energy and protein intake which often barely meets their maintenance requirements, and partly to do with their capacity to perform and survive under the unfavourable environment. Predation is a significant element of this environment, and through natural selection, indigenous breeds have evolved to combat this by being alert to the various threats, by being light-bodied with long legs and are capable of running fast and flying up into trees to escape attack (Alders and Pym, 2009). Being slow moving, meat-type stocks are particularly susceptible to attack from predators. The other form of predation is theft, and commercial breeds are particularly attractive to and easy prey for thieves.

THE ROLE OF WOMEN AND CHILDREN IN SMALL-SCALE FAMILY POULTRY PRODUCTION

An important feature of small-scale family poultry farming in most countries is that women and children often own and manage the small household poultry flock (see plates 9 and 10 in colour plate section). Poultry are an important feature of many female-headed households (Guèye, 2000; Bagnol, 2001). This has very desirable ramifications for the empowerment of women and impacts favourably on household nutrition, particularly that of the children, as well as the use of funds from the sale of chickens or eggs for educational purposes. Where the productivity and profitability of family poultry egg and meat production can be improved through effective low-cost interventions as described later in this chapter, there is an opportunity to overcome the poverty cycle and for the woman and her family to move into other areas of income generation. While moving into commercial poultry production might be a possibility for some, there is limited opportunity for many households to take this path. Apart from anything else, there is a considerably greater degree of risk in small-scale commercial poultry production due to financial exposure and competition with large-scale production. Another reality of such a transition is the likelihood of the enterprise being taken over by the male household member, with the woman/wife being relegated to a worker/assistant role. Such a move is counterproductive to the empowerment of women and has the potential to impact negatively on the welfare and nutrition of the family.

BROODINESS AND ATTRITION RATE IN CHICKS

One very important reason for the widespread retention of indigenous rather than commercial breeds in village production systems worldwide is the capacity of the hens to go broody after a clutch of eggs has been laid. This means that the birds are capable of reproducing themselves, something that the large majority of commercial breeds are not (FAO, 2010). With chickens, the village hen will typically lay a clutch of between eight and 14 eggs and then sit on these to hatch a brood of chicks. She is usually an excellent incubator, and where there are one or more adult males in the flock, hatchability is typically in excess of 80%. So, for a hen laying a clutch of 12 eggs and being allowed to sit on all of these, she will typically hatch out about ten chicks, and it is commonplace to see a mother hen with a brood of ten or more baby chicks (see Plate 4 in colour plate section).

The low annual egg production of indigenous breed village chickens is thus due in no small part to their role as incubators and mothers. Once they have laid the clutch of eggs, which may take from 2 to 4 weeks, they spend the next 3 weeks hatching the eggs and then the following 7 to 8 weeks rearing their chicks to an age when they can more or less fend for themselves, before they recommence laying. They typically produce between three and four batches of chicks per year (Gueye, 1998). Thus over the 15 or so weeks between one batch and the next, the hen is out of production for about 11 weeks, or about 75% of the time. It is thus small wonder that their annual egg production is so low. Having said this, even when these birds are put in laying cages, given *ad libitum* high-quality layer diets and prevented from going broody, for the large majority of indigenous village breeds, their egg production is very much lower than that of commercial layers (FAO, 2010). This is an important consideration where 'commercial' egg production with indigenous breed birds housed in cages or pens is contemplated. Their low productivity needs to be compensated for by a high premium for their eggs, to justify the costs associated with housing and feeding. The critical issue related to the cost of feed in this equation is the relative contribution to feed intake of maintenance and egg production. In the case of modern layer breeds, their feed conversion efficiency is considerably higher than for the large majority of indigenous breeds (FAO, 2010).

In the village scavenging environment, it is rare to see a hen with more than four or five chicks by the time they have reached about 6 weeks of age. The high attrition rate of typically between 50 and 80% is due to a combination of disease, predation, malnutrition and climatic exposure (Farrell *et al.*, 2000). Because of this high rate of loss, it is common practice in many countries to set most of the eggs under the hen in order to produce chicks, most of which die. Statistics (Pym *et al.*, 2006) suggest that, in the absence of effective interventions, few eggs are eaten and typically from each batch of ten or so chicks, one bird may reach adulthood to replace the hen or the rooster and typically only one or two of the growers are eaten or sold.

NEWCASTLE DISEASE

Compounding the effects of young bird mortality, grower and adult mortality is also often high due principally to the effects of disease, with lesser impact of predation and malnutrition. Where there is no effective vaccination programme, Newcastle disease (ND) is often the principal cause of mortality in grower and adult stock. While there are a number of available vaccines that have been demonstrated to be very effective against ND in commercial flocks, the small-scale, semi-scavenging system poses considerable challenges to the effective protection of the flocks: the flocks are small and are usually distributed over a relatively wide area; the birds are free roaming and therefore difficult to catch; the flocks are being constantly regenerated with young stock; and the ability to adequately conserve vaccine during distribution to many villages is often severely limited or absent, which means that traditional heat-labile vaccines cannot be used. To counter some of these problems, heat-tolerant vaccines have been developed, which to some extent overcome the problems associated with the lack of refrigeration (Spradbrow, 1993; Alders and Spradbrow, 2001; Alexander *et al.*, 2004).

Results from a number of projects have demonstrated that small-scale semi-scavenging flocks can be effectively protected against ND using a water-soluble, heat-tolerant vaccine with eye-drop vaccination, employing a vaccination interval of 4 months (Alders *et al.*, 2010). Apart from questions of duration of immunity following vaccination, the suggested frequency of vaccination is required in any case to protect the young stock, which is being produced continuously, and the eye-drop method allows young birds to be vaccinated. There is real risk of injury to young birds through vaccination by injection with oil-based ND vaccines. The logistics of effective vaccination require proper coordination at the village level, with a well-trained vaccination team supplied with appropriately refrigerated viable vaccine and cooperating farmers who have been instructed to keep their birds in their overnight shelters so that they can be easily caught prior to vaccination. By following these procedures, it is possible to obtain effective coverage of the large majority of the village's poultry population (Alders *et al.*, 2010) (see plates 9 and 10 in colour plate section).

OTHER DISEASES

Highly pathogenic avian influenza (HPAI) is clinically indistinguishable from ND and has caused large mortality in village and backyard flocks since the emergence of the H5N1 strain (Capua and Alexander, 2009; Azhar *et al.*, 2010). The detection of HPAI H5N1 was delayed in many countries as the mortalities were initially thought to be due to ND. The detection of HPAI would be facilitated by the effective control of ND (Alders *et al.*, 2010). Appropriate biosecurity practices under backyard and village conditions have been defined but require a high degree of community compliance to be successful (Ahlers *et al.*, 2009).

In the south-eastern region of Africa, fowl pox has emerged as an important problem in village chickens following the control of ND, while in some parts of Asia, fowl cholera is widespread. Duck plague is a serious constraint in South-East Asia. Diseases related to poor nutrition, for example vitamin A deficiency, may have a seasonal appearance in areas where the SFRB is limiting. Internal and external parasitism is also widespread (Ahlers *et al.*, 2009).

MANAGEMENT PROCEDURES TO IMPROVE PRODUCTIVITY AND PROFITABILITY

Results from project work in a number of countries have demonstrated a number of interventions that can be applied to reduce the chick attrition rate from typically between 50 and 80% to less than 25%. These include vaccination against ND, confinement with the mother hen for the first week or two combined with creep feeding, and secure confinement at night with the hen for about 6 weeks (Ahlers *et al.*, 2009; Alders and Pym, 2009; Henning *et al.*, 2009) (see Plates 2 and 5 in colour plate section). Their effects are complementary and additive.

The reduction in attrition rates afforded by such interventions provides an opportunity to reassess the typical practice of setting all, or nearly all, of the eggs under the hen. Where the SFRB is limiting, it is counterproductive to produce large numbers of chickens for which there is inadequate feed, thus requiring feed to be purchased. A more efficient alternative is to either restrict the number of eggs that each hen is allowed to set, or identify certain hens as 'brooder/mothers' and others as 'egg producers' (Cumming, 1992). The latter approach has the advantage of bringing the 'egg producer' hens back into lay sooner, resulting in a higher annual egg production. With either approach, all eggs surplus to requirements for setting are available for consumption by the family or for sale. The former has a very desirable impact on family food security and nutrition, particularly if emphasis is given to providing the children with eggs in their diet as frequently as possible (Ahlers *et al.*, 2009).

The dramatic impact of the increased availability of eggs for consumption or sale from a reduction in chick attrition rate from 70% to 25% can be seen by taking the example of a hen laying 13 eggs in a clutch. In the first scenario, one of these eggs is taken for eating or sale and the remaining 12 are set under the hen. Assuming 80% hatchability, she hatches ten chicks, but with a 70% mortality rate only three of these are alive at 6 weeks of age. In the second scenario, with the above simple interventions, the mortality rate is reduced to 25%. Here, only five eggs need be set under the hen, four of which hatch and three of these are still alive at 6 weeks of age – the same number as in scenario 1. By setting only five eggs, there are now eight eggs available for consumption or sale, an increase of seven eggs (or 700%). This is a truly profound improvement and is achievable with very modest and cost-effective interventions (Henning *et al.*, 2009).

SMALL-SCALE COMMERCIAL PRODUCTION

While not the specific brief of this chapter, it is pertinent to draw a comparison between the above 'safety net' form of poultry production and small-scale commercial poultry ('asset builder flock') production, which is also frequently encountered in village situations. For the purposes of this discussion, the maximum number of birds in these units is assumed to be about 500. The very large majority of the birds in these units are commercial broiler or layer chickens. In some cases, however, the birds are from indigenous breeds, e.g. 'yellow chickens' in China, or crosses, e.g. Sonali birds in Bangladesh, but globally these are a small minority. In the case of broilers, these are typically reared in much the same way as in large-scale production, namely in groups on deep litter, but in much smaller groups and in buildings constructed from local materials and utilizing simple, non-automated equipment in the form of feeders and drinkers constructed from metal or local materials (see Plate 6 in colour plate section). Electricity is rarely available, so ventilation is natural. The high metabolic heat production of broilers as they approach market weight, combined with the typically high ambient temperatures and lack of shed cooling, means that the birds are usually marketed at much earlier ages and lower weights than for broiler flocks in the temperate, developed countries. One means of alleviating this problem in all stock is through the incorporation of genes which either reduce feather cover and in so doing facilitate heat loss, or reduce body size, which increases the ratio of surface area to body mass. Such genes include Naked neck (*Na*), Frizzle (*F*), Scaleless (*Sc*) and sex-linked dwarfism (*dw*) (Cahaner *et al.*, 2008).

Layers may be group-housed in relatively small houses in a similar manner to broilers on deep litter, but with the provision of nest boxes. They may also be housed in cages constructed from wire or wood (e.g. split bamboo) and, similar to broilers, equipment such as feeders and drinkers is simple and non-automated, and constructed of a variety of materials.

A universal feature of this system is the provision of a compounded diet which notionally meets the birds' nutrient requirements. Birds may or may not be fed *ad libitum* depending on a range of factors. In hot climates, in an attempt to reduce heat stress, feed is sometimes restricted during the hottest part of the day to avoid the coincidence of metabolic heat load and peak ambient temperatures. Often, small producers have difficulty in accessing the best-quality feeds and when they do, the price is often significantly higher than paid by the larger-scale operations. It is not uncommon for feed to be of poor quality, and even adulterated, which impacts severely on productivity. Veterinary services, medication and supplements may also be expensive or difficult to access, which also impacts on productivity and profitability. Further, unless they are in a favourable niche market situation, small-scale operators can have significant difficulty in establishing a consistent market for their produce.

MEETING PROJECTED DEMAND FOR POULTRY EGGS AND MEAT

Much of the world's projected increased demand for poultry meat and eggs over the next decade will occur in the developing countries (FAPRI, 2010). It is generally accepted that the increasing demand for poultry meat and eggs from the cities and towns in most developing countries will be largely met by commercial poultry production. Despite the preference for the meat and eggs from indigenous breeds in most countries, the modest production and high price of these products mean that they will have limited impact on this increasing demand. In the interests of social equity and regional development, it is the responsibility of governments to provide an environment which allows for the development of in-country, small- to medium scale poultry production. Such development is not incompatible with support for efficient production of meat and eggs from 'safety net' flocks in the rural areas.

BACKYARD POULTRY PRODUCTION IN 'DEVELOPED' COUNTRIES

Backyard poultry-raising is experiencing a resurgence in many 'developed' countries, including the USA (Alders, 2010). The number of families raising backyard poultry is on the increase due to both a growing enthusiasm for organic poultry products and the economic downturn (Block, 2008; Neuman, 2009; Anonymous, 2011b). Game fowl production is also a significant activity in many US states (Garber et al., 2007). Ornamental poultry breeds have been raised worldwide by fanciers for many years (Damerow, 2010). Backyard production systems vary in accordance with local government regulations, producer preferences, household residential circumstances and climatic conditions (Damerow, 2010; Anonymous, 2011b).

In the USA, backyard poultry flocks have been defined as flocks containing fewer than 1000 birds other than pet birds (i.e. birds not normally kept for food and usually housed in cages in the home, such as parrots, parakeets, finches and canaries) (Garber et al., 2007). Poultry fanciers are most interested in the birds' appearance and place little attention on production characteristics

(Damerow, 2010). The majority of backyard flocks are based on dual-purpose chicken breeds that enable owners to produce both home grown eggs and meat. Backyard poultry are usually owned and maintained by the family in whose homestead they are located. In a small number of cases it is possible to rent a flock, enabling families to decide if raising backyard poultry is feasible prior to making a substantial financial commitment to do so (Anonymous, 2011a).

The type of shelter provided for backyard flocks usually reflects the production system involved:

- no confinement (free range) – seen most often in rural areas;
- confinement to a portable shelter with a fenced foraging area (pastured, range fed, day range) – used on farms with available pasture to periodically move the fence or shelter;
- confinement within a floorless portable shelter – used in family gardens (ark, chicken tractor) and on farms (pastured poultry) with enough land for the shelter to be moved frequently;
- confinement to a permanent building with an outdoor fenced yard (yarding) – the traditional method for housing homestead poultry and other small backyard flocks (see Plate 7 in colour plate section);
- confinement within a permanent building (loose housing) – generally used for raising broilers or breeders or maintaining a flock during cold or wet weather; and
- cage confinement (hutch, ark) – most often used in urban and suburban areas and for show chickens (Damerow, 2010).

The provision of food and water to the backyard poultry varies considerably. In extreme climatic conditions water may need to be either cooled or warmed to ensure that birds have constant access to fresh, clean water. The type of feed provided varies according to the production system, with organic producers needing to meet the regulatory requirements set by their national authorities. The majority of backyard poultry are nourished by a combination of scavenged feed, household leftovers, commercial rations and home mixed rations (Damerow, 2010; Anonymous, 2011b). In peri-urban settings, backyard poultry are on occasions becoming companion rather than production animals as demonstrated by the development of chicken 'nappies' (see Plate 8 in colour plate section) that enable birds to enter their owner's homes without problems associated with random defecation (Anonymous, 2011a).

Backyard poultry owners traditionally report few health problems (Damerow, 2010). However, backyard flocks can encounter problems due to internal and external parasites, infectious disease and intoxication. Healthcare for backyard poultry flocks is often provided by individual owners in combination with agricultural extension services and agricultural suppliers. Since the appearance of HPAI H5N1, many governments are actively working to increase awareness of good biosecurity practices for backyard poultry flocks (Damerow, 2010; Anonymous, 2011b; USDA, 2011).

REFERENCES

Ahlers C., Alders, R.G., Bagnol, B., Cambaza, A.B., Harun, M., Mgomezulu, R., Msami, H., Pym, B., Wegener, P., Wethli, E. and Young, M. (2009) Improving Village Chicken Production: A Manual for Field Workers and Trainers. ACIAR Monograph No. 139. Australian Centre for International Agricultural Research, Canberra; available at: http://aciar.gov.au/publication/mn139 (accessed 2 January 2011).

Aini, I. (1990) Indigenous chicken production in South-East Asia. *World's Poultry Science Journal* 46, 51–57.

Alders, R. (2010) Teaching family poultry health and production in the USA. In *Proceedings of the XIII European Poultry Conference*, Tours, France, 23–27 August 2010. World's Poultry Science Association, Beekbergen, The Netherlands, p. 267.

Alders, R. and Pym, R.A.E. (2009) Village poultry: still important to millions, eight thousand years after domestication. *World's Poultry Science Journal* 65, 181–190.

Alders, R. and Spradbrow, P. (2001) *Controlling Newcastle Disease in Village Chickens: A Field Manual.* ACIAR Monograph No. 82. Australian Centre for International Agricultural Research, Canberra; available at: http://aciar.gov.au/publication/mn082 (accessed 2 January 2011).

Alders, R.G., Bagnol, B. and Young, M.P. (2010) Technically sound and sustainable Newcastle Disease control in village chickens: lessons learnt over fifteen years. *World's Poultry Science Journal* 66, 433–440.

Alexander, D.J., Bell, J.G. and Alders, R.G. (2004) *Technology Review: Newcastle Disease with Special Emphasis on Its Effect on Village Chickens.* FAO Animal Production and Health Paper No. 161. Food and Agriculture Organization of the United Nations, Rome.

Anonymous (2011a) City chicks: Eggs on legs. http://citychicks.com.au/category//rental-packages/ (accessed on 2 January 2011).

Anonymous (2011b) Keeping chickens: What are the rules, regulations and benefits of having poultry in a Brisbane backyard? http://www.ourbrisbane.com/lifestyle/keeping-chickens (accessed 2 January 2010).

Awuni, J.A. (2002) Strategies for the improvement of rural chicken production in Ghana. In: *Characteristics and Parameters of Family Poultry Production in Africa.* International Atomic Energy Agency, Vienna, pp. 33–37.

Azhar, M., Lubis, A.S., Siregar, E.S., Alders, R.G., Brum, F., McGrane, J., Morgan, I. and Roedar, P. (2010) Participatory disease surveillance and response in Indonesia: strengthening veterinary services and empowering communities to prevent and control highly pathogenic avian influenza. *Avian Diseases* 54, 749–753.

Bagnol, B. (2001) The social impact of Newcastle disease control. In: Alders, R.G. and Spradbrow, P.B. (eds) *SADC Planning Workshop on Newcastle Disease Control in Village Chickens.* Proceedings of an International Workshop, Maputo, Mozambique, 6–9 March 2000. ACIAR Proceedings No. 103. Australian Centre for International Agricultural Research, Canberra, pp. 69–75.

Bamhare, C. (2001) Country report: Namibia. In: Alders, R.G. and Spradbrow, P.B. (eds) *SADC Planning Workshop on Newcastle Disease Control in Village Chickens.* Proceedings of an International Workshop, Maputo, Mozambique, 6–9 March 2000. ACIAR Proceedings No. 103. Australian Centre for International Agricultural Research, Canberra, pp. 26–31.

Besbes, B. (2009) Genotype evaluation and breeding of poultry for performance under sub-optimal village conditions. *World's Poultry Science Journal* 65, 260–271.

Block, B. (2008) US city dwellers flock to raising chickens. http://www.worldwatch.org/node/5900 (accessed 30 May 2010).

Cahaner, A., Druyan, S., Hadad, Y., Yadgari, L., Astrachan, N., Kalinowski, A. and Romo, G.

(2008) Breeding broilers for tolerance to stresses. In: *Proceedings of the 23rd World's Poultry Congress*, Brisbane, Australia, 30 June–4 July 2008 (CD ROM). World's Poultry Science Association, Beekbergen, The Netherlands.

Capua, I. and Alexander, D.J. (2009) *Avian Influenza and Newcastle Disease: A Field and Laboratory Manual.* Springer-Verlag Italia, Milan, Italy.

Chitate, F. and Guta, M. (2001) Country report: Zimbabwe. In: Alders, R.G. and Spradbrow, P.B. (eds) *SADC Planning Workshop on Newcastle Disease Control in Village Chickens.* Proceedings of an International Workshop, Maputo, Mozambique, 6–9 March 2000. ACIAR Proceedings No. 103. Australian Centre for International Agricultural Research, Canberra, pp. 46–50.

Cumming, R.B. (1992) Newcastle disease research at the University of New England. In: Spradbrow, P. (ed.) *Newcastle Disease in Village Chickens.* ACIAR Proceedings No. 39. Australian Centre for International Agricultural Research, Canberra, pp. 84–91.

Damerow, G. (2010) *Storey's Guide to Raising Chickens.* Storey Publishing, North Adams, Massachusetts.

Ekue, F.N., Pone, K.D., Mafeni, M.J., Nfi, A.N. and Njoya, J. (2002) Survey of the traditional poultry production system in the Barmenda area, Cameroon. In: *Characteristics and Parameters of Family Poultry Production in Africa.* International Atomic Energy Agency, Vienna, pp. 15–25.

FAO (2008) *Biosecurity for Highly Pathogenic Avian Influenza: Issues and Options.* FAO Animal Health and Production Paper No. 165. Food and Agriculture Organization of the United Nations, Rome.

FAO (2010) *Chicken Genetic Resources Used in Smallholder Production Systems and Opportunities for Their Development.* FAO Smallholder Poultry Production Paper No. 5. Food and Agriculture Organization of the United Nations, Rome.

FAPRI (Food and Agricultural Policy Research Institute) (2010) US and World Agricultural Outlook. http://www.fapri.iastate.edu/outlook/2010/text/Outlook_2010.pdf (accessed 2 January 2011).

Farrell, D.J., Bagust, T.J., Pym, R.A.E. and Sheldon, B.L. (2000) Strategies for improving the production of scavenging chickens. *Asian-Australian Journal Animal Science* 13, Supplement, 79–85.

Garber, L., Hill, G., Rodriguez, J., Gregory, G. and Voelker, L. (2007) Non-commercial poultry industries. surveys of backyard and gamefowl breeder flocks in the United States. *Preventive Veterinary Medicine* 80, 120–128.

Gueye, E.H.F. (1998) Village egg and fowl meat production in Africa. *World's Poultry Science Journal* 54, 73–86.

Guèye, E.F. (2000) The role of family poultry in poverty alleviation, food security and the promotion of gender equality in rural Africa. *Outlook on Agriculture* 29, 129–136.

Henning, J., Morton, J., Pym, R., Hla, T. and Meers, J. (2009) Evaluation of strategies to improve village chicken production: controlled field trials to assess effects of Newcastle disease vaccination and altered chick rearing in Myanmar. *Preventive Veterinary Medicine* 90, 17–30.

Khalafalla, A.I., Awad, S. and Hass, W. (2002) Village poultry production in the Sudan. In: *Characteristics and Parameters of Family Poultry Production in Africa.* International Atomic Energy Agency, Vienna, pp. 87–93.

McLeod, A., Thieme, O. and Mack, S. (2009) Structural changes in the poultry sector: will there be smallholder poultry development in 2030? *World's Poultry Science Journal* 65, 191–199.

Marvin, G. (1984) The cockfight in Andalusia, Spain: images of the truly male. *Anthropological Quarterly* 57, 60–70.

Mavale, A.P. (2001) Country report: Mozambique. In: Alders, R.G. and Spradbrow, P.B. (eds)

SADC Planning Workshop on Newcastle Disease Control in Village Chickens. Proceedings of an International Workshop, Maputo, Mozambique, 6–9 March 2000. ACIAR Proceedings No. 103. Australian Centre for International Agricultural Research, Canberra, pp. 20–25.

Minga, U.M., Katule, A.M., Yongolo, M.G.S. and Mwanjala, T. (1996) The rural chicken industry in Tanzania: does it make sense? In: *Proceedings of Tanzania Veterinary Association Scientific Conference*, vol. 16. Tanzanian Veterinary Association, Dar es Salaam, pp. 25–28.

NAFRI (2005) *Indigenous Chicken and Rural Livelihoods in Lao PDR*. National Agriculture and Forestry Research Institute. Vientiane, Lao PDR.

Njue, S.W., Kasiiti, J.L., Macharia, M.J., Gacheru, S.G. and Mbugua, H.C.W. (2002) Health and management improvements of family poultry production in Africa – survey results from Kenya. In: *Characteristics and Parameters of Family Poultry Production in Africa*. International Atomic Energy Agency, Vienna, pp. 39–45.

Neuman, W. (2009) Keeping their eggs in their backyard nests. *New York Times*, 4 August; available at: http://www.nytimes.com/2009/08/04/business/04chickens.html?_r=1&scp =1&sq=william+neuman+chickens&st=nyt (accessed 30 May 2010).

Pym, R.A.E., Guerne Bleich, E. and Hoffmann, I. (2006) The relative contribution of indigenous chicken breeds to poultry meat and egg production and consumption in the developing countries of Africa and Asia. In: *Proceedings of the 12th European Poultry Conference*, Verona, Italy, 10–14 September 2006 (CD ROM). World's Poultry Science Association, Beekbergen, The Netherlands.

Roberts, J.A. (1992) The scavenging feed resource base in assessments of productivity of scavenging village chickens. In: Spradbrow, P. (ed.) *Newcastle Disease in Village Chickens. Control with Thermostable Oral Vaccines*. Proceedings of an International Workshop held in Kuala Lumpur, Malaysia, 6–10 October 1991. Australian Centre for International Agricultural Research, Canberra, pp. 43–49.

Sonaiya, E.B. (2004) Direct assessment of nutrient resources in free range and scavenging systems. *World's Poultry Science Journal* 60, 523–535.

Sonaiya, E.B., Branckaert, R.D.S. and Gueye, E.F. (1999) Research and development options for family poultry. *First INFPD/FAO Electronic Conference on Family Poultry*, 7 December 1998–5 March 1999. Introductory paper. http://www.fao.org/ag/againfo/ themes/en/infpd/econf_scope.html (accessed 2 January 2011).

Spradbrow, P.B. (1993) Newcastle disease in village chickens. *Poultry Science Review* 5, 57–96.

USDA (US Department of Agriculture) (2011) Biosecurity for Birds. http://www.aphis.usda. gov/animal_health/birdbiosecurity/ (accessed 2 January 2011).

CHAPTER 7

Technology and Programmes for Sustainable Improvement of Village Poultry Production

B. Besbes, O. Thieme, A. Rota, E.F. Guèye and R.G. Alders

ABSTRACT

Although the commercial poultry sub-sector has reached a dominating position globally during the last three decades, village poultry production is still very important in most developing countries. Village poultry makes up more than 80% of the poultry stocks in many of the countries of Africa, Asia Pacific and Latin America. The village poultry sub-sector contributes significantly to food self sufficiency, poverty alleviation and gender empowerment. It is a noticeable source of employment and well-being, especially for disadvantaged groups and in less-favoured areas. Despite many constraints, including high mortality from diseases and poor nutrition, significant improvements can be achieved through well-designed development programmes that endow the different actors dealing with village poultry with the necessary knowledge, skills and resources. Beyond this need for substantial improvement in human and institutional capacity building, planners and policy makers should be sensitized to recognize the significance and potentials of village poultry production. This chapter highlights the importance of village poultry production as a tool for poverty reduction, food security and gender empowerment in developing countries. It identifies the development objectives for such production systems and their contributions to meeting the Millennium Development Goals. It reviews the options, strategies and technologies that have been used to achieve these objectives and draws a few lessons.

INTRODUCTION

Over the past three decades, a 'livestock revolution' (Delgado *et al.*, 1999) has led to a rapid growth in poultry production. This 'revolution' and the rapid increase in demand for meat and other livestock products in developing countries has been fuelled by population growth, income growth and urbanization.

From the 1990s to 2005, consumption of poultry meat in developing countries increased by 35 million tonnes – almost double the increase that occurred in developed countries. This increase has been most evident in East and South-East Asia and in Latin America, particularly in China, India and Brazil (FAO, 2008). The share of the world's poultry meat consumed in developing countries rose from 43 to 54% between 1990 and 2005. Similarly, the proportion of the world's poultry meat produced in developing countries rose from 42 to 57%. It is projected that production and consumption of poultry meat in developing countries will increase by 3.6 and 3.5% per annum, respectively, from 2005 to 2030 (FAO, 2008).

To meet this increasing demand, the poultry sector has undergone major structural changes throughout the world during the past two decades. Poultry production has become more intensive, geographically concentrated, vertically integrated and linked with global supply chains. This has been enabled by technological developments and innovations in all aspects of poultry production: breeding, reproduction, feeding and housing, disease control, transportation, processing, storage and marketing. The poultry industry also took advantage of increased preference for lean meat and processed poultry food products among consumers in developed countries.

Sectoral growth has, so far, mainly benefited large commercial producers, while small-scale poultry producers have been unable to participate in this dynamic market. As a consequence, a growing divide has emerged in the poultry sector: a large-scale commercial sub-sector dominated by industrial vertically integrated companies serving a growing market in the developed world and in many developing countries, and a small-scale sub-sector that is predominant in the least-developed countries. The large-scale commercial sub-sector now accounts for an estimated 72% of global poultry meat production and 61% of global egg production (Steinfeld *et al.*, 2006), with large differences between countries.

In most developing countries, small-scale production systems represent the alternative production systems to the industrial one, and include village and urban and peri-urban production systems. Chicken is generally the dominant species, but turkey, duck and guinea fowl are also important in some regions. A recurrent question concerns the future of small-scale production in general and village production in particular. This raises a series of other questions:

- What are small-scale and village poultry production systems?
- Why should we care about village poultry production?
- What are the development objectives for village poultry production?
- What strategies, options and technologies have contributed to achieving these objectives?
- What are the lessons learned?

This chapter tries to answer these questions based on reviewed literature and the authors' personal experiences. It focuses on village chicken production, but reference to other systems and species is made when appropriate.

SMALL-SCALE AND VILLAGE POULTRY PRODUCTION SYSTEMS

In all developing countries, many people keep small numbers of poultry for home consumption, occasional sales and various socio-cultural uses. This practice was termed 'village poultry' production as it was originally concentrated in villages. The term 'scavenging poultry' was then created to describe the feed supply of this production system, and became almost synonymous with village poultry. Increasing urbanization has made the keeping of small numbers of poultry in urban and peri-urban areas more common. If they are housed all or most of the time, the system is often called 'backyard production'. With the decreasing scavengeable feed resource base in villages and the absence, or very limited availability, of natural feed resources in urban environments, supplementary feeding has become more common. The term 'family poultry' was, therefore, created to describe the variety of small-scale poultry production systems that are found in rural, urban and peri-urban areas of developing countries. Rather than defining the production systems per se, the term is used to describe poultry production that is practised by individual families as a means of obtaining food security, income and gainful employment for women and children (Sonaiya, 1999).

The species of birds that are kept, the management and feeding practices and the common size of flocks may vary among regions and countries. Chickens are kept almost everywhere, while ducks, turkeys and guinea fowl are important in some regions. The level of external inputs for feeding and health is usually low. Birds are generally of a non-descript type and are multiplied by broody hens. However, influenced by commercial poultry production and development programmes, more intensive forms of small-scale poultry production have evolved and this has influenced feed and chick supply and healthcare. For further description, see Chapter 6 (this volume).

Small-scale production systems have been categorized by Guèye (2005) as: (i) traditional scavenging backyard or village systems; (ii) semi-scavenging systems; or (iii) small-scale intensive systems. Common features of all these poultry production systems are that they are usually practised by women based on traditional knowledge and that they contribute an important, but small, proportion of household income. This chapter addresses the state of these small-scale poultry production systems in village environments and therefore uses the term 'village poultry production'.

WHY SHOULD WE CARE ABOUT VILLAGE POULTRY PRODUCTION?

Village poultry makes a substantial contribution to food security and poverty alleviation in many countries around the world (Dolberg, 2008; Alders and Pym, 2009); 80 to 95% of rural households in sub-Saharan Africa and South Asia keep one or more species of poultry (Guèye, 2003). Village poultry makes up about 80% of poultry stocks in low-income food-deficit countries (Guèye, 2003; Pym et al., 2006) and provides high-quality food that improves the nutritional status and health of household members. This is important for rural poor people in developing countries, especially young children and their mothers who do not consume enough animal-based food and suffer high rates of under-nourishment and micronutrient deficiency.

Village poultry also help to diversify incomes and act as a form of household savings and insurance. Poultry are easily and quickly sold for essential household needs such as food, medicines and school fees. Village poultry production is an activity that is generally carried out by women assisted by children (see Plates 9 and 10 in colour plate section) and its promotion can thus contribute to women's empowerment. However, in order to fully understand the significance and potential of village poultry production in supporting smallholder livelihoods, we must also appreciate the socio-cultural and religious functions of poultry in local communities, beyond the strictly economic or nutritional importance of the birds or their products (Kryger et al., 2010).

Village poultry is an integral component of the livelihoods of poor rural households, and is likely to continue to play this role for the foreseeable future (FAO, 2008). However, as figures for the proportion of poultry kept under family-based operations are not available, village poultry production is not visible in national statistics and often not considered by policy makers and development planners. It is therefore essential to estimate properly the relative contributions of the different sources to poultry meat and egg production and consumption (Pym et al., 2006) as well as the impact on employment and income generation.

Despite its importance, there are threats to the future of village poultry production from the rapidly changing structure of the poultry sector, with increasing pressure for biosecurity regulations, such as the closure of open wet markets, or the requirement that all poultry be permanently housed. A careful analysis is therefore needed to identify potentials for development and for the future role of village poultry in the livelihoods of poor rural people.

DEVELOPMENT OBJECTIVES FOR VILLAGE POULTRY PRODUCTION

This section identifies development objectives and their contributions to meeting the Millennium Development Goals (MDGs; Box 7.1) adopted by world leaders in 2000 (UN, 2008) and reiterated during the MDG Review Summit in September 2010. The overall development objective for village poultry production is to increase its contributions to poverty reduction and food security (MDG 1), gender equity and women's empowerment (MDG 3), well-being of the rural population (MDGs 2, 4, 5, 6 and 8) and environmental sustainability (MDG 7) in developing countries (Alders and Pym, 2009). These can be achieved through the following specific development objectives:

- improve village poultry production and productivity in a sustainable manner;
- adapt village poultry production to the changing environment;

- maintain the diversity of local bird types;
- improve access to markets and the supply of poultry and poultry products to remote locations that do not attract commercial poultry producers;
- build the capacity of stakeholders involved in village poultry production; and
- raise awareness and influence livestock policy to promote village poultry production.

Box 7.1. Millennium Development Goals

Goal 1: Eradicate extreme poverty and hunger
Goal 2: Achieve universal primary education
Goal 3: Promote gender equality and empower women
Goal 4: Reduce child mortality
Goal 5: Improve maternal health
Goal 6: Combat HIV/AIDS, malaria and other diseases
Goal 7: Ensure environmental sustainability
Goal 8: Develop a global partnership for development

STRATEGIES, OPTIONS AND TECHNOLOGIES FOR ACHIEVING THE DEVELOPMENT OBJECTIVES

Addressing the development objectives identified above requires the design and implementation of development strategies combining five components. (i) animal health; (ii) management; (iii) breeding; (iv) marketing; and (v) networking. The success of such strategies depends on balanced improvement of all the components, with priority given to the most limiting ones (Nessar and Thieme, 2004). There is evidence that isolated initiatives that aim to develop one component without considering possible limitations in the others often fail to produce sustainable results and impacts. This section reviews the strategy components needed for achieving the development objectives for village poultry production and discusses the options and technologies used to implement them in different regions of the world.

Animal health

Diseases of economic importance that cause high mortality differ according to the species concerned (Table 7.1). In village chickens, the major health constraint in many developing countries is Newcastle disease (ND) (Alexander *et al.*, 2004). In these countries, circulating strains of ND virus are capable of causing 100% mortality in unprotected flocks (Alders and Spradbrow, 2001). Highly pathogenic avian influenza (HPAI) is less widespread than ND but does cause high mortality in chicken and quail flocks when outbreaks occur. In South-East Asia, fowl cholera is recognized as a significant problem in most

Table 7.1. Species variation in susceptibility to diseases causing high mortality in different types of village poultry (Ahlers *et al.*, 2009; Capua and Alexander, 2009).

Species	Highly pathogenic avian influenza	Newcastle disease	Fowl cholera
Chickens	+++	+++	++
Mallard ducks	+/–	Ducklings +/–	++
Pigeons	+	+	++
Quail	+++	–	++

(–) no mortality; (+/–) occasional low mortality; (+) low mortality; (++) medium mortality; (+++) high mortality.

species of village poultry, with duck plague (duck virus enteritis) being the disease that causes major economic losses in ducks (Meers *et al.*, 2004). There is a range of other infectious diseases, such as fowl pox, and internal and external parasites, that affect the health of village poultry (Ahlers *et al.*, 2009). These diseases cause lower mortality and so are not usually ranked by communities as priorities.

The effective control of diseases is an essential first step towards improving village poultry production (Ahlers *et al.*, 2009). To date, the most successful disease control programmes in village poultry have involved community vaccinators (Alders *et al.*, 2010) or poultry workers (Schleiss, 2001). The introduction of thermotolerant ND vaccines that are administered by community vaccinators (see Plate 11 and 12 in colour plate section) has greatly increased flock sizes and contributed significantly to household food security, poverty alleviation and mitigating the effects of HIV/AIDS (Alders and Pym, 2009; Alders *et al.*, 2010; Moreki *et al.*, 2010). Experience gained from ND control activities involving live, thermotolerant ND vaccines has shown that sustainable programmes are composed of five essential elements:

1. appropriate vaccine, vaccine technology and vaccine distribution mechanisms;
2. effective extension materials and methodologies that target veterinary and extension staff as well as community vaccinators and farmers;
3. simple evaluation and monitoring systems for both technical and socio-economic indicators;
4. economic sustainability based on commercialization of the vaccine and vaccination services and marketing of surplus chickens and eggs; and
5. support and coordination by relevant government agencies for the promotion and implementation of vaccination programmes (Copland and Alders, 2005).

The ND control activities initiated by the Southern Africa Newcastle Disease Control Project that was implemented in Malawi, Mozambique and Tanzania continue under the management of national authorities (Alders *et al.*, 2010). Appropriate biosecurity practices for village poultry can be implemented at the household level but, for best results, engagement with the entire village and its leadership is critical (Alders *et al.*, 2007; Ahlers *et al.*, 2009). The Village-based Biosecurity Education and Community Programme in South and West Sulawesi in Indonesia has facilitated the active participation of community members in village-wide biosecurity activities to prevent and control HPAI (FAO, 2011).

Management

The productivity of poultry depends on the management system adopted. Figure 7.1 illustrates the effect of the level of intensification on the productivity of chickens. Village poultry production systems are low-input and low-output production systems. In addition to the diseases described above, they experience a number of constraints such as insufficient nutrition, predation, poor housing and scarce husbandry practices. Together, these factors result in high losses and corresponding low levels of productivity. The introduction of improved technology, particularly 'adapted technology', aims to improve productivity and reduce mortality of village poultry using few management inputs. Unfortunately, systematic studies on factors influencing farmers' decisions to invest in poultry production technologies, such as introduction of improved breeds and associated improved management practices (feeding, housing, health, etc.), are rather limited. There is, however, strong evidence that by adopting a number of 'simple adapted technologies' of proven impact, it is

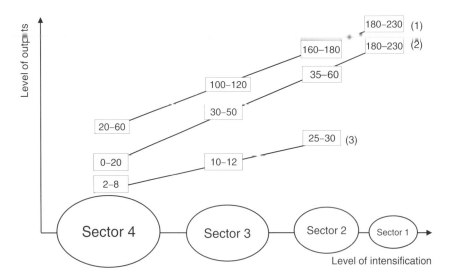

Fig. 7.1. Effect of intensification in production systems on the productivity of chickens in developing countries (adapted from Guèye, 2003): (1) productivity (no. of eggs per year per hen); (2) uses (no. of eggs for consumption, sale or gift per hen): (3) survivability (no. of year-old chickens per hen). Classification of poultry production systems (FAO/OIE, 2007): Sector 1 = industrial and integrated system; Sector 2 = small-scale commercial system; Sector 3 = commercial system; Sector 4 = village or backyard system.

possible to drastically reduce bird mortality (to an acceptable 4–5% in young birds and 1–2% in adults), improve productivity even with native poultry breeds and contribute to doubling the income of US$1 per day on which many less-privileged persons live. Three examples of simple technologies effective in addressing some of the above-listed constraints are presented below.

Protecting baskets for chicks

Many wild and domestic predators can cause extensive losses to rural poultry flocks, especially among birds that are less than a month old (Conroy et al., 2005). Apart from adequate night shelter (chicken coop), a protecting basket to cover a hen with its chicks during the day is an effective adapted technology based on traditional knowledge that can be used to reduce such losses (see plate 5 in colour plate section). Once the birds are older, the basket can be partly raised so that they can leave the basket to feed and drink freely but can still return quickly to the shelter in case of danger.

Improved feeding techniques

Feed cost is an essential consideration in achieving profitable poultry activities. Proper feeding is especially important for young chicks. A combination of the basket method described above with supplementary feeding of the chicks and the hen has been found to improve the production and health of village chickens (Sarkar and Bell, 2006; Henning et al., 2009). Many development programmes have promoted the capacity of local technicians to formulate balanced feeding regimes by mixing locally available feedstuffs. However, the protein source is often imported (and thus costly), unavailable or mixed at sub-standard levels. Alternative feeds are used to supplement existing energy and protein sources (Ravindran, 2010). In Africa and India, it is common practice among backyard poultry farmers to feed termites, other insects and maggots to poultry. Farmers fill a mud pot with old jute sacks, paddy straw, maize husk or any available dried crop stubble and dried cow dung. This mixture is moistened with water and the pot placed upside down in the field. If the inverted pot is opened the next day, it will be full of termites and can be given as feed to the birds.

Low-tech mini-hatcheries

In rural poultry development projects, the availability of healthy day-old chicks (DOCs) at a low cost is one of the key factors in achieving sustainability and economic viability/profitability. As it is difficult to transport DOCs from commercial hatcheries to remote locations, it is crucial to produce them on site. The Microfinance and Technical Support Project (MFTSP), implemented by the International Fund for Agricultural Development (IFAD) and Palli Karma-Sahayak Foundation (PKSF) in Bangladesh, has made it possible for poor women to operate low-tech (functioning without electricity) mini-hatcheries profitably, essentially by addressing management constraints (Nahar et al.,

2006). Rice husk models are now being replaced by more effective and easier-to-manage 'sand models' (see Plate 13 in colour plate section). Hatchability rate is between 85 and 90%.

Genetic improvement

Breeding poultry for improved production and productivity under village conditions implies improving performance, reproductive capacity and liveability or survival. As indigenous or local breeds are dual purpose, improving their performance means increasing both body weight and egg production in a single population. While it is considered to be a hindrance to high egg yield under commercial production, brooding and natural incubation capacity is a desirable trait under village conditions, as it is not always possible to utilize artificial incubation. Survival or longevity under village conditions is an indication of the bird's ability to withstand bacterial or viral infections.

Despite the high number of poultry and poultry breeds kept by smallholders in developing countries, only 18% of countries reported chicken breeding as a priority, and only 14% had structured breeding activities for chickens (FAO, 2007a). Even fewer countries reported structured breeding activities for turkeys (five countries), ducks (eight) and geese (four). These breeding programmes are mainly in Europe (FAO, 2007a). Therefore, one may question whether the genetic improvement of these populations is an option for achieving the development objectives described above. In other words, why not simply make use of the high-yielding commercial hybrids to get more eggs or meat?

Practical experiences have demonstrated that commercial poultry geno-types, both broilers and layers, are not appropriate for low-biosecurity, semi-scavenging, systems as they have: (i) specific feeding requirements; (ii) lost 'broodiness' traits; (iii) reduced immunity to the harsh environment; and (iv) are unable to scavenge a substantial part of their feed from the field. Arguments in favour of national breeding programmes are: (i) adaptation, improvement and conservation of local genetic diversity, in line with countries' commitments under the Global Plan of Action for Animal Genetic Resources adopted in 2007 (FAO, 2007b); (ii) food sovereignty and independence from imports; and (iii) biosecurity and zoosanitary considerations.

A number of projects have been implemented to genetically improve indigenous chicken populations. They have been based on cross-breeding, whether structured or not, between an improved exotic breed and a local breed, with the aim of combining the better production capacity of the former with the latter's adaptability to harsh environments. Three examples are presented below.

In Bangladesh, a cross-breeding programme was conducted by the Directorate of Livestock Services (DLS) to supply birds to villages. Two breeds were used: Fayoumi and Rhode Island Red. The Fayoumi, an Egyptian breed, is rather common in South Asia (Bangladesh, Pakistan and Afghanistan) mainly for egg production. The Rhode Island Red was the original American breed and not the modern, commercial, high-yielding version. DLS either produces the F_1 chickens and transfers them to the rearing farms or transfers the parents to the breeding farms (see Plate 14 in colour plate section). The cross-bred (F_1) chickens, known as Sonali, have proved to be high yielding and profitable (Rahman *et al.*, 1997). However, in order to perform, Sonali birds must be kept under improved management systems. In Afghanistan, under a joint project of the Food and Agriculture Organization of the United Nations (FAO) and the US Agency for International Development (USAID), the average production of eggs from chickens of a similar type increased from an average yearly yield of 125 eggs under traditional conditions to 241 eggs under an improved semi-scavenging poultry production system. This resulted in an average monthly increase in income from eggs, from US$12–13 to US$144.

Another cross-breeding strategy for improving the performance of local populations is the introgression of genetic material. This can be achieved through back-crossing or cockerel exchange programmes. In Uganda, the F_1 (50% Bovan Brown) and the back-crosses (25% Bovan Brown), generated at the Serere Animal and Agricultural Institute (SAARI), were superior to the local chickens in terms of daily gain, but their superiority decreased gradually and vanished for the 25% Bovan cross at 6 months of age – the traditional market age (Sørensen and Ssewannyana, 2003). No report on reproductive capacity and general fitness of the hens with Bovan genes has been obtained, but higher mortality was observed as the project progressed. Cockerel exchange programmes have been practised for many years in some African countries (Alexander *et al.*, 2004). In such programmes, cocks of improved breeds are distributed to smallholders on the condition that all fertile indigenous cocks are removed from the flocks and the exotic cocks are replaced every year. However, several reports have concluded that this type of activity has not changed the characteristics of the basic populations, except for contributing to a larger variation in plumage colour (Sørensen, 2010).

A unique example of transposing industrial breeding structure and logistics to backyard poultry is offered by Kegg Farms Private Ltd, in India, which produced a cross, named the Kuroiler ('Kuroiler' = 'Kegg + Broiler'), from coloured broiler males crossed with Rhode Island Red or White Leghorn × Rhode Island Red females (Khan, 2008). The Kuroiler chicken is bred for the Indian rural market and is supplied to farmers through a network of local suppliers. Eggs are supplied from the parent company to hatcheries, which produce DOCs for sale to 'mother units' kept by village entrepreneurs. They raise the birds to 2 or 3 weeks old in netted houses, vaccinate them and sell them to *pheriwallahs* (small traders) or directly to owners of scavenging flocks in the same village. The latter keep the birds and market them in much the same way as the traditional 'desi' breed. In the first year (1993), the company sold more than one million day-old Kuroiler chicks. In 2005/6, it sold 14 million – an annual growth rate of almost 22% sustained for more than a decade. A field study of Kuroiler production (Ahuja *et al.*, 2008) showed that a large proportion of those raising the birds were landless households or marginal farmers with less than one acre of land. On average, households raising Kuroilers generated more than five times as much from their poultry enterprise as did households that kept no Kuroilers, but this was also the result of better management.

In these examples, cross-breeding has provided significantly higher productivity when the management has improved, but has resulted in a loss or dilution of the indigenous birds' morphological characters and instinct for broodiness, which decreased smallholders' acceptance. The best way to improve the productivity of indigenous chickens without altering any of the morphological characters that are appreciated by the villagers is to select for production traits within a given population. In terms of rate of improvement, this is a slow process compared with cross-breeding with a genetically superior breed, which probably explains why, to our knowledge, no development project has adopted such an approach.

Access to markets

Village poultry birds and eggs are sold in local and urban markets (see Plate 15 in colour plate section), generally through an informal marketing system, with the involvement of numerous intermediaries such as village traders, collectors, wholesalers and retailers (Kryger *et al.*, 2010; Moges *et al.*, 2010). Village poultry products generally fetch 'premium' prices on local markets because of various factors such as freshness of eggs, special taste of free range meat, a particular colour for traditional rituals, etc. In 1995, for example, the average indigenous chicken meat prices in Dakar, Senegal, varied from US$2.5 kg^{-1} to US$3.9 kg^{-1} at markets and supermarkets, respectively (Guèye, 2003). These represented increases of about 13% at markets and 27% at supermarkets in comparison with prices of broilers' meat. Higher prices on the markets may also be linked to a particular event. For example, in Bure, Ethiopia, market prices during festival/holy days compared with ordinary market days show an increase of 19% for matured male chickens, 15% for matured hens, 24% for pullets/cockerels and 1% for eggs (Moges *et al.*, 2010).

The marketing system can be improved by providing technical assistance (e.g. training sessions, veterinary assistance, credits/loans) to small-scale poultry actors (producers and sellers, in particular). Similarly, direct transactions between producers and consumers can be supported by reducing the number of intermediaries. The examples presented below illustrate the marketing strategies envisaged by different projects.

In the 'Bangladesh model' (Sørensen, 2010), efforts have been made to facilitate access to the market and to use small-scale poultry rearing as a source of income generation in the form of a value chain involving eight activities/ actors: (i) key poultry rearers; (ii) model rearers; (iii) pullet rearers; (iv) mini poultry farmers; (v) chick rearers; (vi) mini-hatcherers; (vii) egg collectors; and (viii) poultry workers. All these activities, except for egg collection, are entirely in the hands of women, This approach resulted in a yearly average income ranging from US$60 to US$375 per stakeholder and a gross national product per person estimated at US$450 in the year 2000.

In the 'Kuroiler model', the company supplied 1500 mother units with day-old 'Kuroiler' chicks – directly or through its appointed dealers/suppliers. The mother units are operated by local entrepreneurs who keep anywhere between 300 and 2000 birds at one time. They sell them to vendors (*pheriwallas*) who travel to villages and sell the chicks to households at the price of about Rs20 (US$0.5) per chick. Typically, the mother unit entrepreneur and the *pheriwallas* make a profit of approximately Rs3 per bird. Finally, the rural households make Rs250–300 (US$6.5–7.5) per month as supplementary income. They sell eggs and also the birds for meat through Kegg's distribution channel. From the parent farms to the consuming households and village markets, numerous actors are involved and benefit from this intervention (FAO, 2008).

The 'Rakai chicken model' was launched in Rakai District, Uganda, to improve the livelihoods of smallholder poultry-keeping farmers. A project was implemented from June 2003 to August 2007, with 400 farmers as direct beneficiaries. Because group farmers pass on a 2-month-old hen chick for every hen received, the project has reported indirect benefits for a further 2400 farmers through this 'pass-on' mechanism. Through cross-breeding local chickens with commercial layer and broiler breeds and the use of programmed hatching, the project developed a dual-purpose F_3 bird (the Rakai chicken). Synchronizing hatching ensures that groups of chicks of almost the same age are produced, which facilitates vaccination, improved management and marketing. The external evaluation, conducted in April 2005, revealed that demand from other farmers for products was particularly high, with about two-thirds of both fertilized eggs and improved chicks sold to other farmers from surrounding areas. These other farmers travelled to the project participants' farms to access their supplies; other market outlets included local non-governmental organizations (NGOs) and hotels (Ewbank *et al.*, 2007).

Networking

Several networks have been established to enable exchanges of views, experiences and research and development results among people engaged in small-scale poultry keeping in developing countries (Branckaert and Guèye, 2000; Guèye, 2009; Dolberg, 2010). The FAO encouraged and supported the setting-up of the African Network for Rural Poultry Development (ANRPD) in November 1989. This Information Exchange Network has been renamed the International Network for Family Poultry Development (INFPD, http://www.fao.org/ag/againfo/themes/en/infpd/home.html). Members of INFPD include researchers, policy makers, educators, staff of NGOs and development agencies, aid donors and smallholder farmers. The information collected is disseminated through a newsletter – *Family Poultry Communications* – to around 900 members from 103 countries. In September 2002, INFPD was accepted as the first Global Working Group within the World's Poultry Science Association (WPSA). Since March 2007, INFPD has been collaborating with the *World's Poultry Science Journal* (WPSJ) to publish, twice a year, a new section in WPSJ called 'Small-scale Family Poultry Production'. INFPD/FAO periodically run electronic conferences on family poultry such as 'The Scope and Effect of Family Poultry Research and Development' (from December 1998 to July 1999), 'The Bangladesh Model and Other Experiences in Family Poultry Development' (from May to July 2002) and 'Opportunities of Poultry Breeding Programmes for Family Production in Developing Countries: The

Bird for the Poor' (http://www.fao.org/ag/againfo/themes/en/infpd/home. html). Other networks and organizations with a networking function devoted to smallholder poultry include the Network for Smallholder Poultry Development, established in 1997 and managed by the University of Copenhagen in Frederiksberg, Denmark (http://www.ivs.life.ku.dk/Om-instituttet/IVS%20Development/Network_for_Smallholder_Poultry_ Development.aspx), and the International Rural Poultry Centre (IRPC), a subsidiary entity within the KYEEMA Foundation (http://www.kyeema foundation.org/content/irpc.php).

These networks have been promoting information exchange, supporting human capacity building (through training, education, sensitization and advocacy) and implementing development projects (along with training, research and human capacity development). The fight against ND (e.g. the use of heat-tolerant vaccines to control ND in village poultry flocks in Asia and Southern Africa in the early 1980s and 1990s) and the genetic improvement of local poultry (e.g. 'cock exchange programmes' in West Africa in the early 1970s) have been the key activities around which the various organizations and networks have centred their work.

Finally, the 'South Asia Pro-Poor Livestock Policy Programme' maintains a portal (http://www.sapplpp.org) which includes interesting documents on 'good practices' for village poultry development.

PROSPECTS FOR THE PROPER IMPLEMENTATION OF TECHNOLOGY AND PROGRAMMES

Any development strategy for village poultry production should consider the main motivations and knowledge of the producers, and the opportunities for reliable delivery of improved services and inputs. If selling products is a major motivation for production, then good knowledge of the type of products required and the best time for selling them will be important criteria. Existing structures for knowledge transfer and services can help to introduce new ideas and suitable innovations to village poultry production even if not created specifically for that purpose. Such positive synergies have been found, for example, between programmes for micro-credit and poultry production in Bangladesh (Nahar, 2008). The potential utilization of existing structures for services and input supply created for commercial producers, such as feed outlets and shops that supply vaccines and drugs, will ease the introduction and diffusion of innovations. Benefits for sustainability have been seen when provision of supplies and services is less reliant on public funds and more on demand from producers (Dolberg, 2010). In remote locations where such opportunities do not exist, the introduction of village poultry development activities may be expensive and difficult to sustain without links to broader development programmes.

Lessons learned prove that the likelihood of succeeding with technology transfer is higher when rural poultry development projects promote a comprehensive approach that includes motivation, group organization and intensive 'hands-on' training in poultry management. Using gender-sensitive approaches is critically important (Bagnol, 2009). This must be combined with vaccination, supply of small credit to the target groups, regular supervision and advice. Poultry development projects often target women. An important element of success is the development of tailored intensive training and demonstration packages for women which take into consideration their needs, interests and constraints (e.g. heavy daily workload) and are delivered by female trainers (see Plate 16 in colour plate section).

Another element that plays an important role in effective technology transfer is marketing. It is well recognized that resource-poor people are the least likely to take risks and, as a result, adopt new technologies only once they are sure of an adequate return on their investment in terms of time and money. In simple words, if motivated women or small farmers investing in rural poultry are provided with assistance in marketing their products in a regular and remunerative manner, innovative and adapted technologies are quickly adopted and disseminated within a specific project area. This implies that analysis of market opportunities, development of appropriate business plans, support to value chains and facilitation of access to markets for poultry products are necessary prerequisites for sustainable poultry development.

At the macro level, the potential of village poultry development with technology transfer as a component is not yet recognized. Despite the importance of village poultry in the national economies of developing countries and its role in improving the nutritional status and incomes of many small farmers and landless communities, this sub-sector does not rate highly in the mainstream of national economies and policies due to the lack of measurable indicators of its contribution to macro-economic indices, such as gross domestic product (GDP), and the scarce attention paid to it by agricultural decision makers and policy makers (including livestock specialists).

There is a need for coordinated advocacy and comprehensive strategic plans to bring village poultry development (and other small livestock sectors) back into the agenda of international and national institutions. This would include in particular:

- Raising the awareness of decision makers in national governments and donor agencies about the effectiveness of village poultry as a tool for poverty reduction, food security and women's empowerment. For instance, there is little inclusion of village poultry development action plans in Poverty Reduction Strategy Papers (PRSPs), IFAD's Country Strategic Opportunities Programmes (COSOPs) or other national development plans.
- Developing effective and consistent national pro-poor policies, which are crucial to capitalizing on the opportunities offered by the increasing demand for livestock products and the poverty-focused agendas of several countries.
- Supporting the formation and capacity of institutions for village poultry farmers, which can help to voice their needs and facilitate the provision of services and inputs to farming communities.
- Supporting participatory adaptive research, needed to identify appropriate technologies and models that are pro-poor, culturally acceptable,

economically viable and environmentally sustainable. Institutions that are part of the Consultative Group on International Agricultural Research (CGIAR) and National Agricultural Research Extension Systems (NARES) should include topics for development of village poultry in their respective research programmes.

- Strengthening extension and training programmes for capacity building, especially those that are adapted for women and other under-privileged groups.
- Enabling a market-led approach supported by services and infrastructures that are effective, accessible and of good quality (breeding, veterinary services, credit, processing, marketing, extension/training, etc.).
- Strengthening global networks such as the INFPD.

CONCLUSIONS

Sustainable village poultry development programmes are those that are built on the existing practices and capabilities of the beneficiaries from the local communities. They make efficient use of locally available resources (i.e. farmers' knowledge and practices, feed resources, building materials, equipment). Objectives and activities of village poultry development programmes in the developing world should relate to the perceptions, needs, priorities, interests and suggestions of relevant members of local communities, considering different groups, such as men and women, young and old people, poorer and wealthier families, and members of different socio-cultural groups (Guèye, 2003, 2005). Among a wide range of on-farm and off-farm activities, keeping poultry may compete with other activities, especially for women, and may not always have priority. A well-designed investigation of the socio-cultural and economic environments of potential target communities should therefore be undertaken in order to assess the situation before embarking on a village poultry development programme.

DISCLAIMER

The views expressed in this publication are those of the authors and do not necessarily reflect the views of their organizations.

REFERENCES

Ahlers, C., Alders, R.G., Bagnol, B., Cambaza, A.B., Harun, M., Mgomezulu, R., Msami, H., Pym, B., Wegener, P., Wethli, E. and Young, M. (2009) *Improving Village Chicken Production: A Manual for Field Workers and Trainers.* ACIAR Monograph No. 139. Australian Centre for International Agricultural Research, Canberra.

Ahuja, V., Dhawan, M., Punjabi, M. and Maarse, L. (2008) *Poultry Based Livelihoods of the Rural Poor: Case of Kuroiler in West Bengal.* Study Report, Document 012. South Asia Pro-Poor Livestock Policy Programme, New Delhi; available at: http://sapplpp.org/

informationhub/files/doc012-PoultryBasedLRPKuroiler-updated09Mar31.pdf (accessed 2 January 2011).

Alders, R.G. and Pym, R.A.E. (2009) Village poultry: still important to millions, eight thousand years after domestication. *World's Poultry Science Journal* 65, 181–190.

Alders, R. and Spradbrow, P. (2001) *Controlling Newcastle Disease in Village Chickens: A Field Manual*. ACIAR Monograph No. 82. Australian Centre for International Agricultural Research, Canberra; available at: http://aciar.gov.au/publication/mn082 (accessed 2 January 2011).

Alders, R., de Almeida, A., Cardoso, A., do Karmo, A., Dunn, S., Gusmão, D. and Jong, J. (2007) *Avian Influenza – A Manual for the National Investigation and Response Team for Avian Influenza in Animals*. Ministry of Agriculture, Forestry and Fisheries and Food Agriculture Organization of the United Nations, Dili, Timor Leste.

Alders, R.G., Bagnol, B. and Young, M.P. (2010) Technically sound and sustainable Newcastle disease control in village chickens: lessons learnt over fifteen years. *World's Poultry Science Journal* 66, 433–440.

Alexander, D.J., Bell, J.G. and Alders, R.G. (2004) *Technology Review: Newcastle Disease with Special Emphasis on Its Effect on Village Chickens*. FAO Animal Production and Health Paper No. 161. Food Agriculture Organization of the United Nations, Rome.

Bagnol, B. (2009) Improving village chicken production by employing effective gender-sensitive methodologies. In: Alders, R.G., Spradbrow, P.B. and Young, M.P. (eds) *Village Chickens, Poverty Alleviation and the Sustainable Control of Newcastle Disease*. Proceedings of an International Conference held in Dar es Salaam, Tanzania, 5–7 October 2005. ACIAR Proceedings No. 131. Australian Centre for International Agricultural Research, Canberra, pp. 35–42.

Branckaert, R.D.S. and Guèye, E.F. (2000) FAO's programme for support to family poultry production. In: Dolberg, F. and Petersen, P.H. (eds) *Poultry as a Tool in Poverty Eradication and Promotion of Gender Equality*. Proceedings of a Workshop held at Tune Landboskole, Denmark, 22–26 March 1999. DSR Forlag, Frederiksberg, Denmark, pp. 244–256; available at: http://www.smallstock.info/reference/KVLDK/tune99/24 Branckaert.htm (accessed 14 January 2011).

Capua, I. and Alexander, D.J. (2009) *Avian Influenza and Newcastle Disease: A Field and Laboratory Manual*. Springer-Verlag Italia, Milan, Italy.

Conroy, C., Sparks, N., Chandrasekaran, D., Sharma, A., Shindey, D., Singh, L.R., Natarajan, A. and Anitha, K. (2005) The significance of predation as a constraint in scavenging poultry systems: some findings from India. *Livestock Research for Rural Development* 17(6); available at: http://www.lrrd.org/lrrd17/6/conr17070.htm (accessed 14 January 2011).

Copland, J.W. and Alders, R.G. (2005) The Australian village poultry development programme in Asia and Africa. *World's Poultry Science Journal* 61, 31–37.

Delgado, C., Rosegrant, M., Steinfeld, H., Ehui, S. and Courbois, C. (1999) *Livestock to 2020. The Next Food Revolution*. Food, Agriculture and the Environment Discussion Paper No. 28. International Food Policy Research Institute, Washington, DC/Food and Agriculture Organization of the United Nations, Rome/International Livestock Research Institute, Nairobi.

Dolberg, F. (2008) Poultry production for livelihood improvement and poverty alleviation. In: *Poultry in the 21st Century: Avian Influenza and Beyond*. Proceedings of the International Poultry Conference, Bangkok, Thailand, 5–7 November 2007; available at: http://www.fao.org/ag/againfo/home/events/bangkok2007/docs/part3/3_1. pdf (accessed 31 January 2011).

Dolberg, F. (2010) *Poultry as a Tool in Human Development: Historical Perspective, Main Actors and Priorities*. FAO Smallholder Poultry Production Paper No. 6. Food and Agriculture Organization of the United Nations, Rome; available at: http://www.fao.org/ ag/againfo/themes/documents/poultry/SPP6.pdf (accessed 14 January 2011).

Ewbank, R., Nyang, M., Webo, C. and Roothaert, R. (2007) *Socio-Economic Assessment of Four MATF-Funded Projects*. FARM-Africa Working Paper No. 8. FARM-Africa, London.

FAO (2007a) *The State of the World's Animal Genetic Resources for Food and Agriculture*. Food and Agriculture Organization of the United Nations, Rome.

FAO (2007b) *The Global Plan of Action for Animal Genetic Resources and the Interlaken Declaration*. Food and Agriculture Organization of the United Nations, Rome; available at: http://www.fao.org/docrep/010/a1404e/a1404e00.htm (accessed 14 January 2011).

FAO (2008) *Poultry in the 21st Century: Avian Influenza and Beyond*. Proceedings of the International Poultry Conference, Bangkok, Thailand, 5–7 November 2007. FAO Animal Production and Health Proceedings No. 9. Food and Agriculture Organization of the United Nations, Rome; available at: ftp://ftp.fao.org/docrep/fao/011/i0323e/i0323e.pdf (accessed 14 January 2011).

FAO (2011) Village-based biosecurity: community participation in prevention and control of H5N1 HPAI in South and West Sulawesi, Indonesia. *FAO AIDE News*, Situation Update 73, 17 January 2011; available at: http://www.fao.org/docrep/013/al844e/al844e00.pdf (accessed 14 January 2011).

FAO/OIE (2007) The Global Strategy for Prevention and Control of H5N1 Highly Pathogenic Avian Influenza, revised March 2007. Food and Agriculture Organization of the United Nations and World Organisation for Animal Health in collaboration with World Health Organization. http://www.fao.org/docs/eims/upload/210745/glob_strat_HPAI_apr07_en.pdf (accessed 14 January 2011).

Guèye, E.F. (2003) Poverty alleviation, food security and the well-being of the human population through family poultry in low-income food-deficit countries. *Journal of Food, Agriculture and Environment* 1(2), 12–21.

Guèye, E.F. (2005) Gender aspects in family poultry management systems in developing countries. *World's Poultry Science Journal* 61, 39–46.

Guèye, E.F. (2009) The role of networks in information dissemination to family poultry farmers. *World's Poultry Science Journal* 65, 115–124.

Henning, J., Morton, J., Pym, R., Hla, T. and Meers, J. (2009) Evaluation of strategies to improve village chicken production: controlled field trials to assess effects of Newcastle disease vaccination and altered chick rearing in Myanmar. *Preventive Veterinary Medicine* 90, 17–30.

Khan, A.G. (2008) Indigenous breeds, crossbreds and synthetic hybrids with modified genetic and economic profiles for rural family and small scale poultry farming in India. *World's Poultry Science Journal* 64, 405–415.

Kryger, K.N., Thomsen, K.A., Whyte, M.A. and Dissing, M. (2010) *Smallholder Poultry Production – Livelihoods, Food Security and Sociocultural Significance*. FAO Smallholder Poultry Production Paper No. 4. Food and Agriculture Organization of the United Nations, Rome; available at: http://www.fao.org/ag/againfo/themes/documents/poultry/SPP4.pdf (accessed 14 January 2011).

Meers, J., Spradbrow, P.B. and Tu, T.D. (eds) (2004) *Control of Newcastle Disease and Duck Plague in Village Poultry*. Proceedings of a Workshop held at NAVETCO, Ho Chi Minh City, Vietnam, 18–20 August 2003. ACIAR Proceedings No. 117. Australian Centre for International Agricultural Research, Canberra.

Moges, F., Tegegne, A. and Dessie, T. (2010) *Indigenous Chicken Production and Marketing Systems in Ethiopia: Characteristics and Opportunities for Market-oriented Development*. IPMS of Ethiopian Farmers Project/ILRI Working Paper No. 24. International Livestock Research Institute, Nairobi; available at: http://mahider.ilri.org/bitstream/10568/2685/4/WorkingPaper_24.pdf (accessed 14 January 2011).

Moreki, J.C., Dikeme, R. and Poroga, B. (2010) The role of village poultry in food security and HIV/AIDS mitigation in Chobe District of Botswana. http://www.lrrd.org/lrrd22/3/more22055.htm (accessed 14 January 2011).

Nahar, J. (2008) Poverty reduction in Bangladesh through microfinance and poultry development. In: *Poultry in the 21st Century: Avian Influenza and Beyond.* Proceedings of the International Poultry Conference, Bangkok, Thailand, 5–7 November 2007. FAO Animal Production and Health Proceedings No. 9. Food and Agriculture Organization of the United Nations, Rome; available at: http://www.fao.org/ag/againfo/home/events/bangkok2007/docs/part4/4_2.pdf (accessed 31 January 2011).

Nahar, J., Fattah, K.A., Rajiur Rahman, S.M., Ali, Y., Sarwar, A., Mallorie, E. and Dolberg, F. (2006) The rice husk hatchery in the microfinance and technical support project in Bangladesh. *INFPD Newsletter* 16(2), 27–29.

Nessar, M.H. and Thieme, O. (2004) Family poultry production in Afghanistan. In: *Proceedings of XXII World's Poultry Congress*, Istanbul, Turkey, 8–13 June 2004 (CD-ROM); available at: http://www.fao.org/ag/againfo/themes/en/infpd/documents/papers/2004/7afghan1503.pdf (accessed 20 October 2010).

Pym, R.A.E., Guerne Bleich, E. and Hoffmann, I. (2006) The relative contribution of indigenous chicken breeds to poultry meat and egg production and consumption in the developing countries of Africa and Asia. In: *Proceedings of XII European Poultry Conference*, EPC Verona, Italy, 10–14 September 2006 (CD-ROM); available at: http://www.cabi.org/animalscience/Uploads/File/AnimalScience/additionalFiles/WPSAVerona/10222.pdf (accessed 20 October 2010).

Rahman, M., Sorensen, P., Jensen, H.A. and Dolberg, F. (1997) Exotic hens under semi scavenging conditions in Bangladesh. *Livestock Research for Rural Development* 9(3); available at: http://www.lrrd.org/lrrd9/3/bang931.htm (accessed 4 February 2011).

Ravindran, V. (2010) Poultry feed availability and nutrition in developing countries – Alternative feedstuffs for use in poultry feed formulations. http://www.fao.org/docrep/013/al706e/al706e00.pdf (accessed 31 January 2011).

Sarkar, K. and Bell, J.G. (2006) Potentialities of the indigenous chicken and its role in poverty alleviation and nutrition security for rural households. *INFPD Newsletter* 16(2), 15–26.

Schleiss, K. (2001) The Smallholder Livestock Development Project, Bangladesh. In: Alders, R.G. and Spradbrow, P.B. (eds) *SADC Planning Workshop on Newcastle Disease Control in Village Chickens.* Proceedings of an International Workshop, Maputo, Mozambique, 6–9 March 2000. ACIAR Proceedings No. 103. Australian Centre for International Agricultural Research, Canberra, pp. 140–142.

Sonaiya, E.B. (1999) International Network for Family Poultry Development: origins, activities, objectives and visions. In: Dolberg, F. and Peterson, P.H. (eds) *Poultry as a Tool in Poverty Eradication and Promotion of Gender Equality.* Proceedings of a Workshop held at Tune Landboskole, Denmark, 22–26 March 1999. DSR Forlag, Frederiksberg, Denmark, pp. 39–50; available at: http://www.fao.org/DOCREP/004/AC154E/AC154E02.htm#ch2.2 (accessed 20 October 2011).

Sørensen, P. (2010) *Chicken Genetic Resources Used in Smallholder Production Systems and Opportunities for Their Development.* FAO Smallholder Poultry Production Paper No. 5. Food and Agriculture Organization of the United Nations, Rome; available at: http://www.fao.org/docrep/013/al675e/al675e00.pdf (accessed 14 January 2011).

Sørensen, P. and Ssewannyana, E. (2003) *Progress in SAARI Chicken Breeding Project – Analyses of Growth Capacity.* DANIDA's Agricultural Sector Research Programme and NARO, Kampala, pp. 172–178.

Steinfeld, H., Gerber, P., Wassenaar, T., Castel, V, Rosales, M. and de Haan, C. (2006) *Livestock's Long Shadow. Environmental Issues and Options.* Food and Agriculture Organization of the United Nations, Rome.

UN (2008) *The Millennium Development Goals Report 2008.* United Nations, New York City, New York; available at: http://unstats.un.org/unsd/mdg/resources/static/products/progress2008/mdg_report_2008_en.pdf (accessed 31 January 2011).

CHAPTER 8
Production Systems for Waterfowl

D. Guémené, Z.D. Shi and G. Guy

ABSTRACT

Ducks and geese, which represent up to 7% of world poultry production, are raised for meat production, eggs, foie gras, down and feathers. Asia represents more than 80% of total waterfowl production, with China being the most important, and Europe accounting for around 13%. Traditional rearing conditions differ widely across the world, due to various species and breeds, environmental, cultural backgrounds and production conditions and objectives. Nevertheless, due to growing involvement of Western duck breeding companies, the demand for good production performances and high economic returns, more uniformity has been observed in recent years. However, demands for alternative rearing systems are emerging although, depending upon the country, the significance of alternative systems differs. Indeed, while this trend is exemplified by the ban of some cage systems and the development of free range systems in Europe, it is illustrated by the recent placement of duck layers for table eggs in conventional cages and the placement of an increasing number of geese in lightproof barns under a control photoperiod to get out-of-season production in China. These few examples illustrate the different uses and meanings of the term 'alternative system' in Eastern and Western countries.

INTRODUCTION

At the world scale, commercial waterfowl production accounts for near 7% of the total poultry production (Anonymous, 2010). Ducks and geese represent about 4.1% and 2.6% of the poultry meat production, respectively. Besides meat production, waterfowl are also raised for eggs, foie gras, down and feathers, with many local specialities or niche products. Expressed in number of birds produced, waterfowl represented about 1100 million ducks (5.4%) and 350 million geese (1.7%), respectively, in 2008 (Anonymous, 2010). Asia is by far the main contributor to this production with more than 80% of the total waterfowl production, while Europe accounts for about 13%. In Asia, China is by far the major contributor (87%). In Europe, France is the largest with more than 50% of the local production, but remains a relatively minor producer on a worldwide basis (7%). Production of waterfowl eggs for human consumption is not common in European countries, whereas it represents a significant volume of total egg production in Asia. Foie gras is mostly produced in France (80%), while the biggest single producer (> 40%) of down and feathers is China. Of importance, egg production for duckling and gosling production purposes is the prior step and requires specific rearing and housing conditions for particular duck and goose production systems.

In this chapter, we first present in more detail the species and breeds that are presently used as well as the economic context of the respective production systems. Their biological characteristics are described briefly as this is of importance in determining rearing conditions. Due to their respective importance in world production, we thereafter describe present rearing systems and some recent shifts to alternative production systems. We focus primarily on Asia and Europe, specifically and very largely on China and France. We first present the most common production systems and outline some recent trends in alternative production systems, both in Europe and Asia. Then we give a brief description of the specific rearing practices and housing conditions for egg laying production.

WATERFOWL SPECIES RAISED FOR PRODUCTION PURPOSES

Ducks and geese are both raised for different production purposes, which imply the use of different species and breeds as described below.

Ducks

Worldwide, three genotypes of domestic ducks, two species and their intercross are used for production purposes (Fig. 8.1). Numerous breeds representing over 90% of the ducks raised are of the common duck species (*Anas platyrhynchos*) and share the wild mallard duck as a common ancestor. The second species is the Muscovy duck (*Cairina moschata*) originating from South America. The last genotype is the mule duck and is the interspecific and sterile hybrid resulting from the cross-breeding between a Muscovy drake and a common duck female (Guémené and Guy, 2004; Brun *et al.*, 2005). These different genotypes vary greatly in their behavioural and physiological characteristics, and require specific rearing and housing conditions.

Common ducks

All breeds of domestic or common ducks originate from domestication of the wild mallard (*A. platyrhynchos*) about 2000 years ago (Conseil de l'Europe, 1999a). The Pekin duck, a heavy-body common duck, is the most widespread. The name of this breed originates from the famous 'Pekin duck' Chinese delicacy, known for its crispy skin when cooked. Most of the modern heavy strains exhibit white plumage and orange to yellow bills and legs, but locally

(a)   (b)

(c)   (d)

Fig. 8.1. (a) Pekin duck (*Anas platyrhynchos*), (b, c) white and coloured Muscovy duck (*Cairina moschata*) and (d) their intercross, the mule duck. (Image (a) courtesy of Grimaud Frères Sélection; images (b, c, d) courtesy of D. Guémené.)

several breeds appear in various forms or colours. Of much lower importance in volume, other common ducks that have coloured plumage are raised for meat production in various European countries. Among them, we can list the Aylesbury, Rouen, Cayuga and Pomeranian ducks (Ashton and Ashton, 2001).

Although they are raised mostly for meat production, some specific strains of common ducks are devoted to table egg production or are dual purpose in Asia. Although overall world duck egg production is unknown, it is estimated to represent about 7% and 15% (over 4 billion per annum) of total fowl eggs produced in India and China, respectively. Specific breeds exhibiting lower body weight and laying up to 350 eggs per reproductive cycle (Sauveur, 1988; Sauveur and De Carville, 1990) are present in this geographical area. Among them, the Indian runner, the Tsaiya, the Shao-Xing and the Khaki-Campbell ducks are prolific layers (Ashton and Ashton, 2001); this last being the best producer. Individual egg production of almost an egg a day in this breed for well over 12 months has been recorded and flock averages in excess of 300 eggs per duck per year are not uncommon. Although common duck breeds raised for meat production are far less productive, with less that 250 eggs laid per reproductive cycle (Sauveur, 1988; Sauveur and De Carville, 1990), these genotypes are used as female lines for the production of mule ducks.

The biological characteristics of the wild mallard duck were briefly described in the recommendation concerning domestic ducks (Conseil de l'Europe, 1999a). Under wild conditions, mallard ducks fly, swim and walk efficiently, but are largely aquatic. Being omnivorous, feeding for example on seeds, plants, insects and worms, they feed by foraging on land or by dabbling their beak along the water. Ducks perform complex behavioural sequences of feeding, bathing, preening and sleeping that are repeated a number of times throughout the day (Reiter, 1997; Guémené *et al.*, 2006).

Muscovy ducks

The Muscovy duck (*C. moschata*) used nowadays for meat production was domesticated by the Colombian and Peruvian Indians and then introduced to Europe by the Spanish and Portuguese in the 16th century. This species has supplanted the common duck in France due to its high yield of a reddish, lean and flavoured meat. It is also quite popular in other countries such as Italy, Egypt and to a lesser extent Spain, Malaysia and countries of South and North America.

This species is not used for egg production for human consumption due to its short laying cycle duration and low laying intensity compared with the common duck. Females will generally lay less than 200 eggs in two consecutive reproductive cycles of about 24 weeks separated by a moulting period of 12 weeks (Sauveur, 1988; Sauveur and De Carville, 1990). Originally, it was also used for foie gras production in France, but mule ducks, which benefit from hybrid vigour, became the bird of choice after artificial insemination techniques were optimized.

The biological characteristics of wild Muscovy ducks have also been briefly described (Conseil de l'Europe, 1999b). In the wild, the Muscovy duck lived in marshy tropical forests, but its robustness has enabled it to adapt to different climates and habitats. Muscovy ducks perch, fly, swim and walk efficiently and are omnivorous. Their diurnal rhythm of activity and anti-predator response is comparable to that of mallard ducks. Muscovy ducks are so-called 'mute ducks' and their vocalization is in the form of hissing. The early growth and development of the Muscovy duck is relatively slow compared with its Pekin counterpart and it generally takes a minimum of 10–11 weeks to achieve market weight, compared with 6–7 weeks for Pekin ducks. Muscovy ducks are sexually dimorphic, males being up to 45% heavier than females when they reach market weight. The difference between male and female growth rates makes single sex rearing a common practice. They can also be raised in mixed pens, with females removed for slaughter at a younger age than males (10 versus 12 weeks), but males can negatively affect female growth rates through competition for food. Wild Muscovy ducks, especially the males, are more aggressive than mallard ducks and fight frequently using their claws, wings and beaks, particularly for chasing off intruders. Under commercial production, feather pecking, injuries and cannibalism are important welfare problems (Rodenburg *et al.*, 2005).

Mule ducks

Cross-breeding between the two previously described species, i.e. a Muscovy drake and a common female duck, results in the production of a hybrid called the mule duck. The reverse intercross can also occur and produces the hinny duck, which has similar biological and behavioural characteristics to the mule duck (Arnaud *et al.*, 2008). The common female duck being a much better layer than the Muscovy one, this last cross is not of commercial use. Both intercrosses produce a sterile hybrid because of the difference in chromosome sizes and numbers between the two parent species (Brun *et al.*, 2005). Mule ducks are nowadays exclusively produced using artificial insemination, due to poor fertility rates after natural mating. It shows little sexual dimorphism and is able to flourish in cooler conditions than the Muscovy duck (Conseil de l'Europe, 1999b). The male mule duck is used for the production of foie gras, and currently accounts for more than 97% of the production in France. For this reason it also represents more than 50% of duck meat production in this country. It is also kept for meat production in Asia, mainly Taiwan, because of its flavoured and lean red meat (Raud and Faure, 1994).

Geese

Geese are thought to be the first poultry to be domesticated and breeds of geese currently used for commercial purposes originate from two species (Fig. 8.2): the greylag goose (*Anser anser*), which is considered to be the ancestor of domestic goose breeds of European origin; and the swan goose (*Anser cygnoïdes*), the ancestor of breeds of Asiatic origin. Geese belonging to the *A. cygnoïdes* species can be differentiated by the presence of a large knob on the front of the head, the one in the gander being even more prominent. Geese are mainly raised in China and the *A. cygnoïdes* breeds are consequently the most commonly used worldwide. Geese are kept primarily for the production of meat, down and feathers but also for ornamental purposes, and at least historically also to keep watch for intruders. In China, the Huoyang geese are valuable layers for table egg production, with production over 120 eggs per annum. In Europe, different breeds originating from the *A. anser* species have been developed. Among these are the grey Landaise goose for foie gras production, and the white Italian, the Embden and the Pilgrin for meat production. However, European goose production is nowadays very limited relative to the total amount of poultry meat produced (<3%) and has declined sharply by 20% in the last decade. Compared with other poultry species, geese have a long lifespan and are often used for seasonal egg production for 3 to 5 years, with the best performances being obtained in the second year (Sauveur, 1988). The biological characteristics of wild geese were briefly described in the European Council recommendation (Conseil de l'Europe, 1999c). Geese are gregarious birds which, in the wild, congregate in large flocks that stay together except during the breeding season where they develop monogamous bonds and disperse into pairs. Geese eat a variety of food items including different small invertebrates when foraging and were originally mainly kept on marginal grazing lands. Water is an important factor in their grooming behaviour and wild geese almost exclusively mate in open water. Although swimming water seems to encourage domestic geese to mate, they can mate satisfactorily without it. Domestic geese may be mated with four to six females to one gander, and it is important that these flocks are established as soon as possible. Communication by a variety of vocalizations is an important part of their behaviour. Wild geese migrate over long distances and they walk and run efficiently. The ability to fly is reduced in many domestic breeds especially in the heavy ones. Domestic geese are good walkers and if necessary they can range over a huge distance (5 km or more) to find their food and come back to the resting area in the evening.

(a)   (b)

Fig. 8.2. (a) The swan goose (*Anser cygnoïdes*) of Asiatic origin and (b) the greylag goose (*Anser anser*) of European origin. (Image (a) courtesy of INRA-UEPFG; image (b) courtesy of D. Guémené.)

GEOGRAPHICAL DISTRIBUTION OF WATERFOWL PRODUCTION

Although waterfowl account for less than 7% of the total world poultry production, it is not negligible with approximately 6 million tonnes produced in 2008. However, waterfowl production is significant only in a few countries. Indeed, some 83% and 95% of world duck and geese meat, respectively, is produced in Asia, with China dominating the output for both species (Table 8.1). Thus China represents 80.7% of the total Asian and 66.6% of the world production of ducks. Moreover, due to the difficulty in evaluating non-commercial production, duck production may be much higher in China than figures of the Food and Agriculture Organization of the United Nations (Anonymous, 2010) suggest. Duck meat has traditionally played an important role in Chinese culture and the increase in production in China is averaging 3.8% per annum. Some other Asian countries such as India, Indonesia and Bangladesh are also significant contributors to total duck production.

While Europe is the second largest duck-producing region, this output accounts only for about 12% of the world total and its share of production is decreasing, with the notable exception of France, representing nearly 54% of the European production (Table 8.1). Interestingly, French production differs from other countries in terms of species and breeds used. Indeed, most of the production in France originates from mule ducks and Muscovy ducks, with 60 and 40% for the meat produced and the reverse for the number of ducklings raised to reach these volumes, respectively, while Pekin ducks are by far the most common elsewhere in the world. Ducks are produced in America in equivalent proportions in the south and the central-northern part, for a total of 3.0% of world production. Africa produces about 1.5%, mainly in Egypt. Ducks are also present in Oceania but account for only 0.3% of the world production.

Table 8.1. World duck and goose production in 2008 by geographical region and for the most important production countries (Anonymous, 2010).

Zone	Country	Duck production		Goose production	
		'000 tonnes	%	'000 tonnes	%
World		3780		2377	
Asia		3122	82.6	2250	94.7
	China	2518	66.6	2238	94.0
	Malaysia	111	2.9	–	–
Europe		459	12.1	70	2.9
	France	249	6.6	2.5	0.1
	Germany	61	1.6	–	–
	Hungary	51	1.3	–	–
Africa		57	1.5	55	2.3
America		112	3.0	2	<0.1
Oceania		12	0.3	0.1	<0.01

Regarding goose production, China's leadership is even more important than in duck production. Thus, China contributes 99% of the Asian production and 94% of the 2.5 million tonnes of goose meat consumed annually worldwide. Europe is the second largest producer but represents a very small percentage (<3%), mainly in the eastern countries, with Ukraine, Romania, Poland and Hungary being the most significantly involved in this production. African countries account for 2.3% of the production, with Egypt being again the most important contributor. America and Oceania have very limited goose production.

Foie gras can be produced from specific strains of both duck (mainly mule duck) and geese (*A. anser*) species. French production accounted for about 80% of world production with about 19,000 tonnes produced in 2009; France is by far the world leader in both foie gras production and consumption. Duck foie gras represents the greater part of the national production (98%). In Europe, the other countries with significant foie gras production are Hungary and Bulgaria, while minor production originates from Spain and Belgium. In 2009, the Hungarian production was about 2550 tonnes and the major part was represented by goose foie gras (1800 tonnes). The Bulgarian production was about 2300 tonnes and solely represented by duck foie gras.

Down and feathers are by-products from waterfowl production, being collected live or after slaughter. According to old data, 67,000 tonnes of feathers and down from all waterfowl species were traded at an international level in 1994 (Buckland and Guy, 2002), of which about 30% were collected from geese. Nowadays, China is reported to produce 100,000 tonnes annually (Wang, 2008) and once again is the major world producer, followed by Taiwan, Thailand and Hungary.

REARING PRACTICES AND HOUSING SYSTEMS

Overall context in Europe and Asia

Regardless of the species and breeds kept, or production purpose, there are some basic requirements in waterfowl farming practices that apply everywhere:

- Protection, especially for ducklings and goslings, from extreme weather conditions and predators.
- Access to a clean, dry, sheltered area. Although waterfowl can spend part of their time outdoors, on ponds or in wet areas, they require a clean, dry, sheltered area where they can retreat, rest, clean and preen their feathers. This allows them to waterproof their plumage, which protects their skin from injury and helps keep their body warm.
- Clean drinking water, i.e. water that is free from pathogens and toxins harmful to ducks. Water for swimming is not essential, but can be beneficial in areas where temperatures are high.
- Diet that covers nutritional needs.
- Adequate light regimen, especially for layers.
- Protection from disease established and maintained by a biosecurity programme that will prevent the introduction of diseases into the premises where waterfowl are kept.

Taking these requirements on board, the context differs greatly between Europe and Asia. In Europe, legislation established to promote animal welfare, which is relevant in the context of rearing practices and housing conditions, takes these requirements into consideration. A general convention covering all domestic species (Council of Europe, 1976) and a comparable directive (98/58/EC; Anonymous, 1998) have been adopted by the Standing Committee of the European Convention (European Council) and the European Union (EU), respectively. Both of these texts apply to all domestic species and therefore to domestic birds. Furthermore, in 1999, the Standing Committee of the European Council adopted three specific recommendations devoted to waterfowl. These recommendations concern the domestic duck (*A. platyrhynchos*) (T-AP (94) 3; Conseil de l'Europe, 1999a), the Muscovy duck (*C. moschata*) and the mule duck and their hybrids (T-AP (95) 20; Conseil de l'Europe, 1999b), and the domestic goose (*Anser* sp.) (T-AP (95) 5; Conseil de l'Europe, 1999c). In accordance with Article 9, Paragraph 3 of the Convention, these recommendations came into force in December 1999. These require-ments shall apply for new accommodation or when existing ones are replaced, from 31 December 2004. All accommodation shall fulfil these requirements by 31 December 2010; however, a delay of 5 years for fulfilling these requirements has been requested by France, especially regarding the ban on individual cages.

The most important dispositions are the following. The use of completely slatted floors and of individual cages is forbidden. At present, the production of foie gras can be carried out only where it is current practice and then only in accordance with domestic legislation. Countries such as Italy and Greece, which incidentally were not producing any before the legislation came into effect, and Poland have officially banned foie gras production. Feed restriction strategies, ahemeral rhythms and split photoperiods are forbidden. Further, usage of certain water facilities (pipette, nipple drinker) might be restricted, if not forbidden. Mutilations shall be prohibited; therefore beak and claw trimming are not allowed for domestic ducks and geese, and only tolerated under severe restrictions for Muscovy and mule ducks. Moreover, due to these driving forces, alternative systems will go towards less intensive systems with generalization of floor rearing and often access to a free range area in Europe.

In China, income growth, associated with population urbanization, has been a major factor driving the demand for duck meat. This growing demand has resulted in a move away from traditional backyard or smallholder flocks to large-scale commercial systems. Trends in food processing, increased specialization and food safety concerns, especially public health issues, have also provided a stimulus to intensive production. Indeed, since 2005, to promote intensive waterfowl production, the Ministry of Agriculture has indicated that traditional farming methods need to change. Moreover, the growing presence on the Chinese market of international companies of Western origin also contributes greatly to the move to occidental types of rearing practices and housing systems. This standardization is further illustrated by the fact that the UK's duck breeding company Cherry Valley, the world market leader, was taken over by Thailand's Bangkok Ranch Group, Thai Co. Company, in 2010. The move to alternative systems in China is thus undoubtedly going towards more intensive systems, although traditional ones are still presently in use.

Rearing practices and housing systems for ducks

Duck production systems in Europe

In Europe, large differences exist in housing and management between and within species, with husbandry systems ranging from intensive confined houses to free range production systems (Raud and Faure, 1994; Rodenburg *et al.*, 2005) or one after the other in the specific case of mule ducks (Guémené and Guy, 2004). The type of management system depends on a variety of factors such as the availability of funding, amount and cost of labour, technology, sanitary regulation and also the market for which the ducks are destined. Breeding stocks and to a lesser extent table ducklings are normally housed in intensive, closed, lightproof accommodations, which are comparable in some respects to the buildings provided for other types of poultry. Nevertheless, ducks drink, spill over and excrete more water than chickens or turkeys which makes it more difficult to maintain litter floors in a dry condition. In order to achieve this, drinkers are located over slatted plastic or wire flooring drained to an effluent disposal system (Merlet *et al.*, 2010). Ducks are even often kept on a fully slatted floor made of wood, metal or plastic material, instead of straw or wood shavings, in order to improve hygiene and reduce risk of pathological problems. Detailed descriptions are provided in Tables 8.2 and 8.3, adapted from Rodenburg *et al.* (2005).

For Muscovy ducks, slatted flooring was the most common system, usually without litter after a certain age. However, from 1 January 2011 an entirely

Table 8.2. Typical stocking rates and densities of ducks, according to genotype, rearing system and country in Europe. (Adapted from Rodenburg *et al.*, 2005.)

Genotype	System	Country	Stocking rate (birds m^{-2})	Stocking density (kg m^{-2})[a]
Muscovy	Conventional	Germany	9 (no litter)	35
			5 (litter)	19
		France	13 (no litter)	52
Muscovy	Free range	France	9	28
Mule	Rearing	France	4	16
Mule	Overfeeding	France	10	60
Pekin	Conventional	Germany	6	20
		UK	7 (litter)	22
			8 (no litter)	25
		Netherlands	8	25
		France	15	46
Pekin	Free range	France	8	35
Pekin	Organic	UK	0.25–0.50[b]	
		Germany	6	20

[a]Stocking density means the maximum density at any period of the fattening period.
[b]2500 ducks ha^{-1}, but 5000 ducks ha^{-1} on well-grassed outdoor runs.

Table 8.3. Characteristics of conventional, free range and organic systems for Pekin ducks, Muscovy ducks and mule ducks. (Adapted from Rodenburg et al., 2005.)

	Pekin Conventional	Pekin Organic	Muscovy Conventional	Muscovy Free range	Mule Rearing	Mule Foie gras
Floor	Straw/slatted/ partly slatted	Straw	Slatted	Slatted	Straw	Cage
Flock size	3,000–13,000	3,000	3,000–10,000	3,000–10,000	2,500	600
Stocking density (kg m^{-2})	25	20	40	28	1	60
Final body weight (kg)	3	3	4	3	4	6–7
Drinkers	Nipple/bell/ trough	Nipple/ bell	Nipple/bell/ trough	Nipple/trough	Nipple/ trough	Water trough
Outdoor run	No	Yes	No	Yes	Yes	No
Open water	No	Yes	No	No	No	No
Beak trimming	No	No	Yes	Yes	No	No
Claw trimming	No	No	Yes	Yes	No	No

slatted floor was forbidden and the floor must be covered with suitable material, according to the Council of Europe's Recommendation (Conseil de l'Europe, 1999a). This Recommendation has not been transcribed yet into European regulation and/or national regulation; nevertheless all European countries have ratified the European Convention for the Protection of Animals kept for Farming Purposes (Council of Europe, 1976) and should have done so since then (Guémené and Faure, 2004). Muscovy ducks are kept in groups of about 3000 to 10,000 birds. The stocking rate is about 5–13 birds m^{-2} (19 to 52 kg m^{-2}). In Germany, group size in conventional systems is generally smaller. Sexes are kept separately, but most often in the same house, in France and Germany. As the females are slaughtered at a younger age than the males (at around 10 weeks instead of 12 weeks or later), the full barn is available for the males by the end of the rearing period. The theoretical stocking density of 52 kg m^{-2} is thus never reached and in fact corresponds to the total cumulative live body weight produced in the barn. Female mule ducks are reared under similar conditions to conventional Muscovy ducks or they are killed immediately after hatching at the hatchery, using welfare-recommended methods. In recent years, increasing numbers of day-old female Muscovy ducklings are eliminated as well. This is because drakes, which are heavier, are better adapted for marketing as meat parts, which have become more popular than whole duck carcasses. Ducks produced in France, under the Label Rouge code of practice, are kept at a lower stocking density and group size, and the birds have an access to an outdoor run. This free range system represents about 2% of the French production.

In Europe, Pekin ducks are mainly kept on deep litter systems with straw or wood shavings. Pekin ducks in conventional systems are kept in large groups of about 3000 to 13,000 birds with a stocking rate of about 6–15 birds m^{-2}, partly depending on the flooring (Table 8.2). Although representing a negligible proportion at present, some Pekin ducks are raised in organic farms where group sizes and stocking densities are smaller (3000 birds with 8 birds m^{-2} in Germany, 0.25–0.50 m^{-2} in the UK). In conventional production system, Pekin ducks have neither access to an outdoor run or to open water. Access to an outdoor run and open water is available for ducks for free range production systems in the UK (2500–5000 birds ha^{-1}; 2–4 m^2 per duckling), Germany (organic production) and France (Label Rouge: 2 m^2 per duckling). For environmental and sanitary reasons, it is prohibited to keep ducks in free range systems in the Netherlands. For reproduction purposes, Pekin ducks were typically also often kept on slatted floors or on partly slatted floors in the water supply areas, at least in Germany and France (Rodenburg *et al.*, 2005). More details thereafter are provided in Table 8.3. However, as for other duck species, fully slatted floors should no longer be used in the EU because the Council of Europe's Recommendation (Conseil de l'Europe, 1999a) states that whole slatted floors are forbidden from 31 December 2010 and that the floor must be covered with suitable material.

In France, half of the duck meat produced is from male mule ducks which are kept for foie gras production purposes and raised under specific rearing conditions (Fig. 8.3). Three rearing phases can be distinguished in mule duck production: (i) the rearing or growing period (from hatching to 11 weeks of age); (ii) the preparatory period (during 1 week); and (iii) the force feeding period (during 10 to 12 days) (Guémené and Guy, 2004). During the rearing period the ducks are raised in collective floor pens on straw and they have access to a free range area as soon as biologically possible. Flock sizes are about 2500 ducklings raised at low stocking densities (3 to 5 m^2 per duckling). After a period of food restriction, ducks will progressively receive an increasing amount of food once daily. At the end of this preparatory period, until now ducks have most often been placed in individual cages for the force feeding period. Nevertheless, with the adopted recommendation (Conseil de l'Europe, 1999b), individual cages are to be banned and increasing numbers of male mule ducks are now being reared in collective conditions during the force feeding period. As a consequence, alternative systems are presently under development. In these systems, ducks are kept in slatted collective cages set on platforms (3–10 ducks per pen) or floor pens (12–15 ducks per pen), where they are fed twice daily from the day after transfer for 10–12 days, and slaughtered immediately thereafter.

Duck production systems in Asia

Duck production systems in Asia are generally differentiated into three systems combining spatial and economic criteria: (i) small-scale free ranging family production; (ii) commercial medium-size integrated systems; and (iii) fully confined ones. In this chapter, we present two integrated systems and a confined one.

TRADITIONAL INTEGRATED FISH AND DUCK PRODUCTION. In southern provinces of China, both meat and layer duck productions are still mostly conducted using the integrated 'fish–duck' system (Fig. 8.4). Such integrated production

(a)

(b)

Fig. 8.3. Mule ducks in open barn rearing systems in France: (a) inside view and (b) outside view. (Image (a) courtesy of G. Guy; image (b) courtesy of INRA-UEPFG.)

(a)

(b)

Fig. 8.4. (a) Goose and (b) duck breeding rearing systems in integrated 'fish–bird' production systems in China. Ducks (b) have access to a slatted floor that extends the available surface above the pond. (Images courtesy of Z. Shi.)

systems are also popular in Eastern Europe. It consists of producing waterfowl and fish simultaneously on the same site. Fish raised under these conditions include common carp, silver carp and tilapias, while the ducks concerned are Pekin ducks. Husbandry conditions during the starting period do not differ from classical ones worldwide. At 4 to 6 weeks of age, depending on the climate, the ducks will go out for extensive rearing, sometimes with access to shelter. Since local duck egg laying breeds do not seem to be very much influenced by photoperiod, housing for egg type ducks is essentially similar to that used for meat ducks or geese, i.e. having a shed littered with rice straw or rice husks, a yard with feeders and additional drinkers, and a fenced area on water. For laying purposes, collective laying nests are provided under the shed. In some areas having hot summer months, the fish pond owners, who may differ from the duck owners, will install aerators on the water, to improve water quality by oxygenating the water to prevent fish deaths. Alternatively, although access to water is not strictly necessary for ducks, one can think that provision of water is good for their welfare by reducing heat stress and allowing bathing. Farmers noticed some benefits in setting such a system of production for the ducks, with skin and feathers being cleaner than in usual rearing systems (Varadi, 1995). However, although it has been thought to be an efficient system for waste recycling, it has not been thoroughly investigated scientifically yet. In any case, as the availability of freshwater resources is decreasing worldwide, there is a need for further research in order to optimize this type of combined production system.

RICE–DUCK INTEGRATED PRODUCTION. This type of production can be subdivided in two categories depending on whether the ducks have access to the rice field after transplanting or only after harvesting (Fig. 8.5). In China, ducklings are released into the paddy fields after their rearing period, i.e. at 7 to 10 days of age, 10 days after the rice seedlings are transplanted. In comparison, in India, the ducks generally only have access to the paddy fields after harvesting. These systems concern both duck raised for meat and egg productions (Jalaludeen *et al.*, 2004). In order to be set up, the ducklings have to be hatched at periods which take into account the availability of paddy fields for pasture. Ducks are raised in a nursery after hatching using a simple shed made out of local material (bamboo, coconut leaves, etc.). Under tropical conditions, no heating is required and they are fed with various diets using local resources (cooked rice, dried fish, etc.). In India, the whole flocks can be moved from one place to another over long distances using trucks to reach the recently harvested field areas under this integrated system (Jalaludeen *et al.*, 2004). On the paddy fields, ducklings forage all day long and can find paddy grain, weeds, insects, snails, frogs, worms, fish, crabs, crawfish and aquatic plants, while their droppings have the role of fertilizing the soil for the next crop. Ducks consume available resources that will otherwise mostly be lost and are consequently produced at lower feed cost. Furthermore, ducks raised in paddy fields play a key role in the control of pests, by eating for example snails and mosquito larvae, which reduces the intermediary hosts of certain parasites.

(a)

(b)

Fig. 8.5. (a, b) Ducks raised in rice paddy fields in China. A small shed is provided as shelter (b), and the paddy field is surrounded by nets. (Images courtesy of Z. Shi.)

These integrated rearing systems can be subdivided further into two sub-types. One is the open range extensive rearing system where birds have full-time access to the paddy fields and a river. The second is the semi-intensive system in which a shelter is provided for night-time. In China, a paddy field plot of 5000 m² is surrounded by a plastic fence (1 m high) and a shelter is built close to the free range foraging areas. This last semi-confined system is well adapted for meat production, but also for egg production. In India, such a practice is for egg production only. In addition to keeping the birds protected during the night, it restricts the area they can use at the time of laying, so that eggs can be easily collected. In practice, a second enclosure of nylon mesh net is placed around the ducks within the shelter for the night, so that ducks benefit from two forms of protection from predators. Early in the morning, the farmer removes the inner circle and the ducks move to the outer circle where they can lay eggs on clean litter.

Depending on the food they are able to find and the objective of the producer, the ducks can be provided with additional food. In India, if any, the complementary diet is generally a rough one, made out of local resources considered as waste or at least which do not compete with human requirements (shellfish waste, paddy chaff, rice bran, palm pith, oyster shell). On the other hand, in China ducks are generally fed near the shelter with complete feed once daily providing up to half of their nutritional requirement. Regarding egg production, the local breeds of ducks have the potential to produce about 180 to 200 eggs per annum. However, in regard to the amount and quality of the food provided, the effective production is generally much lower. The average body weight of meat ducks does not exceed 1.5 kg at 20 weeks, even if food is provided in sufficient amount. Some attempts to use non-local strains with higher levels of performance have been recognized as unsuccessful because of their low resistance to disease and their lower adaptability to the local climatic conditions. The slow growth of indigenous ducks is thus a better choice under these rearing conditions. Even if the 'rice–duck integrated production' system is rapidly expanding throughout China, the total duck production from this system is still far smaller than from conventional duck-raising systems (Fig. 8.6).

Recent alternatives in conventional duck production systems

In northern China, as water is not as abundant as in the south, duck rearing is mostly carried out inside poultry houses or sheds (Fig. 8.7). Partly because of the wide extension of occidental duck breeds, duck houses are nowadays designed and built in similar styles and with standard equipment similar to those seen in developed Western countries. This is especially true for breeding ducks. In subtropical areas, water spraying sprinklers are normally installed in duck houses and air renewed through exhausting fans placed on the side of walls.

More recent innovations in housing and rearing for duck meat production in China are the use of slatted platforms and biological beddings. High slatted platform rearing was developed to maintain hygiene and health status by eliminating direct contact with faeces. The slatted platform, placed at around 80 cm in height, is made of plastic-coated wire mesh. Automatic chain feeders

(a)

(b)

Fig. 8.6. (a) Duck and (b) goose rearing systems with access to a ditch. (Images courtesy of Z. Shi.)

(a)

(b)

Fig. 8.7. Ducks in confined rearing systems: above (a) biological bedding in China and (b) normal litter in Taiwan. (Image (a) courtesy of Z. Shi; image (b) courtesy of G. Guy.)

and nipple-cup drinkers are installed over the slatted floor and ducks fed *ad libitum*. Galvanized iron bathing tubs (1.5 m long × 0.7 m wide × 0.4 m deep) can be provided during hot summer months. The water will contribute to reducing heat stress and to maintaining feed intake at the required level. Faeces drop through the floor down to the ground where they can be scraped away.

Alternatively, the use of biological bedding has recently been introduced in China and is nowadays widely used in commercial duck farms. Biological bedding results from pre-treating litter materials such as straw, wood shavings or sawdust with probiotic microorganisms. The bedding should be at least 20 cm deep right from the start, so that microorganism growth and fermentation can proceed following placement of ducklings. Although it remains to be scientifically validated, it has been claimed that raising ducks on this biological bedding can save up to 10% of feed cost, without any loss of duck production. Moreover, the major bonus comes from the energy saving, due to the heat produced by microbial fermentation, which keeps young ducks warm in winter. Another attractive point of biological bedding is that it is spread on to crop fields after use, transferring faecal nutrients to crop fields. Within these systems, feeders and bell-shaped drinkers are placed alongside one of the walls, above a slatted covered ditch.

Cages for duck layers

Although breeding ducks are mostly reared on floors inside lightproof barns during the reproductive period, in recent years increasing numbers of duck layers have been placed in conventional cages for the production of table eggs in China (Fig. 8.8g and h). This is especially the case in areas where open water resources are limited and pollution concerns prohibit raising ducks on waterways. Not taking into account possible welfare concerns, the advantages of cages are higher efficiency of space utilization, high numbers of ducks managed per human capita, cleaner eggs with much less *Salmonella* contamination, higher egg production with less feed required, easier and better disease management, no use of litter and reduced waste discharge (Chen and Tao, 2007).

The laying duck houses are similar in structure to the laying hen ones. Inside the house (typically 90 m long, 8 m wide) are placed four rows of three-level stair step cages (Californian cage models) leaving the space between rows and wall sides as walkways. Each cage is square shaped (40 cm × 40 cm), with the front and back heights set at 40 and 30 cm, respectively. One cage holds two or three ducks, depending upon the size of duck breed used. Under this setup, a house of 720 m^2 can hold up to 7200 ducks (Zhang *et al.*, 2006). Concrete ditches (200 to 220 cm wide, 30 to 40 cm deep) are built under the rows of cages to collect faeces. Both natural and mechanical ventilation systems are required for cage layer duck barns. In summer time, water-pad cooling ventilation is required to remove the large amount of gases and excess moisture from the house. This is achieved by installing water pads of 14 to 15 m^2 and two to four large exhaust fans on the walls at opposite ends. Air inlets at south–north side walls should be laid out for natural ventilation.

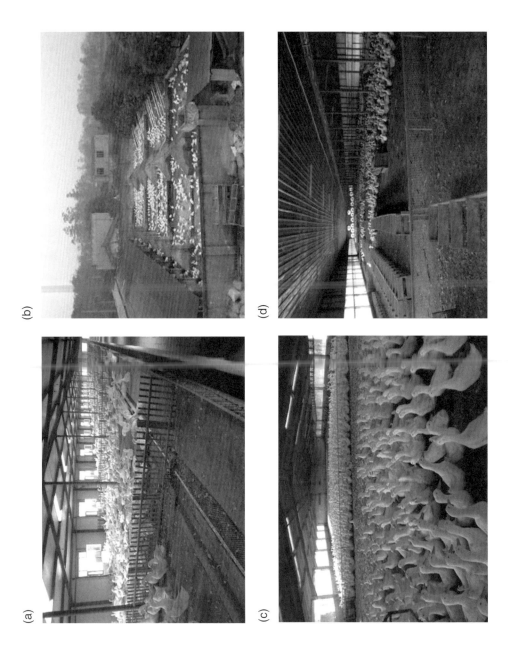

(a)

(b)

(c)

(d)

Fig. 8.8. Worldwide rearing systems for duck breeding: (a) northern China; (b) southern China; (c) Portugal; (d) Malaysia, (e) Thailand; (f) Egypt; (g, h) China. (Images (a–f) courtesy of Y. Le Pottier; images (g, h) courtesy of Z. Shi.)

Rearing practices and housing systems for geese

As geese production worldwide is located mainly in China, here we essentially refer to raising conditions in this country (Figs 8.4 and 8.6). In developed countries such as France and East European countries, housing for geese is similar to that for other types of poultry such as ducks. As for ducks, geese production systems can differentiate into different systems among which integrated and confined systems are described below.

Traditional and modern confined systems

Originally, most Chinese goose houses were built with simple locally produced materials and naturally ventilated. These houses were built with bamboo or wooden supporting frames, covered with an inner layer of asphaltic felt and an outer layer of fir bark. Although primitive and cheap, fir bark is good in preventing both rain and sun damage, besides giving certain heat insulation properties. Fresh air inlets are built at the base of the walls, which also stop sunlight entering the house. The roof ridge is designed in a double layer structure, leaving a gap of 15 cm between the top and lower layers to allow exit of warm moist air. For efficient air exchange, no less than 4 m vertical distances between air inlets and outlets should be reached. In addition, both air inlets and outlets are built to prevent illumination by sunlight without stopping air flow. Walls are also built with large sized windows that are covered with black curtains made of cloth or plastic sheets. Closing the curtains during the daytime prevents sunlight entering into the house. During the night when solar illumination is no longer present, automatic or manual opening of curtains allows free air exchange.

In recent years, more and more goose houses have been equipped with fully automatic ventilation systems. Such automatically ventilated houses can be built with brick walls and tile roofs, or with rigid plastic, fibreglass-covered polystyrene foam boards, in the southern provinces of China. In this type of house, air outlets equipped with electric fans are installed on the side walls. Air inlets can be installed either at the base of walls, or at the opposite end walls of the fans, and be fitted with water pads for cooling the influx air. This type of negative pressure tunnel ventilation system, conferred by the cooling water pads and large fans, is especially useful for out-of-season egg laying in summer months in the subtropical areas of Guangdong and Taiwan, where heat stress can become a problem. With mechanical ventilation, the house can accommodate four or five geese per square metre, which should be decreased to three in naturally ventilated houses.

Inside the house, light devices are evenly placed, to provide artificial or additional illumination, for controlling daily photoperiod. Both incandescent lamps and fluorescent tubes can be used but for energy efficiency reasons, compact fluorescent lights are being increasingly used. In non-confined barns, drinkers and feeders are sometimes placed inside, but most often outside. In order to limit water spillage and maintain dry floors, it is preferable to use bell-shaped drinkers with automatic level adjusters.

Houses used for housing laying geese flocks are very similar with the addition of extra equipment. Thus, one or two collective nesting areas littered with rice straw representing a total area of 15 m^2 for a flock of 1000 geese are typically provided. Another enclosed area of 4 to 5 m^2 is also necessary for confining incubating geese, which would stop broodiness after 7 to 10 days in confinement. This confining area can also be connected to an outside yard extending on to the water, so the confined incubating geese are exposed to the same daily photoperiod as the rest of the flock. The yard is paved with bricks or laid with concrete. The size of the yard is normally 1.5 to 2 times that of the house. For a 1000 bird flock, a yard with 300 to 400 m^2 is adequate. Shade should also be created above the yard, to avoid direct exposure of the geese to sun during the hot summer months.

During the rearing period up to 10 days of age, goslings are raised in small slatted pens (1 m × 2 m). The floor normally consists of wire mesh material but half of the surface is covered to provide a littered resting area protected from the wire mesh gaps, as well as for holding creep feed before goslings learn to feed from the feeders. Infra-red lamps must be provided during cold seasons from autumn to early spring, to protect goslings from hypothermia. Housing for growing meat geese or goslings of greater than 10 days of age is much simpler than those for breeding geese as they do not require facilities for artificial lighting and ventilation. A simple shed, sometimes only with a roof, a yard and a pond will be sufficient to meet geese requirements. The space needed within the house and on the yard is also smaller than those for breeding geese. As for breeding geese, shading is also provided on the yard during summer months.

Traditional integrated systems

In the Chinese traditional integrated 'fish–geese' production system, housing consists of a barn or shed, a yard and a pond. However, as with all domestic breeds, geese are seasonal breeders; thus goose housing must be built in order to be able to control the photoperiod (using lightproof barns) and ventilation (dynamic system), in order to get out-of season reproduction. In this integrated system, ponds not only enable fish production but are also the site for mating and preening, besides providing drinking water for the geese. However, side effects, due to poor water quality, on goose health and production performances have been frequently observed over the past years in many parts of China. Therefore pond water quality management is critical to out-of-season egg laying in summer and, in order to avoid or reduce geese from drinking water from the pond, bell drinkers should also be provided in the yard. In practice, it is also recommended to replace pond water once every week. Otherwise, in circumstances where clean water is hard to source, reducing the stocking density to as low as 0.5 birds m^{-2} water surface is recommended. Additionally, aerators can be installed which, when used in conjunction with supplementation of photosynthetic bacteria or probiotic microorganisms, will substantially reduce proliferation of pathogenic bacteria growth and endotoxins (Shi *et al.*, unpublished data). In this integrated system, the yard area is generally restricted to limit the labour needed for cleaning faeces. In order to compensate, a closed extension with wire mesh floor supported by concrete posts, placed above the pond, can be provided. Apart from this, an area on water is also provided as a water playground for the goslings.

In situations lacking a sufficient water surface, a bathing ditch is provided along one edge of the yard. In practice, a flock of 1000 geese requires a ditch of 25 m length, 2 m width and 0.5 m depth. Water is changed every 1 or 2 days. This goose production system is normally combined with crop or vegetable production, such as rice, and discharged waste water is used to irrigate green crops or vegetables. In the most up-to-date goose houses seen in Taiwan, temperature is maintained below 30°C in summer by using water-pad and tunnel ventilation. In these systems, a bathing ditch is inserted inside the barn (Cheng *et al.*, 2001). Access to water, which limits heat stress, allows good laying performance in geese, even in the hot summer months, without negative impacts on egg fertility and hatchability (Li, 2008). These conditions would all contribute to increase the total number of goslings produced.

CONCLUSIONS

Different species, and breeds among species, account for worldwide multi-purpose waterfowl production such as meat, eggs for human consumption, foie gras, down and feathers. Rearing systems differ greatly between countries depending upon species and breeds, production purposes, as well as different environmental conditions and cultural backgrounds. Until recently, the driving forces were basically an intensification trend in the occidental countries, but welfare and sustainability concerns have impacted them deeply with a strong demand for a move to less intensive alternative systems. On the other hand, fish–waterfowl or rice–waterfowl integrated systems were among the traditional systems in Asian countries. However, the move to 'alternative systems' in Asia undoubtedly means going towards more intensive systems, although traditional ones are still present on a large scale. In those countries, income growth, associated with an increasing urbanization, is a major factor driving the demand for meat all year long. Due to increases in the sanitary problems encountered in traditional backyard production, a move away to more intensive large-scale confined commercial systems is being observed, a move which is further promoted by local governments since the Second World War in Europe, albeit for other reasons. The growing presence on the Chinese market of international companies of occidental origin also greatly contributes to the move to occidental types of intensive rearing practices and housing systems, as well as the use of breeds from occidental origins at least in duck species. Interesting is thus the apparent opposite interpretation of what an alternative system is, depending upon which part of the world one lives in and what the resulting driving forces are.

ACKNOWLEDGEMENTS

We are very grateful to the two reviewers for their contribution in improving the manuscript and to those who provided us with relevant information and images for the figures.

REFERENCES

Anonymous (1998) Directive 98/58/CE du Conseil du 20-09-98 concernant la protection des animaux dans les élevages. *Journal Officiel des Communautés Européennes* L 221, 08/08/1998, 23–27.

Anonymous (2010) Asia dominates world waterfowl production. http://www.thepoultrysite. com/articles/1633/asia-dominates-world-waterfowl-production (accessed 8 April 2011).

Arnaud, I., Mignon-Grasteau, S., Larzul, C., Guy, G., Faure, J.-M. and Guémené, D. (2008) Behavioural and physiological fear responses in ducks: genetic cross effects. *Animal* 2, 1518–1525.

Ashton, C. and Ashton, M. (2001) *The Domestic Duck*. The Crowood Press, Marlborough, UK.

Brun, J.-M., Richard, M.-M., Marie-Etancelin, C., Rouvier, R. and Larzul, C. (2005) Le canard mulard: déterminisme génétique d'un hybride intergénérique. *INRA Productions Animales* 18, 295–308.

Buckland, R. and Guy, G. (2002) *Goose Production*. FAO Animal Production and Health Paper No. 154. Food and Agriculture Organization of the United Nations, Rome.

Chen, Y.C. and Tao, Z.R. (2007) Comparisons between battery cage and floor bed layer duck raisings. *China Poultry* 29(12), 29–30 (in Chinese).

Cheng, G.F., Hsu, J.C. and Lei, P.K. (2001) Design of an environmental-control closed type geese breeder house. *Journal of Agricultural Machineries* 10, 99–116 (in Chinese with an English abstract).

Conseil de l'Europe (1999a) Comité Permanent de la Convention Européenne sur la protection des animaux dans les élevages. Recommandations concernant les canards domestiques (*Anas platyrhynchos*). T-AP 94 (3): 13 pp

Conseil de l'Europe (1999b) Comité Permanent de la Convention Européenne sur la protection des animaux dans les élevages. Recommandations concernant les canards de barbarie (*Cairina moschata*) et les hybrides de canards de barbarie (*Cairina moschata*) et de canards domestiques (*Anas platyrhynchos*). T-AP 95 (20): 16 pp.

Conseil de l'Europe (1999c) Comité Permanent de la Convention Européenne sur la protection des animaux dans les élevages. Recommandations concernant les oies domestiques (*Anser anser* F. *domesticus*, *Anser cygnoides* F. *domesticus*) et leurs croisements. T-AP 95 (5): 14 pp.

Council of Europe (1976) European Convention for the Protection of Animals kept for Farming Purposes. *Official Journal* L 323, 17/11/1978, pp. 14–22.

Guémené, D. and Faure, J.M. (2004) Productions avicoles: bien-être et législation européenne. *INRA Productions Animales* 17, 59–68.

Guémené, D. and Guy, G. (2004) The past, present, and future of force-feeding and 'foie gras' production. *World's Poultry Science Journal* 60, 210–222.

Guémené, D., Guy, G., Noirault, J., Destombes, N. and Faure, J.-M. (2006) Rearing conditions during the force-feeding period in male mule ducks and their impact upon stress and welfare. *Animal Research* 55, 443–458.

Jalaludeen, A., Peethambaran, P.A., Leo, J. and Manomohan, C.B. (2004) *Duck Production in Kerala*. NATP on Ducks Centre for Advanced Studies in Poultry Science, Kerala Agriculture University, Kerala, India.

Li, H.-Z. (2008) Analysis of use of environmental-control goose breeder house. *Rural Animal Husbandry Technologies* 5, 12 (in Chinese).

Merlet, F., Rousset, N., Ponchant, P. and Corson, M. (2010) Evaluation de l'impact carbone en production de canards à rôtir: une approche de terrain des systèmes d'élevage sur caillebotis et sur litière. *TeMA* 13, 31–38.

Raud, H. and Faure, J.M. (1994) Welfare of ducks in intensive units. *Revue Scientifique et Technique (International Office of Epizootics)* 13, 119–129.

Reiter, K. (1997) Das Verhalten von Enten (*Anas platyrhynchos* f. *domestica*). *Archiv fur Geflugelkunde* 61, 149–161.

Rodenburg, T.B., Bracke, M.B.M., Berck, J., Cooper, J., Faure, J.-M., Guémené, D., Guy, G., Harlander, A., Jones, T., Knierim, U., Kuhnt, K., Pingel, H., Reiter, K., Servière, J. and Ruis, M.A.W. (2005) Welfare of ducks in European duck husbandry systems. *World's Poultry Science Journal* 61, 633–646.

Sauveur, B. (1988) *Reproduction des volailles et production d'oeufs*. INRA Editions, Paris.

Sauveur, B. and De Carville, H. (1990) *Le canard de Barbarie*. INRA Editions, Paris.

Varadi, L. (1995) Ecological aspects in integrated duck and fish production. In: *Proceedings of the 10th European Symposium on Waterfowl*, Halle, Germany, 22–31 March 1995. Offset Köhler KG, Giesen, Germany, pp. 8–19.

Wang, D.Z. (2008) Chinese feather and down industry development and the international trade outlook. *Guide to Chinese Poultry* 25(14), 12–13 (in Chinese).

Zhang, X.E., Gong, S.M., Jia, W.L., Zhu, Z.M. and Lou, L.F. (2006) Keys of egg duck raising in battery cages. *Hangzhou Agricultural Science and Technology* 2, 34–35 (in Chinese).

CHAPTER 9

Game Bird Breeding, Brooding and Rearing – Health and Welfare

T. Pennycott, C. Deeming and M. McMillan

ABSTRACT

Large numbers of game birds, mainly pheasants (*Phasianus colchicus*) and red-legged partridges (*Alectoris rufa*), are released for sporting purposes in the UK. The released birds are derived from eggs produced by captive breeding flocks and incubated artificially. This chapter considers the systems of housing of adult pheasants and partridges during the breeding season, including group size, stocking density and the use of sight barriers as a means of environmental enrichment and refuge provision. The conditions and constraints of artificial incubation are explored. It is concluded that radical changes to the conditions under which pheasants and partridges are bred will require experimentation and research. Further development of the game industry will also require a switch to modern incubation equipment designed for poultry but adapted to hold pheasant and partridge eggs. The chapter then goes on to describe some of the current methods of rearing pheasants and partridges from day-old to release, and relates husbandry practices to the requirements of the new Codes of Practice for the Welfare of Gamebirds Reared for Sporting Purposes. In general most of the recommendations in the welfare codes are being complied with. Areas that require further attention by game bird rearers include improved environmental enrichment in some systems; the best use of artificial lighting; a re-evaluation of the use of bits; improved cleaning and disinfection of crates and vehicles used to transport birds to release pens; greater involvement of local veterinary practices; preparation of flock health plans; and improved biosecurity.

INTRODUCTION

Considerable numbers of game birds, mainly pheasants (*Phasianus colchicus*) and red-legged partridges (*Alectoris rufa*), are reared and released in the UK each year. The birds are released to be shot for sporting purposes several months later but many of the carcasses also enter the human food chain. The actual number of birds reared and released is not known with certainty, but one estimate (FAWC, 2008) suggests that 30–35 million pheasants and 5–10 million red-legged partridges are released each year for sporting purposes. These birds are derived from eggs laid by captive pheasants and partridges kept in breeding accommodation and incubated artificially. Pheasants and red-legged partridges are neither fully domesticated nor fully wild, and are not truly indigenous to the UK. Nevertheless, they will be expected to cope with a period of intensive rearing followed by several months in a semi-wild environment, where feed and water will still be provided but the birds will be exposed to hazards including predators, collisions with motor vehicles and adverse weather.

The qualities needed in a bird being reared for release into the semi-wild are very different from those of broiler chickens being reared intensively for meat production. The game birds will need to leave the protection of their rearing accommodation and adapt to a completely different environment. Their feathering must be sufficiently good that they will be able to withstand variations in temperature and showers of rain, feed and water will be provided in a different way, and the birds will be exposed to a number of hazards including avian and mammalian predators. Similarly the requirements for breeding pheasants and red-legged partridges differ from those of domesticated poultry such as chickens and turkeys.

Concerns about various aspects of game bird breeding and rearing were expressed by the Farm Animal Welfare Council (FAWC, 2008) and after consultation and revision the Department for Environment Food and Rural Affairs (Defra) produced a Code of Practice for the Welfare of Gamebirds Reared for Sporting Purposes in England (Defra, 2010) that came into effect in January 2011. This code is based on a voluntary code produced earlier by the Game Farmers' Association and is designed to help game bird rearers cater for the welfare needs of their birds as required by the Animal Welfare Act 2006. An almost identical welfare code was produced by the Scottish Government for game birds reared in Scotland (Scottish Government, 2011), where the equivalent legislation is the Animal Health and Welfare Act (Scotland) 2006. Hereafter these codes will be referred to as the 'game bird welfare codes'.

The Animal Welfare Act 2006 states that an animal's needs include:

- its need for a suitable environment;
- its need for a suitable diet;
- its need to be able to exhibit normal behaviour patterns;
- any need it has to be housed with, or apart from, other animals; and
- its need to be protected from pain, injury and disease.

The game bird welfare codes state that birds must therefore:

- 'Have an environment appropriate to their species, age and the purpose for which they are being kept, including adequate heating, lighting, shelter, ventilation and resting areas.'
- 'Have ready access to fresh water and an appropriate diet to maintain growth, health and vigour.'
- 'Be provided with appropriate space and facilities to ensure the avoidance of stress and the exhibition of normal behaviour pattern.'

- 'Be provided with company of their own kind as appropriate for the species concerned.'
- 'Be adequately protected from pain, suffering, injury or disease. Should any of these occur a rapid response is required, including diagnosis, remedial action and, where applicable, the correct use of medication.'

Numerous different systems are used in the UK to breed, rear and release game birds for sporting purposes, but very little scientific research on the welfare of UK game birds has been published in peer-reviewed journals (FAWC, 2008). Flocks of breeding pheasants vary from less than ten birds to over 300. Breeding red-legged partridges are usually kept in pairs. Pheasant and partridge chicks are usually reared fairly intensively for the first few months, are 'hardened off' and are then released into the semi-wild under controlled conditions. Pheasants are usually released into large pens of several hectares and gradually leave these release pens into the semi-wild. Red-legged partridges are usually released from smaller pens.

This chapter considers the housing conditions of adult pheasants and red-legged partridges during the breeding season, and describes the equipment used to incubate the hatching eggs produced. The initial rearing stages of the young birds before being moved to the release pens are then discussed. To find out more about the rearing systems in use in Scotland, in-depth questionnaires were completed on 12 sites rearing pheasants from day-old and 11 sites rearing red-legged partridges from day-old. Summaries of the findings from the questionnaires are presented, with further discussion about how the rearing practices meet the requirements of the Animal Welfare Acts and the game bird welfare codes.

BREEDING PHEASANTS

The breeding of pheasants by game farmers was reviewed by Deeming (2009) and it was concluded that there was little consensus within the UK regarding the best way to keep this species. Although some breeding birds are kept in metal floored pens raised above the ground these are relatively uncommon in the UK. The majority of breeding pheasants are kept in outdoor pens laid out in grass fields with wire and board sides and mesh roof coverings. Some larger pens lack the mesh roof because the birds have a brail fitted to one wing to restrict flying. There is great variation between the size of the pen and the size of the flock (Game Conservancy Trust, 1993), with group sizes ranging from seven females with one male in a small pen (9 m × 3 m – 3.375 m^2 per bird) through to pens containing 332 birds at 8:1 sex ratio (50 m × 25 m – 1.255 m^2 per bird). Indeed the ideal sex ratio for pheasants is far from clear although Deeming and Wadland (2002) showed that keeping birds in flocks of 250 hens at an 8:1 sex ratio provided better fertility than a 12:1 ratio. A 7:1 ratio in smaller flocks (56 hens) produced even higher fertility (Deeming et al., 2011b). Although space requirements of 4.5 to 6 m^2 per bird are recommended (Game Conservancy Trust, 1993) this is not always achieved in commercial practice – birds described by Deeming et al. (2011a) were kept at 2.74 m^2 per bird.

The main problem is that no research exists that would allow us to allocate a value for the optimum space requirements of pheasants.

A consideration of alternative housing systems for pheasants is rather difficult given the variety of pen types that are currently used (Game Conservancy Trust, 1993). There are almost no scientific studies that examine variations in different housing conditions on pheasant welfare, behaviour or reproductive performance. This is despite the fact that pheasants and partridges are kept in pens, often in the former case in large groups, which are not natural breeding conditions. From what is known about how housing impacts on the welfare of other birds, e.g. poultry, it is anticipated that there is tremendous potential for improvements in pheasant and partridge welfare and productivity. The lack of research could be largely due to the economic structure of the game industry in the UK. Although game farm operations are larger than operations run by game keepers, they are still small relative to average poultry operations. This means that there is rarely any financial support for research and the relatively small membership of organizations such as the Game Farmers' Association means that, despite its economic importance, large sums of money are unavailable for research into pheasant breeding.

However, pheasant welfare is increasingly becoming important (Butler and Davis, 2010) and the effects of a change in pen layout were recently investigated by Deeming *et al.* (2011a,b). Working on a commercial game farm in the UK, the presence of sight barriers was tested as a form of environmental enrichment that would be beneficial for breeding pheasants. Straw bales and metal sheets were placed in pens in a >–< shape in the middle of the pen in order to prevent birds from seeing the whole of the breeding pen at all times, which was the case for the control pens. It was postulated that these slight barriers would provide refuges for pheasants wishing to escape unwanted attention from birds in the same pen, whether to avoid aggression or to seek some privacy for courtship and copulation. As a result there would be an improvement in bird welfare through reduced aggression and mortality accompanied by higher and more persistent fertility (Deeming *et al.*, 2011a,b).

The breeding pheasants were in their first breeding season and kept in groups of 56 hens with eight cocks in pens measuring 13.2 m × 13.2 m. The walls were wooden boards up to 60 cm with a further 180 cm of wire mesh above, the floors were grass and the roofs were plastic mesh. Birds had spectacles and brails fitted prior to the breeding season. There were 11 trial pens with barriers measuring 60 cm high and 11 control open pens, and birds were monitored over the 10-week laying season. Observations of general behaviours and social behaviours were recorded three or four times over the laying season. Eggs were collected on a per pen basis and set once a week. Eggs were candled each set week and clears were opened to determine whether the eggs were fertile or not. Data were also collected for embryonic mortality and hatchability (Deeming *et al.*, 2011a,b).

Bird mortality was not significantly affected by the presence of the barriers but plumage scores at the middle and end of the laying season were significantly (P<0.05) better in the barrier pens (Deeming *et al.*, 2011a). The main difference in the behaviour time budgets was the extent of perching, which was increased in barrier pens as a direct result of the increased opportunity to perch on the straw bales. In open pens the birds could only perch on top of metal half-barrels used as nest sites. For social behaviours aggression was significantly ($P<0.05$) reduced in barrier pens with both males and females pecking other birds to a very much less extent.

Measures of reproductive performance were also affected by the presence of the barriers (Deeming *et al.*, 2011b). Egg production was unaffected by the presence of the barriers and the numbers of rejected eggs, while they increased over the course of the laying season, were unaffected by the presence or absence of the barriers. By contrast, true fertility, expressed as a percentage of incubated eggs, was higher in barrier pens by 1% during week 3 of laying; and, although fertility in general decreased over time, the barrier pens had higher fertility from weeks 3–10 and at week 10 the eggs from open pens had 4% fewer fertile eggs. Hatchability was similarly higher in barrier pens from weeks 3–10, averaging 3% higher. This difference was not a function of embryonic mortality, which was unaffected by the presence of the barriers, but rather was associated with a higher average fertility of 2.7%.

It would seem that providing pheasants with a more heterogeneous environment that provides refuges for birds does have positive impacts on measures of welfare, behaviour and reproductive performance (Deeming *et al.*, 2011a,b). These results were comparable to those of Leone and Estévez (2008), who used a series of panels running the length of broiler breeder houses and demonstrated an elevation in fertility that persisted for longer during the second half of the laying period. In addition, egg production and hatchability were increased. The studies by Deeming *et al.* (2011a,b) do suggest that there is scope for investigating different environmental conditions used for brooding pheasants that would have benefits for welfare and productivity. This could include the effects of differing group sizes or bird densities on aggression or reproductive performance, or developing a better understanding of the reproductive behaviour of captive pheasants under different conditions.

BREEDING PARTRIDGES

There is very little published with respect to appropriate conditions for keeping of breeding grey or red-legged (French) partridges (*Perdix perdix* and *A. rufa*, respectively) other than those suggested by the Game Conservancy Trust (1993). These species are smaller than pheasants but males are more intolerant of potential rivals. This means that these species are generally kept in pairs in small wire cages, although red-legged partridges can be kept in communal groups at a 3:1 sex ratio in a 6 m × 12 m pen (Game Conservancy Trust, 1993). It is unfortunate that there are few scientific studies that have examined the breeding biology of partridges (see review by Deeming, 2009) and no reports have been published that attempted to see the effects of the existing methods for keeping either species of partridge on their behaviour or welfare. There is a pressing need for such research so that alternative housing systems can be developed and assessed in an effective way.

ARTIFICIAL INCUBATION

Despite the total size of the UK game industry, most game operations for either pheasants or partridges are small relative to most poultry producers. This is confounded by the short laying season of only 10 weeks. The most noticeable economic effect of this is that hatchery operations run by most game farms are typically starved of investment. Although the incubation phase of an operation can often be crucial in determining the productivity, and profitability, of a rearing operation, through necessity most UK game operations are reliant on old equipment.

Artificial incubation requires a setter that serves to incubate the eggs from setting through to transfer around 3 days prior to hatch, and a hatcher where the eggs spend these last 3 days allowing the chicks to hatch. The economic realities of game farming have meant that many game farms are operating hatcheries containing equipment that was designed and first used in the 1960s. This causes numerous problems that all have negative impacts upon hatchability. The age of most machines means that they are relatively simple to operate and fix but this is reliant on an understanding of the machines in the first instance. Experience has shown that this is often not the case – for instance, many Western 'Turkeybator' incubators are being operated at a heaters-off temperature of 100°F despite the fact that the original machine was designed to operate at 99.1°F. This difference causes the machines to continually heat and the eggs achieve very high temperatures that are deleterious to their survival. Moreover, it is becoming increasingly difficult to service and maintain these old machines. The lack of understanding of how old machines operate means that fundamental errors can be made in their operation, which can have adverse effects on results.

Larger game operations have invested in more modern equipment but the design of such equipment often leaves a lot to be desired. Designers of some types of incubators appear to have no understanding of the basic principles of artificial incubation and construct these machines with inefficient fan systems which results in inadequate provision of fresh air. Most importantly these machines have water cooling systems that are wholly inadequate for their role. It seems that game farmers purchase this equipment because it is priced within their budgets but they are all too often disappointed when the results are lower than their expectations of what they perceive as modern, tailor-made incubation equipment. Only the largest, most progressive game operations have invested in modern incubation equipment that is routinely sold to the poultry industry and this is typically reflected in their higher hatchability figures.

Another problem with many game operations is that they take no account of the room environment that the machines have to operate in. Farmers seem unaware of the idea that the room conditions can affect the operation of the setter or hatcher. Rooms most often lack adequate ventilation with the machines drawing in air from the same space into which the exhaust air is vented. This can reduce the oxygen levels of the room, which can reduce hatchability in the machines because even a 1–2% drop in oxygen in the air entering the machine can increase lethal hypoxia in eggs with low conductance eggshells. Simple things like fitting a fan to extract stale air from the room so that it can be replaced by fresh air can have positive impacts on results.

The consequence of old or poorly designed incubation equipment kept in the incorrect room environment is that hatchability is generally well below that expected from eggs in the poultry industry. Deeming (2009) found that one measure of mean hatchability for UK farm operations was 67.3% but ranged from 49 to 87% of eggs set. Average hatchability values of 64–66% have been reported in Turkey (Demirel and Kirikçi, 2008) and 62–65% and 67–71% in the UK (Deeming and Wadland, 2002; Deeming et al., 2011b). Such values are 10–15% lower than would be expected for the breeding life of broiler breeders. Much of the losses were due to high infertility, but mortality post 4 days of incubation, and particularly during the last 4 days, was crucial in determining success rates (Deeming and Wadland, 2001; Deeming et al., 2011b). There are no studies of hatchability in modern incubation equipment but it is likely that values will be higher yet still not reach those achieved by broiler eggs. This is partly due to problems associated with egg selection. In scientific studies, by the end of a 10-week season around 20% of pheasant eggs laid are being rejected for setting (Deeming and Wadland, 2002; Deeming et al., 2011b). Unfortunately, most game farmers are not as stringent in egg selection and many eggs that are not of sufficient quality for incubation are set. Mortality during the first 4 days of development is typically 4–5% and appears to be uniform between different species including pheasants, partridges, poultry and ratites (Deeming, 2009).

However, reasons for embryonic mortality are not well documented in poultry in general and even less so in game birds. Although pheasant eggs rarely addle (D.C. Deeming, personal observation), the presence of microbes inside the shell can still be a problem. Deeming et al. (2002) compared the incidence of in ovo yolk sac infection of dead-in-shell, unpipped pheasant eggs within 36 h of hatching. The presence of bacteria in the yolk sac contents of unhatched embryos was tested for eggs collected from conventional grass floored pens and from wire floored cages. Bacteria were observed in 60% of the sample of eggs from conventional pens, which was significantly higher than in eggs from wire floored pens (24%), but these values were very much higher in than broiler eggs collected from nests or the floor (10% and 30%, respectively). There is a pressing need for further research to identify why game bird embryos die, the results of which would be invaluable in developing better incubation practice that will maximize hatchability.

Radical changes to the conditions under which pheasants and partridges are bred will require experimentation and research. The use for pheasants of wire floored cages raised above the ground appears not to be widespread at the present time but this system has not been scientifically evaluated in terms of welfare or productivity. Whether it improves productivity but at a cost of welfare is yet to be investigated. Similarly, the use of more open conditions for partridges is rare but if effective systems could be developed this could represent a source of real improvements in welfare for these birds. Unfortunately, there seems little interest in the industry to investigate such alternatives and financial resources to support such scientific research are difficult to obtain. Future legislation aimed at improving game bird welfare may be needed as a stimulus to start such research programmes.

Future development of the game industry would ideally also involve a switch to modern incubation equipment designed for poultry but adapted to hold pheasant and partridge eggs. Incubator designers have made incredible improvements to their machines such that modern commercial incubators, which should be used by game farmers, are able to produce high hatchability. Research into such alternative systems would be of general use and value to the game industry as a whole.

REARING FROM DAY-OLD TO RELEASE

Source of birds

Twelve pheasant-rearing sites were studied after the 2009 rearing season had finished. These sites placed a total of over 500,000 pheasants, with individual sites rearing from 5500 to over 100,000 birds annually. Almost half (46%) of the pheasants were imported as day-old chicks from France, and another 19% were imported from France as hatching eggs. The remaining 35% of pheasants (on six of the 12 sites) were derived from eggs produced and hatched in the UK.

Eleven sites rearing red-legged partridges (some of which also reared pheasants) were also studied after the 2009 rearing season. Nearly 200,000 birds were reared on these sites, with individual sites rearing between 2000 and 60,000 birds in 2009. Similar to the pheasants, 50% of the birds were imported from France as day-olds and 18% were imported from France as hatching eggs. Ten per cent of the birds (one site) were imported from Spain as day-old chicks and the remaining 22% of red-legged partridges were derived from eggs produced and hatched in the UK.

Overall in this survey about 65% of pheasants and 78% of partridges were imported as hatching eggs or day-old chicks. FAWC (2008) estimated that in the UK about 50% of pheasants and up to 90% of partridges were imported, comparable to the figures obtained in this small survey. Taking pheasants and red-legged partridges together, UK-produced birds constituted only 30% of the birds placed on the sites studied. This could potentially pose problems if certain infectious diseases were present in the imported eggs or chicks, and welfare problems if the birds were in transit for extended periods. However with the exception of birds coming from Spain (for which journey time was not known), those interviewed confirmed that the chicks were in transit for less than 24 h. Such heavy reliance on imported birds also renders the UK game bird industry vulnerable to interruptions in supply in the event of import restrictions should a notifiable disease be confirmed in the country of origin.

Housing systems – pen sizes and stocking densities

Many different housing systems were used, sometimes on the same site. Ten of the 12 sites rearing pheasants and nine of 11 sites rearing partridges used systems of heated brooder huts giving access to unheated nursery/shelter pens and large netted grass runs. This system could be moved to fresh ground after being used for a few years. Typically the day-old chicks were confined to the brooder area for approximately a week, after which access was given to the shelter pens. Depending on the weather, the birds were allowed into the grass runs when 2–3 weeks of age. To begin with the birds would be moved back into the shelter pens and brooding area at night but eventually would spend the night in the grass runs. The size of the brooder huts varied from 2.5 m × 2.5 m to 7.4 m × 6.0 m, each housed between 150 and 800 pheasants or 350 to 1000 partridges. Stocking density in the brooder huts varied from 18 to 64 birds m^{-2} for pheasants and from 32 to 80 birds m^{-2} for partridges.

Four sites rearing pheasants and three rearing partridges had fixed sheds with no access to grass runs. Such sheds were stocked at 20–33 birds m^{-2} (pheasants) and 22–47 birds m^{-2} (partridges). On one site partridges were reared on elevated wire floored runs, with access to solid floored brooding areas.

The game bird welfare codes require that accommodation provides 'appropriate size, stocking densities and facilities, including appropriate environmental enrichment, to ensure good health and welfare' but do not specify actual stocking densities. Wise (1993) suggests that stocking densities up to 70 pheasants m^{-2} brooding area are acceptable if there is access to a shelter pen and a grass run providing 0.2 m^{2} per bird. In their 1983 booklet *Red-Legged Partridges*, the Game and Wildlife Conservation Trust (then The Game Conservancy) suggest stocking densities of 31–44 birds m^{-2} for red-legged partridges (Game Conservancy, 1983) but this advice may be out of date. The pheasant stocking densities described in the questionnaires are likely to comply with the welfare codes, but more research may be needed to establish suitable stocking densities for partridges.

Seven pheasant sites surveyed rarely or never reused accommodation in the same rearing season, two sites sometimes reused accommodation and three sites (including the two largest pheasant sites) frequently reused accommodation. A similar situation was found with partridges – nine sites rarely or never reused accommodation, but two sites frequently reused accommodation. The welfare codes require that houses be cleaned and disinfected between different batches of birds.

Group sizes and environmental enrichment

Overall, 44% of the pheasants placed were reared in groups of 200 to 400, 14% in groups of 400 to 1000 and 42% of the birds in groups of over 1000. Birds in the latter category were kept in larger fixed sheds on three sites. Pheasants remained in their accommodation until sale or release around 6–7 weeks of age. By contrast, partridges tended to be housed in larger groups, especially in fixed shed accommodation. Forty-two per cent of partridges on the sites interviewed were kept in groups of 400–1000, 54% in groups over 1000 birds and only 4% in groups under 400.

The original draft game bird welfare codes suggested that 'batch sizes should be kept within easily managed limits. Where possible, groups should be small enough so that if a disease problem arises the group can be isolated.' However the final versions made no mention of batch size. Wise (1993) concluded that 500 pheasants was a reasonable group and that numbers up to 1000 could be manageable. The welfare codes do not stipulate maximum batch sizes, but future research may be needed to see if groups of over 1000 birds (42% of the pheasants in the survey, 54% of the partridges in the survey) prevent the expression of normal behaviour and result in stress. The game bird welfare codes also suggest that rearing pens should provide perches, hiding places and environmental enrichment to minimize aggressive behaviour within the flock. Those systems with access to grass runs undoubtedly provide hiding places and enrichment, but none of the pheasant or partridge sites studied provided perches.

Heating and bedding

Heat in the brooding areas was provided by gas brooders on 11 of the 12 sites rearing pheasants and on all 11 sites rearing partridges. On one pheasant site (rearing the smallest number of birds) heat was provided by 'electric hens' – raised heated elements under which the young birds brood. A combination of gas brooders and electric hens was used on one partridge site. As the birds grew older the amount of heating was gradually reduced and was typically removed around 5 weeks of age. If functioning properly the systems used should comply with the welfare codes' requirement that the birds have adequate heating.

The commonest bedding used in the pheasant and partridge brooder areas was wood shavings or chopped cardboard but some pheasant and partridge sites used chopped straw for bedding. Such bedding complies with the welfare codes, which require that litter be clean, dry, non-toxic and tangle-free.

Feeders and drinkers

Proprietary feed was provided on all sites. Initially the feed was presented as crumbs, frequently provided in shallow trays, egg trays or cardboard paper. The crumbs were then followed by 'mini-pellets' and then larger grower pellets. Drinking water was provided by automatically filled 'mini-master' automatic drinkers or hanging bell drinkers connected to header tanks. Several pheasant and partridge sites additionally provided drinking water in small manually filled drinkers in the first week of life. On some sites access to drinking water was through nipples in the base of hanging reservoir drinkers or via nipples on extended drinker lines. These arrangements fulfil the requirements of the welfare codes, that the birds have ready access to fresh water and an appropriate diet.

Ventilation and lighting

Natural ventilation was employed on nine pheasant sites, three sites had automatically controlled fans and one site had manually controlled fans. Similar systems were found on partridge sites, nine of which had natural ventilation and two had automatically controlled fans. Of those sites with fans, most had some form of automatic back-up in the event of a power failure, as required by the game bird welfare codes. Six of the 12 pheasant sites employed artificial lighting, controlled either manually or automatically, as did four of the 11 partridge sites. The remaining sites used natural lighting supplemented by the light from the gas brooders. The welfare codes require a dimming facility (to allow the birds to prepare for darkness) and a minimum continuous period of night-time darkness of 6 h in every 24 h (to allow the birds to rest). Some but not all of the pheasant/partridge sites employing artificial lighting had dimmers or provided periods of darkness.

Use of bits

Small C-shaped devices termed 'bits' are commonly used by the game bird industry. They are clipped into the nostrils (without penetrating the nasal septum) and lie between the upper and lower beaks to prevent the tips of the beaks from coming together, reducing feather pulling. 'Bumpa Bits' are plastic bits with an additional loop of plastic in front of the upper mandible (FAWC, 2008). Small bits were used routinely on all 12 sites rearing pheasants, typically applied between 16 and 24 days of age. All bits fitted were plastic, with no metal bits used, and all bits were removed prior to release. One site (the smallest) used Bumpa Bits and one site (the largest) routinely beak trimmed the birds prior to release (in case the birds needed to be retained in release pens for a prolonged period). To reduce possible detrimental effects of bitting, seven sites administered electrolytes or multivitamins around the time of bitting. Reducing pecking injuries and mortality during rearing were the main reasons offered for bitting, and two sites also considered that bitting improved feather quality at release. In contrast, bits were used routinely on only one of the 11 partridge sites, inserted around 18–20 days and removed prior to release.

FAWC (2008) recommended that plastic bits should be permitted in young pheasants as long as their use could be justified and monitored. Butler and Davis (2010) reported that bits were an effective way of reducing welfare problems caused by feather pecking and cannibalism. However detrimental effects were also identified such as nostril damage and beak deformity, and in the future alternative ways of rearing pheasants without the need for bitting may be required. The welfare codes permit the use of bits in young pheasants for short periods (3–7 weeks) provided the practice can be justified and closely monitored, and require that bits be fitted/removed by trained and experienced stockmen.

In contrast to the use of standard bits, the welfare codes recommend that beak trimming and Bumpa Bits only be used in exceptional circumstances – in the current survey only one site trimmed their birds' beaks and only one used Bumpa Bits, suggesting that most sites could comply with these recommendations.

Only one partridge site used bits, implying that partridges can successfully be reared without using bits, but bits are currently widely used in young pheasants. The causes of feather pecking in pheasants have not been studied in great detail (Butler and Davis, 2010) but such work should be carried out to see if pheasants could be reared without the need to use such management tools in the future.

Wing clipping

Clipping the outer primary feathers to restrict flight is permitted by the welfare codes, provided the blood quills (growing feathers) are not cut. Approximately 45% of the pheasants sold or released by the sites interviewed had some primary wing feathers clipped at the time of sale/release at 6–7 weeks, to delay their ability to fly out of the release pens by a few weeks. A different situation was encountered on the partridge-rearing sites, where wing clipping was carried out on only a small percentage of the birds produced on one of the 11 sites. This reflects differences in the methods of releasing partridges and pheasants.

Transport to release pens

Pheasants and partridges were typically transported to release pens in plastic crates with straw bedding, although some partridges were transported in wooden crates with bedding. Approximately a third of the pheasants were transported less than 20 miles, a third 20–50 miles and a third were transported over 50 miles. Partridges tended to be transported for shorter distances. Not all sites routinely cleaned and disinfected crates between loads of birds, potentially posing a biosecurity risk. The game bird welfare codes require that all boxes, crates and vehicles, where appropriate, be thoroughly cleansed and disinfected between loads.

Health

Pheasant mortality to 6–7 weeks varied between rearing sites. Five sites expected 5–10% mortality and four sites 10–15% mortality. One site anticipated over 15% mortality but two sites expected less than 5% mortality. Partridge mortality to release was similar – three sites expected less than 5% mortality, seven sites 5–10% mortality and one site 10–15%. Mortality was recorded daily on nine pheasant sites and nine partridge sites, but only five pheasant sites and three partridge sites retained mortality records for 3 years. The welfare codes require that records of mortality and post-mortem reports be kept for a minimum period of 3 years, and game bird rearers should be encouraged to do so.

The main sources of veterinary advice were given as local veterinary practices (four pheasant sites and five partridge sites), specialist poultry veterinary practices in England/Wales (four pheasant sites and four partridge sites) and veterinary advisors of SAC (Scottish Agricultural College) or VLA (Veterinary Laboratories Agency) (four pheasant sites and two partridge sites). Greater involvement of local veterinary practices may help to ensure that the birds are adequately protected from pain, suffering, injury or disease, as required by the welfare codes that state that 'any bird suffering from ill health or injury must receive immediate attention, including where appropriate the attendance of a veterinary surgeon.'

Only one of the 12 pheasant sites had a flock health plan, and only two of the partridge sites. FAWC (2008) recommended that game bird-rearing sites should adopt flock health and welfare plans, prepared in conjunction with a veterinary surgeon and regularly reviewed. This recommendation was repeated in the game bird welfare codes. The findings from this small survey suggest that the game bird industry has some way to go before this requirement is met.

Eight of the 12 pheasant sites had disinfectant foot dips at the site entrance, and 11 had foot dips at some or all of the pens. Seven used hand sanitizers. Seven pheasant sites strongly discouraged visitors but none provided boots or protective clothing for visitors and none kept a formal record of visitors. Similar biosecurity precautions were in place on the partridge sites. The welfare codes state that good biosecurity is essential to prevent disease, and the findings from this survey suggest that most game bird rearers agree with this. However none of the pheasant or partridge sites kept a record of visitors.

Rearing – conclusions

Perhaps the most significant point to emerge from this brief review of game bird rearing is the lack of scientific evidence available to back up many of the recommendations made in some text books and the game bird welfare codes. Nevertheless, the results of this small survey suggest that in general most of the recommendations in the welfare codes are already being complied with. Areas that require further attention by game bird rearers include improved environmental enrichment in some systems (e.g. those without access to grass runs); the best use of artificial lighting; a re-evaluation of the use of bits; improved cleaning and disinfection of crates and vehicles used to transport birds to release pens; greater involvement of local veterinary practices; preparation of flock health plans; and improved biosecurity.

REFERENCES

Butler, D.A. and Davis, C. (2010) Effects of plastic bits on the condition and behaviour of captive-reared pheasants. *The Veterinary Record* 166, 398–401.

Deeming, D.C. (2009) Ratites, game birds and minor poultry species. In: Hocking, P.M. (ed.) *Biology of Breeding Poultry*. CABI Publishing, Wallingford, UK, pp. 284–304.

Deeming, D.C. and Wadland, D. (2001) Observations on the patterns of embryonic mortality over the laying season of pheasant. *British Poultry Science* 42, 580–584.

Deeming, D.C. and Wadland, D. (2002) Influence of mating sex ratio in commercial pheasant flocks on bird health and the production, fertility, and hatchability of eggs. *British Poultry Science* 43, 16–23.

Deeming, D.C., Clyburn, V., Williams, K. and Dixon, R.A. (2002) *In ovo* microbial contamination of the yolk sac of unhatched broiler and pheasant embryos. *Avian and Poultry Biology Reviews* 13, 240–241.

Deeming, D.C., Hodges, H.R. and Cooper, J.J. (2011a) Effect of sight barriers in pens of breeding pheasants: I. Behaviour and welfare. *British Poultry Science* 52, 403–414.

Deeming, D.C., Hodges, H.R. and Cooper, J.J. (2011b) Effect of sight barriers in pens of breeding pheasants: II. Reproductive parameters. *British Poultry Science* 52, 415–422.

Defra (2010) Code of Practice for the Welfare of Gamebirds Reared for Sporting Purposes. http://www.defra.gov.uk/publications/files/pb13356-game-birds-100720.pdf (accessed 2 December 2010).

Demirel, Ş. And Kirikçi, K. (2008) Effect of different egg storage time on some egg quality characteristics and hatchability of pheasants (*Phasianus colchicus*). *Poultry Science* 88, 440–444.

FAWC (2008) *Opinion on the Welfare of Farmed Gamebirds*. Farm Animal Welfare Council, London.

Game Conservancy (1983) *Game Conservancy Booklet 18 (Red-legged partridges)*. The Game Conservancy, Fordingbridge, UK, p. 10.

Game Conservancy Trust (1993) *Egg Production and Incubation*. Game Conservancy Trust, Fordingbridge, UK.

Leone, E.H. and Estévez, I. (2008) Economic and welfare benefits of environmental enrichment for broiler breeders. *Poultry Science* 87, 14–21.

Scottish Government (2011) Code of Practice for the Welfare of Gamebirds Reared for Sporting Purposes. http://www.scotland.gov.uk/Publications/2011/03/03123557/0 (accessed 3 May 2011).

Wise, D.R. (1993) *Pheasant Health and Welfare*. Piggot Printers Ltd, Cambridge, UK.

CHAPTER 10
Housing and Management of Layer Breeders in Rearing and Production

H.-H. Thiele

ABSTRACT

Housing and management of layer breeders has to be done in an optimal way otherwise farmers waste their genetic potential and high economic value. A good start secured by optimal brooding conditions, excellent feed quality and appropriate management in the early life of chicks is a prerequisite. The development of sufficient eating capacities during the later rearing period and a fine-tuned light stimulation ensures a good start into the production phase. Furthermore they have to be adjusted to the different housing systems for layer breeders and via a fitting vaccination schedule prepared to react to the different disease challenges in their production environment. Once in production, the nutrient requirements of the birds have to be secured by a phased feeding programme. Good hatching egg quality can be achieved when avoiding floor eggs and by appropriate egg handling.

INTRODUCTION

Parent stock are bred to produce high-performance layers for profitable egg production. The source lines are carefully selected and each parent flock represents a significant economic investment. To maximize return on investment, good management practices are required. Egg quality at the parent and commercial level depends on a combination of genetic potential and non-genetic factors (health, nutrition, light, temperature, air quality, technical environment). It pays to control the non-genetic variables in order to help the birds express their genetic potential. This chapter contains management recommendations based on comprehensive experience which should help to achieve good performance in layer breeder flocks.

REARING YOUNG BREEDERS

Pullets and cockerels of layer breeder flocks should be reared in deep litter or perchery systems comparable to the production unit, whereas birds destined for veranda or breeder cage systems should be reared in rearing cages. The more closely the growing facility resembles the future production system, the easier it will be for the pullets to settle down in their new environment after transfer to the laying house. Most breeding companies recommend to rear males and females together from day-old.

Deep litter systems

Floor rearing systems for chicks and pullets should consist of a well littered, climate controlled, illuminated shed which, as well as feeders and drinkers, also provides slightly raised roosting places. Chicks learn to fly up to rails or perches at an early age. If perching or flying is learnt too late it can result in reduced mobility of individual hens in the future breeder house. Rails or perches should therefore be available to chicks before 6 weeks of age. Mounting feeders and drinkers on or alongside the perches is a very effective preparation for the production phase. Floor rearing systems with a droppings pit on to which feeders and drinkers are mounted are particularly effective for familiarizing the birds with the design of the breeder house. An important aspect of floor rearing is to develop immunity against coccidiosis. It is recommended to vaccinate the birds as the most reliable method to achieve this goal and coccidiostats should never be given in the feed when pullets are vaccinated.

Perchery system

Percheries can accommodate more birds per square metre of floor area than deep litter systems because the total amount of usable space is greater. Multi-tiered perchery systems of different designs are currently offered by several manufacturers, with appropriate management recommendations. The levels are furnished with plastic or wooden slats and feature manure belt ventilation. Feeders and drinkers are usually located only on the bottom and middle levels. The top level is used by the breeders at night as a roosting area. This natural behaviour can be reinforced by using the lighting system to simulate sunset. This involves turning off the light in a stepwise sequence, starting with the bottom and middle levels and finally the top level. In the morning the birds should go to the two lower levels for feeding. By moving between the resting zone and the other levels the breeder pullets get physical exercise and familiarize themselves with the perchery environment. Staggered feeding on the lower tiers promotes flexibility of movement.

Litter

The type and quality of the litter are especially important for young chicks. Straw must be clean and free of mould. Wheat straw is preferable to barley or oat straw. Barley straw contains awn residues which can cause injury to chicks, and oat straw does not absorb sufficient moisture. To reduce dust formation the straw should not be chopped but should be put down as long straw. Splicing improves moisture absorbency. Long straw has the added advantage of encouraging the chicks to forage. This stimulates the birds' natural investigative and feeding behaviours, thus reducing the risk of feather pecking. Wood shavings are good litter material provided they are dust-free and come from softwood varieties that have not been chemically treated; minimum particle sizes of 1 cm are recommended. Chicks must on no account ingest fine particles as these, when combined with water, swell up in the oesophagus, causing ill health and reduced feed intake.

Litter should be put down after heating the shed, when the floor has reached the correct temperature. Significant differences between floor and room temperature when litter is spread too soon change the dew point. The litter then becomes wet from below and produces sticky litter.

House climate

Environmental conditions affect the well-being and performance of the birds. Important environmental factors are temperature, humidity and the level of toxic gases in the air. The optimal temperature depends on the age of the birds. Table 10.1 is a guide to the correct temperature at bird level. The birds' behaviour is the best indicator for correct temperature. Temperatures should always be reduced gradually to avoid sudden changes. The best temperature for optimal feed conversion in the production period is 22–24°C at bird level at a relative humidity of at least 40–45%.

Table 10.1. Recommended house temperatures for breeder layers at various ages[a] (Lohmann Tierzucht, 2009).

Age	Temperature (°C)
1–2 days	36–35
3–4 days	34–33
5–7 days	32–31
2nd week	29–28
3rd week	27–26
4th week	24–22
5th week	20–18
6th week	18–20

[a]Recommended ambient temperatures (at bird level) at a relative humidity of at least 40–45%.

If the ventilation system is used to regulate temperature, take care that the necessary fresh air is supplied. The air quality should meet the minimum requirements given in Table 10.2.

Table 10.2. Minimum requirements for air quality for indoor-reared breeding layer chickens (Lohmann Tierzucht, 2004).

Air component	Desirable level
O_2	>20%
CO_2	<0.3%
CO	<40 ppm
NH_3	<20 ppm
H_2S	<5 ppm

HOUSING OF DAY-OLD CHICKS

Placement of chicks in floor systems

It is advisable to place the chicks close to the watering and feeding facilities in the building as soon as possible after being delivered to the rearing house. If an even temperature distribution within the house cannot be guaranteed or if radiant heaters are used, the use of chick guards or similar devices for keeping the chicks together restricts the chicks to those areas where the temperature is optimal and where feeders and drinkers are located. This also provides a draught-free and comfortable microclimate for the chicks during the first two to three days after hatching. The shed can also be furnished with chick feeding bowls to ensure a better feed intake in the first few days. Both standard feeders and these additional chick bowls should be filled with a layer of about 1 cm of coarse starter feed. As soon as the chicks are able to eat from standard feeders the bowls should be gradually removed. If the chicks are housed in sheds equipped with dropping pits it is advisable to place narrow strips of thin, corrugated cardboard over the slats (40–50 cm wide) on which drinkers, feeding lines and the chick bowls used for the first week are placed. Chick guards or similar devices are again very useful for keeping the chicks close to water, feed and heat sources during the first few days of life.

Placement of chicks in perchery systems

Depending on the system, the chicks are placed either on the middle or bottom level of the perchery where they remain up to about days 14 to 21. Feed and water are provided by so that the birds become fully accustomed to their environment. From 3 to 4 weeks of age the 'training tiers' are opened. The birds can then move freely throughout the building and learn to jump and fly. Percheries that provide feed and water on all tiers can be operated similar to a battery system by confining the chicks during their first few weeks of life. This may be convenient for the pullet producer but is less suitable for training the chicks to move around the system. The tiers should be opened as early as possible in these systems and chick movement within the house stimulated by staggered feeding on the different tiers. Here, too, it is essential that take-off, landing and flying should be mastered by 6 weeks of age. During the first few days of having access to all parts of the house the chicks should be closely watched. Disorientated birds have to be moved manually and trained by the attendants.

Breeders which will later be moved to production percheries where they have to fly on to perches for feeding should ideally be familiarized with this type of perch while still in the growing facility. The flock should be moved to the breeder house well before the proposed start of production. The birds are then better able to find their way around the different areas (feeding, scratching, roosting). By eliminating stress during the period of adaptation to perchery systems, existing nest boxes are more readily accepted and the daily feed intake is more likely to keep up with the birds' growing requirement at the onset of production.

Chick behaviour

The behaviour of the chicks is the best indicator of their well-being. If the chicks are evenly spread and moving around on the cage or pen floor, then the temperature and ventilation are correct. If the chicks are crowding together in some areas or avoiding others, either the temperature might be too low or there is a draught. If the chicks are lying on the cage floor with outspread wings and gasping for air, the temperature is too high.

The body temperature of chicks after they achieve homeothermy is between 40.0 and 41.0°C (Hill, 2001). This information can be combined with the behaviour of the chicks to adjust house temperatures in an optimal way. Using modern ear thermometers from human medicine, the body temperature of chicks even as young as 1 day old can be easily measured. It is important that chicks are sampled from different parts of the barn. It is advisable to proceed in a normal manner to which the birds are familiar e.g. when weighing chicks or pullets to check for uniformity. Average body (rectal) temperatures can be used to adjust the house temperatures to achieve optimal chick temperatures. A big difference in the actual temperature compared with the ideal given temperature, because either the air distribution and humidity level (heat transfer capacity of the air) are too low or the house was not pre-warmed in time, could lead to a drop in body temperature of the chicks causing them to experience cold stress and increase the risk of death.

FEEDING DURING REARING

The recommended feed schedule for the rearing period of layer breeders should be based on four diets (Table 10.3). The starter is a diet with a high nutrient density, based on a feed formulation including raw materials of excellent quality and digestibility. This feed is supposed to be used until the chicks have reached the body weight target at 3 weeks of age. It is followed by a traditional grower feed based on a slightly lower energy concentration compared with the starter feed. The grower feed should be fed until the chicks have reached body weight target at 8 weeks of age. A developer feed should then be fed. A low nutrient density with good structure and a crude fibre content of up to 4–5% in this feed should be used to develop eating capacities.

Table 10.3. Example composition of rearing feeds[a] (recommended nutrient levels per kilogram of feed for different daily feeds).

Nutrient	Starter[b] 1–3 weeks	Grower 1–8 weeks	Developer 9–16 weeks	Pre-layer week 17 – 5% prod.
Minimum metabolizable energy				
kcal	2900	2750–2800	2750–2800	2750–2800
MJ	12.00	11.40	11.40	11.40
Crude protein (%)	20.00	18.50	14.50	17.50
Methionine (%)	0.48	0.40	0.34	0.36
Digestible methionine (%)	0.39	0.33	0.28	0.29
Methionine + cystine (%)	0.83	0.70	0.60	0.68
Digestible methionine + cystine (%)	0.68	0.57	0.50	0.56
Lysine	1.20	1.00	0.65	0.85
Digestible lysine (%)	0.98	0.82	0.53	0.70
Valine (%)	0.89	0.75	0.53	0.64
Digestible valine (%)	0.76	0.64	0.46	0.55
Tryptophan	0.23	0.21	0.16	0.20
Digestible tryptophan (%)	0.19	0.17	0.13	0.16
Threonine	0.80	0.70	0.50	0.60
Digestible threonine (%)	0.65	0.57	0.40	0.49
Isoleucine	0.83	0.75	0.60	0.74
Digestible isoleucine (%)	0.68	0.62	0.50	0.61
Calcium	1.05	1.00	0.90	2.00
Total phosphorus (%)	0.75	0.70	0.58	0.65
Available phosphorus (%)	0.48	0.45	0.37	0.45
Sodium (%)	0.18	0.17	0.16	0.16
Chlorine (%)	0.20	0.19	0.16	0.16
Linoleic acid (%)	2.00	1.40	1.00	1.00

[a]The basis for switching between diet types is the hens' body weight development. The correct time for changing the diet is determined not by age but by body weight. Chicks and pullets should therefore be weighed at regular intervals.
[b]Chick starter should be fed if the standard body weight is not reached by feeding chick grower or if the daily feed intake is expected to be low.

The use of a pre-layer feed for parent stock has several advantages:

- The pre-layer feed gives a better uniformity due to the higher protein and amino acid content in the critical period around the development of sexual maturity. Individual males and females with weight below standard are able to show compensatory growth.
- The pre-layer feed has higher calcium content than the developer ration and improves the shell quality of early maturing hens at later ages.
- The pre-layer feed supplies additional available phosphorus in the critical period of hormonal changes.
- The pre-layer feed prevents excessive high initial egg weight due to its low linoleic acid content.

Parent stock flocks should be fed *ad libitum* during the growing period. Feed density and quality as described before will influence the body weight and feed consumption.

HOUSING SYSTEMS FOR BREEDING STOCK

Floor systems

Birds kept on the floor during production must also be reared on the floor. The optimal bird density depends on management conditions and to what extent climate can be controlled. A stocking rate in the range of 6–8 birds m^{-2} can be taken as a general guide. Floor housing of parent stock flocks can vary considerably in design and layout depending on the type of building. The classic form consists of 80–90 cm high dropping pits covered with wooden, wire mesh or plastic slats, which take up two-thirds of the floor space. Feeders, drinkers and laying nests should be positioned on top of the dropping pit and the drinkers should be mounted at a distance of 30–50 cm directly in front of the entrance to the nest. A littered scratching area of sand, straw, wood shavings or other materials gives the hens room for moving about, scratching and dust bathing. The littered scratching area takes up about one-third of the total floor space, but can be replaced completely by perforated flooring. Rails or other elevated perching facilities should be provided as resting places for the breeders.

Percheries

Percheries are systems where the birds can roam on several levels. The levels are covered with wooden, wire mesh or plastic slats and can have manure belt ventilation installed beneath them in some systems. Feeding and watering facilities are usually located on the lower tiers. The upper tiers usually serve as resting areas for the birds. Depending on the perchery type, the laying nests are either within the system or outside the perchery. A stocking density of up to 14 birds per square metre of floor area should not be exceeded in this housing system. Manufacturers now supply a wide range of perchery types where layer breeders can be kept successfully and achieve high production (see Chapter 12, this volume). Before deciding on which system to use, the farmer should look at the existing construction and select an installation that can be readily adapted to the existing building. When constructing a new facility the building and the perchery installation should ideally be designed together. If the perchery where the young breeders were raised is similar to the type installed in the subsequent breeder house, familiarization problems can be minimized. This aspect should be considered when establishing a perchery system for layer parent stock.

Laying nests

Laying nests should be designed and positioned in such a way that they are easily accessible to the hens, preferably in a central location in the room. It is recommended to keep the entrance to the nest well lit whereas the interior should be darkened. Pullets should not be allowed access to the nests too early, only just before the onset of lay (at about 10 days before start of egg production). This enhances the attractiveness of the nest and improves nest acceptance. During lay the nests should be opened 2–3 h before the start of the daylight period and closed 2–3 h before the end of the daylight period. Closing the nests at night prevents soiling and broodiness. Close-out prevents the hens from roosting in the nests overnight and also makes the nest less attractive to mites. Tilting floors have proved effective for close-out. They also help keep the floor of the nest box clean.

Housing of young breeders

It is advisable that young breeders are transferred to laying houses in good time before the anticipated onset of production. The recommended age is 18 or 20 weeks. Moving the cockerels to the breeder house one or two days earlier can improve their dominance in the new environment. A good early mating behaviour of the males and a good early fertility are secured in this way. The move from the rearing to the production facility should be handled with care but speedily as capture and transportation are stressful to the birds. Gentle rehousing and careful adaptation of the flock to the new surroundings are crucial for good production results.

After transfer the birds should be dispersed evenly across the building. Especially in floor and perchery systems they should be placed close to feeders and drinkers. Water and feed must be available immediately. On arrival in the new quarters the light should be left on so that the birds can find their way around.

Room temperatures should be within a comfort range for the birds. If the building is too cold the breeders may be inactive and not drink or eat. They should not be disturbed during the first 24 h after the move. Inspection of the stock should therefore only be carried out in an emergency. The attendants should be calm and quiet and always wear the same clothing. Nervous attendants cause stress among the newly housed pullets.

Management during the early days

During the first few days after housing it is important to stimulate a sufficiently high feed intake. The birds should be encouraged to increase their food consumption as quickly as possible. Some ways to achieve this are:

* providing an attractive meal-type ration with good structure (see Table 10.4);
* running the feeding lines more frequently;
* feeding when the trough is empty;
* lighting of feeding installations;
* moistening the feed;
* use of skimmed milk powder or whey-fat concentrate added to the feed; and
* vitamin supplements.

Breeder pullets must on no account lose weight after rehousing. They should continue to gain weight, or at least maintain their body weight. Partially closing the scratching area (leaving the birds a minimum amount of space) and manually moving disorientated hens back on to the dropping pit or into the system have also proved effective in floor and perchery systems. Where nests are used, the light sources should be placed in such a way that the entrances to the nests are well lit.

Table 10.4. Recommended particle size distribution for chick starter, grower and layer feed[a] (Lohmann Tierzucht, 2009).

Sieve size (mm)	Overall proportion (%)	Sieve size interval (mm)	Proportion in interval (%)
0.5	19	0.00–0.5	19
1.0	40	0.51–1.0	21
1.5	75	1.01–1.5	35
2.0	90	1.51–2.0	15
2.5	100	>2.0	10[a]

[a]Individual particles not bigger than 3 mm in chick superstarter and starter diets and 5 mm in grower and layer diets.

LIGHTING

The lighting programme controls onset of lay and affects the performance during the production period. So, within certain limits, performance can be adapted to farm-specific requirements by adjusting the lighting scheme. It is easiest to follow the lighting programme in closed houses. In this case the hours of light and light intensity can be adjusted to changing needs.

Intermittent lighting programme in rearing for day-old chicks

When the day-old chicks arrive on the farm, they have been intensively handled in the hatchery and often had a long transport to their final destination. Common practice, in the first 2 or 3 days after arrival, is to give them 23 h light and 1 h darkness (to make sure that chicks do not panic at power failures) to help them recover and to provide the chicks enough time to eat and to drink. In practice it can be observed that after arrival and housing some chicks continue to sleep, others are looking for feed and water. The activity of the flock will always be irregular. Especially in this phase, farmers have difficulties interpreting the chick's behaviour and their condition. There is a practically proven principal that splitting the day into phases of resting and activity using a special designed intermittent lighting programme can be used to achieve the target, which is to synchronize the activities of the chicks (Fig. 10.1). The stockperson gets a better impression of the condition of the flock and the birds are encouraged to search for water and feed by the behaviour of the flock. It is therefore advisable to give chicks a rest after they arrive at the rearing farm and then start with a periodic schedule of 4 h of light and 2 h of darkness for their first week of life.

The usage of the lighting programme brings about advantages as follows:

- the chicks are resting or sleeping at the same time, which means that the behaviour of the chicks will be synchronized,
- the weak chicks will be stimulated by stronger ones to move as well as to eat and drink;
- condition and well-being of the birds is easier; and
- mortality will decrease.

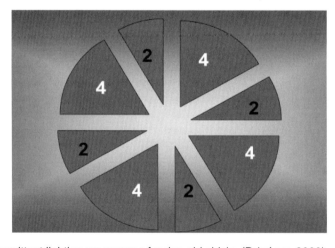

Fig. 10.1. Intermittent lighting programme for day-old chicks (Drinóczy, 2000).

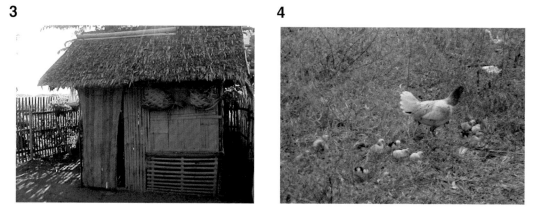

Plate 1. A mixed species semi-scavenging family poultry flock in Lao PDR.
Plate 2. Small-scale family pig and poultry housing in Lao PDR.
Plate 3. Building with woven nest boxes attached to walls under eaves in The Philippines.
Plate 4. A hen with 11 young chicks about 7 days old in Cambodia.

Plate 5. Coop used to confine chicks with hen, with creep feeder (placed inside coop) photographed in Myanmar.
Plate 6. Small-scale broiler production in Bhutan.
Plate 7. This permanent backyard cage contains an insulated coop inside to enable the hens to survive freezing temperatures during the Minnesota winter in the northern USA. (Image courtesy of L. Halcon)
Plate 8. A chicken 'nappy' designed for chickens that are able to roam in backyards and inside their owners' homes. (Image courtesy of I. Dimock)

Plate 9. Poultry are cared for by children in Mauritania. (Image courtesy of S.Issa)
Plate 10. Poultry are cared for by a woman in Afghanistan.
Plate 11. A community vaccinator (right) administers a thermotolerant Newcastle disease vaccine via eye drop to a chicken for a small fee in Gaza Province, Mozambique. (Image courtesy of Kyeema)
Plate 12. The vaccinator, Sr Eduardo Fernando Mondlane, with his vaccination record book showing how many chickens were vaccinated per house and the amount paid. (Image courtesy of Kyeema)
Plate 13. 'Sand mini-hatchery' in Bangladesh functioning with three petrol lamps. (Image courtesy of A. Rota)

Plate 14. Poultry breeding farm in Bangladesh. (Image courtesy of A. Rota)
Plate 15. Birds are offered live for sale to urban consumers at the live bird market of Freetown, Sierra Leone.
They are collected from various regions of the country and neighbouring countries (e.g. Guinea). Around
250–300 local chickens are kept in crates to prevent them from escaping. (Image courtesy of E.F. Guèye)
Plate 16. Training of village women in Afghanistan by a female instructor.

Lighting programmes for layers

The lighting programme (day length and light intensity) to which a flock of breeders is subjected during the growing and production phase is a key factor in determining the onset of sexual maturity and egg production. Lighting programmes for pullets kept in windowless barns can be designed so as to guarantee optimal growth and efficient preparation for the laying period, largely independent of the season.

The 'golden rule' to follow in designing lighting programmes for layers is that they should never experience an increase in day length until the planned light stimulation starts and never experience a decrease in day length during the production cycle (Fig. 10.2). Following this principle, the day length is gradually reduced after placement of the day-old chicks in the rearing farm; after the minimum is reached, a phase of constant day length follows; and finally light hours are gradually increased to stimulate the onset of lay.

The so-called 'step down' procedure in the early days of the chicks' life can be used to make the pullets more sensitive to light. After reaching 10 to 8 h per day, the birds are kept on constant day length for some weeks. The length of the day during this constant period determines the step-down and the following step-up programme and is of minor importance for the pullets' photoperiodic sensitivity to light. The more time the birds have during this constant phase, the more they will eat and grow. In situations where farmers have difficulties to achieve the target body weights, a longer constant day can help to improve pullet quality. Any step-up procedure or increase in day length when birds get to an age of 14 to 15 weeks will stimulate sexual maturation. A quick step-up will induce an earlier onset of egg production, while a slow step-up will delay the onset of lay. The combination of quick step-down and quick step-up lighting is most effective for achieving early onset of lay; slow step-down and slow

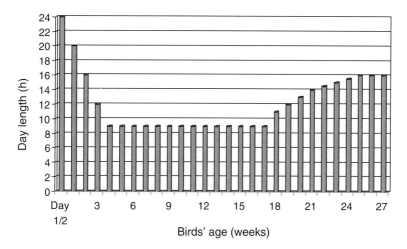

Fig. 10.2. Example of a lighting programme for brown layer breeders; white layer breeders would be very similar (Lohmann Tierzucht, 2009).

step-up will delay it. Many scientific trials and practical experience with different strains of layers have confirmed that the number of eggs and egg weight can easily be influenced utilizing this management tool. If a producer wants early egg production, high total egg number and a moderate egg weight, he should use the quick step-down/step-up variant. To get fewer, but larger eggs, a slow step-down/step-up variant should be chosen. However, layer breeders should never be exposed to the quick step-up/step-down programme, because small eggs at the beginning of the laying period cannot be used as hatching eggs and therefore are undesirable.

Our experience shows that day length should first be increased in the afternoon hours, followed by further increments in the morning hours. Increasing the day length by 2 h in the initial step-up will not only stimulate them more quickly into lay, but also offers two additional hours to eat. This can be taken into consideration when designing lighting programmes for special flocks or housing conditions.

Lighting programme for open houses

A controlled photostimulation of hens should not be abandoned as a management tool in houses with windows. The rearing unit should either be blacked-out or the windows should have a facility for blocking out daylight to maintain the lighting programme. Shutters can be synchronized with the lighting programme and must be seen as very valuable tools.

If the flocks are placed in open houses or if windows, ventilation shafts and other openings cannot be light proofed to keep out natural daylight, this needs to be taken into account when designing the lighting programme. If flocks are moved to production facilities whose windows cannot be blacked-out or where natural light can seep through ventilation shafts ('extraneous light') the lighting programme must be adjusted to match the natural day length at the time the flock is moved and must be kept constant throughout the rearing phase. It is important to distinguish between breeders from a lightproof growing facility or breeders reared with blacked-out windows (with a proper lighting programme) and breeders fully exposed to natural daylight throughout the growing period. When birds unaware of the natural day length during the growing period are moved to open breeder facilities it is essential to prevent stress due to excessive light stimulation by an abrupt lengthening of the day. The number of daylight hours in a naturally lit breeder shed should not be greater than 2–3 h longer than what the birds had in the rearing shed. For example, if the natural day is about 14 h at transfer (18–20 weeks of age), a day length of 12 or 11 h daylight is appropriate in the rearing shed. In the case of birds reared in open housing, premature stimulation of the pullets can only be prevented if the natural day length at the time of the proposed light stimulation of the flock is taken into account when planning the stepwise reduction of light hours in the early growing period. In some extreme cases this would be about 14 h constant day length up to 18–20 weeks of age. In open housing the lighting programme during the spring and summer months is determined by the increasing natural day length, which peaks at about 17 h of daylight in Central Europe. When the natural day shortens from July onwards, the 17 h day length should be kept constant until the end of the laying period.

FLOOR EGGS

The incidence of floor eggs can be reduced by incorporating the following experiences into the design of the breeder house and the management of young flocks:

- Laying nests should be readily accessible to the hens and positioned in a central location in the room.
- The entire building should be well lit – dark corners and excessively littered scratching areas should be avoided.
- Draughty nests disturb the hens during egg laying and should therefore be avoided.
- The entrance to the nest must be clearly visible to the hens.
- Additional lighting of the interior of the nest can improve nest acceptance at the onset of lay.
- Litter depth should not exceed 2 cm at the onset of the laying period. Light-coloured litter material is preferable to dark material.
- Feeders and drinkers should not be more than 2 to 3 m away from the nest area.
- The provision of drinking water in the vicinity of the nest entices the hens to this area.
- Feeders and drinkers should be positioned in such a way that they do not create attractive areas for egg laying.
- If nest boxes are mounted on the dropping pits the perforated floors should have a gradient of about 7° towards the nest. This increases the hens' motivation to deposit eggs in the nest.
- If walkable surfaces are installed in front of the nests these should incorporate barriers every 2 m to stop the hens from parading in front of the nests and blocking access.
- Pullets should not be moved to the production facility before 18–20 weeks of age.
- The laying nests should be opened 10 to 14 days before the onset of lay.
- Hens should not be disturbed while laying eggs – avoid feeding at this time if possible.
- Do not carry out flock inspections during the main morning laying period.
- Floor eggs should be collected quickly, if necessary several times a day.
- If floor eggs still occur, increasing the day length by adding an extra hour of light at the start of the day is often an effective remedy.

FEEDING DURING LAY

For maximum hatching egg production and optimum hatchability a phased feeding programme is recommended. Since the requirements for specific nutrients like essential amino acids, calcium, available phosphorus and linoleic acid change with age, a programme with at least two phases is recommended (Tables 10.5 and 10.6).

Tables 10.5 and 10.6 with recommended nutrient levels per kilogram are based on an average daily feed consumption of 115 g, which can be expected with a feed containing 11.4 MJ = 2720 kcal metabolizable energy per kilogram at an in-house temperature of 22°C and where birds have good feather quality. If feed consumption is greater or less than this the nutrient specifications should be modified to maintain intakes of specific nutrients.

Table 10.5. Example of a phase 1 (bird age approximately 20–50 weeks) breeder feed (Lohmann Tierzucht, 2009). Recommended nutrient levels per kilogram of feed for different daily feed consumption levels.

Nutrient	Requirement per hen (g day^{-1})	Daily feed consumption			
		105 g	110 g	115 g	120 g
Protein (%)	19.20	18.29	17.45	16.70	16.00
Calcium (%)	4.10	3.90	3.73	3.57	3.42
Total phosphorus[a] (%)	0.64	0.61	0.58	0.56	0.53
Available phosphorus (%)	0.44	0.42	0.40	0.38	0.37
Sodium (%)	0.17	0.16	0.15	0.15	0.14
Chloride (%)	0.17	0.16	0.15	0.15	0.14
Lysine (%)	0.87	0.83	0.79	0.76	0.73
Digestible lysine (%)	0.71	0.68	0.65	0.62	0.59
Methionine (%)	0.44	0.42	0.40	0.38	0.37
Digestible methionine (%)	0.36	0.34	0.33	0.31	0.30
Methionine + cystine (%)	0.80	0.76	0.73	0.70	0.67
Digestible methionine + cystine (%)	0.66	0.62	0.60	0.57	0.55
Valine (%)	0.69	0.66	0.63	0.60	0.58
Digestible valine (%)	0.59	0.57	0.54	0.52	0.49
Tryptophan (%)	0.21	0.20	0.19	0.18	0.18
Digestible tryptophan (%)	0.17	0.16	0.15	0.15	0.14
Threonine (%)	0.64	0.61	0.58	0.56	0.53
Digestible threonine (%)	0.52	0.49	0.47	0.45	0.43
Isoleucine (%)	0.66	0.63	0.60	0.57	0.55
Digestible isoleucine (%)	0.55	0.52	0.50	0.48	0.46
Linoleic acid (%)	2.00	1.90	1.82	1.74	1.67

[a]Without phytase.

Feed consumption

The level of feed intake in the production period is mainly affected by:

- In-house temperature – low temperature increases the maintenance requirement for energy.
- Condition of feathering – poor condition due to management mistakes or malnutrition increases the maintenance requirement for energy.
- Feed texture – coarse texture increases whereas fine texture decreases feed intake.
- Energy level – the higher the energy level, the lower the feed intake and vice versa.

Table 10.6. Example of a phase 2 (bird age approximately >50 weeks) breeder feed (Lohmann Tierzucht, 2009). Recommended nutrient levels per kilogram of feed for different daily feed consumption levels.

Nutrient	Requirement per hen (g day^{-1})	Daily feed consumption			
		105 g	110 g	115 g	120 g
Protein (%)	18.40	17.52	16.73	16.00	15.33
Calcium (%)	4.30	4.10	3.91	3.74	3.58
Total phosphorus[a] (%)	0.54	0.51	0.49	0.47	0.45
Available phosphorus (%)	0.38	0.36	0.35	0.33	0.32
Sodium (%)	0.17	0.16	0.15	0.15	0.14
Chloride (%)	0.17	0.16	0.15	0.15	0.14
Lysine (%)	0.85	0.81	0.77	0.74	0.71
Digestible lysine (%)	0.70	0.66	0.63	0.61	0.58
Methionine (%)	0.40	0.38	0.36	0.35	0.33
Digestible methionine (%)	0.33	0.31	0.30	0.29	0.27
Methionine + cystine (%)	0.74	0.70	0.67	0.64	0.62
Digestible methionine + cystine (%)	0.61	0.58	0.55	0.53	0.51
Valine (%)	0.68	0.65	0.62	0.59	0.57
Digestible valine (%)	0.58	0.56	0.53	0.51	0.49
Tryptophan (%)	0.20	0.19	0.18	0.17	0.17
Digestible tryptophan (%)	0.16	0.15	0.15	0.14	0.14
Threonine (%)	0.60	0.57	0.55	0.52	0.50
Digestible threonine (%)	0.49	0.46	0.44	0.42	0.41
Isoleucine (%)	0.65	0.62	0.59	0.57	0.54
Digestible isoleucine (%)	0.54	0.51	0.49	0.47	0.45
Linoleic acid (%)	1.60	1.52	1.45	1.39	1.33

[a]Without phytase.

Micronutrients

The supplementation of parent stock feed with micronutrients like essential vitamins, trace elements and substances like antioxidants and organic acids is essential for maximum hatching egg production and hatchability. By adding these micronutrients in suitable quantities, varying contents in the raw materials are compensated and the correct supply to the parent stock is safeguarded. Some typical recommendations are given in Table 10.7.

A remark on vitamin C: vitamin C is synthesized by poultry normally. This vitamin is not considered essential, but in some circumstances, like during heat stress or in a hot climate, it may be important/beneficial to add 100–200 mg per kilogram of complete feed during the production period.

Table 10.7. Example for micronutrient specifications of a breeder diet (Lohmann Tierzucht, 2009).

Supplement (per kg feed)	Starter/Grower	Developer	Pre-lay/Layer 1 + 2
Vitamin A (IU)	12,000	12,000	15,000
Vitamin D$_3$ (IU)	2,500	2,500	3,000
Vitamin E (mg)	20–30[a]	20 30[a]	50–100[a]
Vitamin K$_3$ (mg)	3[b]	3[b]	5[b]
Vitamin B$_1$ (mg)	2	2	4
Vitamin B$_2$ (mg)	8	6	10
Vitamin B$_6$ (mg)	4	4	6
Vitamin B$_{12}$ (µg)	20	20	30
Pantothenic acid (mg)	10	10	20
Nicotinic acid (mg)	30	30	50
Folic acid (mg)	1	1	2
Biotin (µg)	100	100	200
Choline (mg)	300	300	400
Antioxidant (g)	100–150	100–150	100–150
Coccidiostat	as required	as required	–
Manganese[a] (mg)	100	100	100
Zinc[c] (mg)	60	60	60
Iron (mg)	40	40	40
Copper[c] (mg)	5	5	10
Iodine (mg)	1	1	1
Selenium[c] (mg)	0.3	0.3	0.3

[a]According to fat addition.
[b]Double in the case of heat-treated feed.
[c]So-called 'organic sources' should be considered with higher bioavailability.

MAINTENANCE OF HYGIENE, BIOSECURITY AND HEALTH

General recommendations for hygiene and biosecurity

- Set up the farm at a safe distance from other poultry houses and fence in.
- Keep birds of only one age group on the farm.
- Keep no other poultry on the farm.
- Allow no visitors to enter the farm.
- Wear only the farm's own protective clothing within the farm area.
- Provide the farm's own protective clothing for veterinarians, service and maintenance workers, and consultants.
- Disinfect boots before entering the houses.
- Use bulk feed if possible. Do not allow the truck driver to enter the houses.
- Safeguard the houses against wild birds and vermin. Keep rats and mice under constant control.
- Dispose of dead birds hygienically. Follow local laws and regulations.

Insect and parasite control

If necessary, use a suitable insecticide immediately after the birds have been removed in order to kill the insects before they hide in walls and parts of the equipment. Use a contact insecticide before warming up the house to control remaining insects.

Roundworms and threadworms occur in hens and are transmitted via the droppings. If worm infestation is suspected a bulk faecal sample should be taken and sent for analysis to a veterinary laboratory. If necessary the flock may have to be de-wormed. Red poultry mites are a major problem in alternative production systems. They damage health and reduce the productivity of flocks. Heavy infestation can also cause high mortalities (by transmitting diseases). Infestation causes distress in the flock (feather pecking, cannibalism, depressed production). Continuous monitoring of the flock is therefore advisable. Common hiding places of mites are:

- in the corners of nest boxes;
- under the nest box covers;
- on the feet of feeding chains, trough connectors;
- on crossbars of perches;
- on dropping box trays;
- in corners of walls; and
- inside the perches (hollow tubes).

Mites should be controlled with insecticides or other suitable chemicals. These should be applied in the evening as mites are active during the night. It is important that the treatment reaches all hiding places of the mites. More important than the amount of chemical applied is its thorough and even distribution. The mite and beetle treatment should begin as soon as the flock has been depopulated, while the laying house is still warm. Otherwise the pests crawl away and hide in inaccessible areas of the breeder house.

Vaccination

Vaccination is an important way of preventing disease. Different regional epidemic situations require suitably adapted vaccination programmes and should be guided by the local veterinarian or poultry health service.

Vaccines are applied in different ways. Individual vaccinations, like injections and eye drops, are very effective and generally well tolerated but also very labour intensive. Drinking water vaccinations are not labour intensive but must be carried out with the greatest care to be effective. The water used for preparing the vaccine solution must not contain any disinfectants. During the growing period the birds should be without water for approximately 2 h prior to vaccination (less during hot weather). The amount of vaccine solution should be calculated to be completely consumed within 2–4 h. When vaccinating with live vaccines, add 2 g of skimmed milk powder per litre of water in order to protect the virus titre. Spray vaccinations are not labour intensive and are highly effective, but may occasionally have side effects. For chicks up to the age of 3 weeks apply only using a coarse spray and always use distilled water for vaccine dilution and application. Only healthy flocks should be vaccinated. The expiry date of the vaccine should be checked before use and the vaccine must not be used after the expiry date. Records of all vaccinations and vaccines including their serial numbers should be kept. Applying vitamins in the drinking water in the first two to three days after vaccination can help to reduce stress and prevent undesired reactions. How far this is necessary depends on the specific situation on each farm.

An example vaccination programme is given in Table 10.8.

Cleaning and disinfection

As soon as the hens have been moved out it is advisable to treat walls and ceilings with insecticides while the building is still warm. Then all portable equipment (drinkers, feeders) should be taken outside. Litter and droppings must be removed as far away from the building as possible (>1 km). Prior to the cleaning operation (24 h) the entire interior of the building, including walls, ceilings and the remaining furniture, should be soaked. Fat- and protein-dissolving substances should be used for this purpose. The room should then be cleaned with pressure washers, starting with the ceiling and working down towards the floor. Special attention should be paid to ventilation elements, pipe work, edges and tops of beams. The room should be well lit during the cleaning operation so that dirty deposits are clearly visible. After washing, all surfaces

Table 10.8. Example of a vaccination schedule (Lohmann Tierzucht, 2009).

Disease	Occurrence Worldwide	Locally	Age when given	Methods of application[a]
Marek	X		Day 1 – hatchery	
Coccidiosis	X		As recommended by manufacturer	
Newcastle	X		Number of vaccinations according to disease pressure	DW–SP–ED– ND–IM–BD
Gumboro	X		Two life vaccinations at days 21 and 35	DW
Avian encephalomyelitis	X		One vaccination between weeks 13 and 16	DW
Chicken anaemia virus		X	One vaccination between weeks 13 and 16	DW
Infectious bronchitis	X		Number of vaccinations according to disease pressure	DW–BD–SP
Fowl pox		X	One vaccination between weeks 3 and 10	WW
Pasteurellosis	X		Two vaccinations, approx. at weeks 8 and 14	SC
Infectious coryza	X		Two vaccinations, approx. at weeks 8 and 14	SC
Infectious laryngotracheitis		X	One vaccination between weeks 13 and 16	ED
Mycoplasmosis	X		One vaccination between weeks 8 and 18	IM
Egg drop syndrome		X	One vaccination between weeks 12 and 16	IM
Escherichia coli		X	One vaccination between weeks 12 and 16	IM

[a]DW, drinking water; ED, eye drop; SP, spray; SC, subcutaneous injection; ND, nose drop; BD, beak dipping; IM, intramuscular injection; WW, wing web.

and equipment should be rinsed with clean water. Wood chips or similar should be removed from the outdoor area adjacent to the breeder house and replaced at the same time as the litter. The furniture that was taken outside and the external walls of the building including any concrete surfaces should be washed down. Dirty drinkers are potential hazard sources and must therefore be cleaned and disinfected. Drinker lines should be thoroughly flushed out after disinfection to avoid residues in drinkers. Any leftover feed should be removed from the farm. All parts of the feeding installation and the feed silo should be thoroughly cleaned, washed and disinfected.

MANAGING HATCHING EGGS

Handling of hatching eggs

- Collect hatching eggs frequently.
- Keep floor eggs separately – many will already be contaminated internally.
- If floor eggs are to be used, set in separate incubators (large hatchery) or set at the bottom of setter and hatch trolleys where exploding eggs will cause less damage.
- Take out heavily soiled eggs; do not send them to the hatchery.
- Do not wash hatching eggs.
- Store eggs in a clean egg store. If the farm store does not have a controlled temperature, transfer as soon as possible to the hatchery.
- Store eggs at 22°C if setting within 4 days, or 17–20°C if storing for 5–12 days.
- Older eggs will have markedly lower hatchability.

Hatching egg disinfection

- Fog or spray eggs with a modern disinfectant after collection, then place in store. The most popular method is fogging as it is safe, the fog reaches all of the eggs and the eggs do not get wet.
- Vaporization requires less investment in equipment, but chemicals that can be used in a safe manner are not available everywhere.
- Follow manufacturers' instructions carefully.
- If desired, fog eggs in egg store once a day, but this should not be necessary if the store is regularly cleaned.
- Eggs should be fogged again before pre-warming and setting.
- Several manufacturers are producing modern disinfectants suitable for use in hatcheries.
- A fogging machine is a good investment as there is no wetting of the eggs and the fog will reach all of the eggs.
- Spraying can be carried out using a small droplet size but the spray will not reach all eggs without a fan to aid circulation.
- Formaldehyde (Formalin) is no longer recommended as it is harmful to the embryo, increasing early embryonic death, and it is hazardous to human health. There are modern chemicals available that have the same effectiveness.
- Eggs can be disinfected on the breeder farm, in the hatchery or both. Disinfection on the breeder farm reduces the microbiological load as soon as possible, but keep in mind that this cannot exclude the risk that floor eggs or dirty eggs may have already been contaminated. In the hatchery eggs are usually disinfected after grading/traying or before putting in the storage room.

CONCLUSIONS

Parent stock is bred to produce high-performance layers for profitable egg production and represents a significant economic investment. A combination of the genetic potential and the adjustment or fine tuning of the non-genetic variables helps to achieve a good performance in the layer breeder flocks, a good hatchability and assures a profitable business.

REFERENCES

Drinóczy J. (2000) Management of Lohmann parent stock flocks. Presentation at the *Lohmann Training Course on Poultry Farm and Hatchery Management*, Kitzingen, Germany, 15 April 2000.

Hill, D. (2001) The crucial first 48 hours in the life of a chick. Presented at the *Virginia Poultry Health and Management Seminar*, Roanoke, Virginia, 2001; available at: http://www.docstoc.com/docs/74391148/Chick-first-48-hourscanada-DH-3 (accessed 12 October 2011).

Lohmann Tierzucht (2004) *Management Guide for Layers in Deep Litter, Aviary and Free-Range Systems*. Lohmann Tierzucht GmbH Veterinary Laboratory, Cuxhaven, Germany.

Lohmann Tierzucht (2009) *Parent Stock Management Guide – Lohmann Brown, Lohmann LSL*. Lohmann Tierzucht GmbH, Cuxhaven, Germany.

CHAPTER 11
Furnished Cages for Laying Hens

H.A. Elson and R. Tauson

ABSTRACT

Furnished cages (FCs) were conceived over 30 years ago when the welfare deficiencies of barren conventional ones were realized. Their use was intended to enhance hens' behavioural repertoire and welfare without the disadvantages of non-cage and extensive housing. Since then their design has been refined and improved, resulting in much improved performance and hen welfare. Group size has been an important consideration, especially in relation to variation in damaging pecking in differing genotypes with or without beak treatment. Regulations on the latter vary from country to country and have affected design, group size and management. The trend has been to move from furnished cages for small groups of hens (FCSs), used mainly in Scandinavia, to furnished cages for medium/large groups (FCMs/FCLs) subsequently developed in other countries. The three group sizes have generally performed well under good management. Interventions such as beak trimming and controlled light intensity are most often applied in FCLs and to brown genotypes. Large-scale studies, in which performance and welfare have been compared across all currently available systems, show that they are at least as good in FCs as in any other system, and probably superior. Council Directive 1999/74/EC, which required the demise of all conventional cages in the European Union by 2012, has accelerated the move into FCs and it seems likely that the majority of laying hens in Europe will be housed in them for the foreseeable future, with the aim of enhancing laying hen welfare. However, FCs have potential for further improvement; this chapter suggests some possible developments.

TERMINOLOGY

During the European Union (EU) LAYWEL project on welfare implications of changes on production systems (Blokhuis *et al.*, 2007), the partners involved agreed and adopted terminology already in use by some poultry scientists to describe the systems studied. This differs in some respects from that used by policy makers and legislators and is felt to be more appropriate. EU Council Directive 99/74/EC (1999) (CD 99/74/EC), for example, used the term 'enriched cages', whereas LAYWEL defined them as 'furnished cages'. The latter is more accurate since it describes the system precisely, i.e. as a cage containing furniture such as perches, nest boxes and litter area (see Fig. 11.1), rather than using a term (enriched) that may be a matter of opinion.

Thus LAYWEL used accurate descriptive terms: conventional laying cages (CCs), furnished laying cages (FCs) and non-cage systems (NCs) – called 'alternative systems' in CD 99/74/EC. LAYWEL went further with regard to FCs, categorizing them according to colony size as small furnished cages (FCSs) for up to 14 hens (see Fig. 11.2), medium size furnished cages (FCMs) for 15 to 30 hens and large furnished cages (FCLs) (see Figs 11.3 and 11.4) for over 30 hens (Fiks-van Niekerk and Elson, 2005). FCLs are often described by the UK egg industry as 'enriched colony cages' and currently they each house up to about 90 hens.

In view of the above, this chapter has the title 'Furnished Cages for Laying Hens' and uses the abbreviations for the agreed terms quoted above throughout.

Fig. 11.1. Swedish-made small furnished cage (FCS) with eight white hens showing perch, feed trough, litter box (top right) and nest box (lower right). Hens can be seen sitting, standing, perching, feeding, wing stretching and entering litter box. Nesting was over for the day but note the position of eggs awaiting collection in the cradle in front of nest box. Overall size of cage is 120 cm wide × 50 cm deep approximately. (Image courtesy of I. Pamlenyi.)

Fig. 11.2. UK-made small furnished cage (FCS) with eight brown hens. Same size as, and similar design to, the FCS shown in Fig. 11.1.

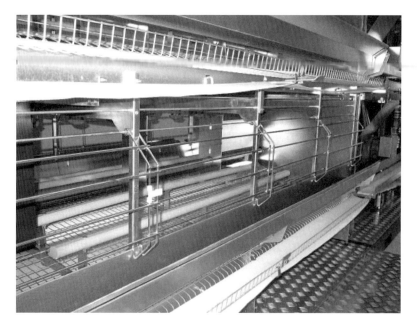

Fig. 11.3. Swedish-made large furnished cage (FCL) for 40–44 hens showing perches, front and centre feed troughs, centre nipple drinkers with drip cups, litter mat (far right) and part of nest box (near left). Note bowed cage front to increase space and improve access to front feed trough. Overall size 242 cm wide × 127–137 cm deep approximately. (Image courtesy of I. Pamlenyi.)

Fig. 11.4. German-made large furnished cage (FCL) with 60 hens. Similar design to that shown in Fig. 11.3 but without centre feed trough. Overall size 362 cm wide × 125 cm deep approximately.

BACKGROUND

Early in the last century European egg production was based on small flocks on general farms. Later farms became more specialized and flock sizes larger. Endoparasitic infestations and certain disease problems brought about the use of wire mesh floor systems for large flocks but these were beset with problems of hysteria, feather pecking and cannibalism (Prip, 1976). Early experiments with wire floored cages for laying hens were carried out in the mid-1920s in the USA and such cages came into commercial use there during the 1930s (Bell, 1995). Their use soon spread to Europe; CCs were in widespread use in the UK by the 1960s and by the early 1980s 95% of the laying flock was housed in them (Elson, 2004).

The book *Animal Machines: The New Factory Farming Industry* (Harrison, 1964) was highly critical of CCs on animal welfare grounds. As a result aspects of cage layout and design were studied during the 1970s and 1980s (e.g. Hughes and Black, 1976; Tauson, 1985). Design modifications and improvements to enhance bird welfare were described by Elson (1990).

French researchers studied the effects of increasing cage height and fitting elevated perches; however vent cannibalism, which led to increased mortality, was a problem (Moinard *et al.*, 1998). Some UK researchers went further and suggested more radical changes to CCs to allow hens to display a wider behavioural repertoire. They involved the use of cages with greater height to allow the inclusion of nest boxes as well as perches and litter; these were the forerunners of FCs.

Bareham (1976) called these 'experimental cages' and Elson (1976) described them as 'get-away cages' because they allowed timid hens to escape to elevated perches. Both designs were for ten to 15 hens, had perches and feeders at two levels and, mainly because of this, had hygienic and inspection problems and in some cases higher mortality rates (Abrahamsson et al., 1995). Although they were subsequently studied at many applied research centres throughout Europe in slightly modified designs, this concept was not completely successful and therefore not introduced commercially.

During the 1980s efforts were made to enlarge and enrich existing laying cages, keeping their original height, by removing some partitions especially the rear ones and adding either a perch or a nest; it was soon found that such cages worked best if both were added. In the late 1980s newly designed modified enriched cages appeared in the UK with a cage height only slightly more than CCs, perches at one level just above the floor and nest and litter boxes. An early version was the Edinburgh modified cage (Appleby and Hughes, 1995; Appleby, 1998). This was further modified and improved in studies at Gleadthorpe (Appleby et al., 2002) and at the Swedish University of Agricultural Sciences (e.g. Abrahamsson et al., 1996). Breeder colony cages were also furnished and adapted for use by laying hens and compared with other FCLs in studies during the 1990s in the Netherlands (Fiks-van Niekerk et al., 2001). The metamorphosis in cage design has been concurrent with studies that established nesting, perching, foraging and dust bathing as behavioural needs of hens, for which any laying hen housing system should make provision (see review by Weeks and Nicol, 2006).

The outcome of this research and development was that FCSs for eight to ten hens came into commercial use in Scandinavia, first in Sweden where there were about 2 million hens in such cages by 2002, i.e. about 40% of the national flock. Subsequently FCLs were developed and installed on farms commercially in a few other European countries; by the end of 2010 about 6 million hens were housed in this system in the UK and many more FCLs have been installed there since then.

Recently interest in FCs has emerged in the USA, stimulated by the activity of the Humane Society of the United States (HSUS). This organization exposed the poor welfare of hens in small barren cages there and prompted some States to phase out all cages, e.g. California by 2015 (HSUS, 2010). Missions to Europe and exchanges of information have been taking place, especially on the performance and welfare of hens in FCLs, but the American Humane Association was against all cages including FCs until recently. However, in June 2010 its American Humane Certified (AHC) accreditation scheme announced that in its animal welfare programme 'it now accepts enriched colony cages (FCLs) as humane systems for housing laying hens'. AHC explained that 'its decision to endorse FCs came following an extensive scientific review of the behaviour and welfare of hens housed in such systems in Europe, where CCs are scheduled to be banned completely in 2012'. The HSUS still rejects FCs, because it does not accept that any form of cage allows the display of sufficient behaviours (World Poultry, 2010).

Meanwhile, a study at Purdue University indicated that their early results 'suggest that furnished cages may be a favorable alternative [to CCs] for housing laying hens' (Pohle and Cheng, 2009). Also in the USA, a Socially Sustainable Egg Production project was set up to understand the issues involved in the possible change from CCs to FCs and NCs (Swanson *et al.*, 2011) and how such a changes would impact the US egg industry (Mench *et al.*, 2011). Part of this project involved a study of hen welfare in different housing systems and led to the conclusion that no single system is ideal from a hen welfare perspective (Lay *et al.*, 2011).

Interest and research in furnished laying cages of different designs and group sizes has also recently developed in Canada (Jendral and Rathgeber, 2009).

Research and Development of Furnished Cages

Most of the early applied research was carried out in FCSs for eight to ten hens per cage. Egg production and feed conversion were generally as good as in CCs and in some cases better (e.g. Abrahamsson *et al.*, 1995). In the better designed models, some of which were further modified and developed in the course of the studies, nest usage was high (over 90% of eggs laid in nest boxes, usually lined with artificial turf) and perch use was about 30% during the day and 75 to 90% at night. Negligible numbers of eggs were laid in the litter boxes, which were generally mounted over the nest boxes, if access to them was controlled, and the small proportion of eggs laid on the wire mesh floors was readily collected as in CCs. Few hens slept in the continuously open nest boxes. Eggs were usually collected clean and, after crack reduction measures ('egg-savers') had been applied, egg quality was also good (Wall and Tauson, 2002).

During the period 1998–2002, when most FCSs were installed in Sweden, mortality was generally similar to or sometimes lower than that of hens in CCs in neighbouring Denmark – a mean of about 3–5% in disease-free flocks by the end of the laying year (slightly higher in brown genotypes and lower in white ones). All the flocks had intact beaks as beak trimming is not permitted in Sweden. A report on mortality, production and use of facilities in the first 53 flocks in FCSs in Sweden during this period was published by Tauson and Holm (2005). No mortality due to hysteria and smothering (commonly experienced in NCs) was observed. In general, scores for plumage condition, peck wounds, deviated keel bone and bumble foot were better or much better than the accepted levels set by the Swedish Board of Agriculture. In some cases scores were worse which led to requests from the Board to alter details in the particular model, e.g. perch location or design, which were carried out before the model was allowed to be introduced commercially.

A 3-year study on FCSs in the UK during the late 1990s produced similar results and led to the conclusion that: 'The results were very good even with 8 birds per cage, but broadly support the provisions for area and facilities made in the EC Directive. They also support the principle behind the Directive that furnished cages provide an acceptable way of protecting the welfare of laying hens' (Appleby *et al.*, 2002).

Studies were also carried out in FCs in Sweden, where the maximum group size allowed is 16 hens, using similar designs as for eight hens per cage but with back-to-back cages modified to have partially or completely open backs. The partially open ones were connected by pop holes (Wall *et al.*, 2004). The purpose of the pop holes was similar to the former get-away cages, i.e. low ranked birds were given the possibility to escape from aggressive individuals and thus lower the potential risk of pecking, but also to offer the hens some stimuli and occupation by changing their position in the cage during the day.

These cages formed FCMs accommodating 16 hens. The results were similar with both back types and to those of FCSs (Wall *et al.*, 2004). Generally the genotypes used showed little aggression, which may have eliminated the expected positive effect. However, the use of the pop holes worked well and many hens changed their position within the whole group (traced by transponders) and were evenly spread in both parts of the cage.

Partly because it was thought that controlled litter boxes and the mechanized delivery of litter to them would be difficult to achieve satisfactorily in large commercial installations, a range of FC sizes was manufactured and tested in Germany for groups of ten, 20, 40 and 60 hens; feed was delivered as 'litter' by auger conveyors on to litter mats on the cage floors. Variable levels of mortality, similar to those expected in CCs, were reported following a study of such FCs by Weitzenbürger *et al.* (2005). Their main finding was that mortality was much higher in FCs when the pullets had been floor reared rather than being reared in cages. Experience with commercial-scale use of these cages, particularly FCLs, shows that beak trimmed flocks housed in them have good feather cover, high egg output and low mortality, although manurial contamination of mats has been a problem in some designs; they quickly came into widespread commercial use on large units in some countries e.g. the UK.

A commercial-scale test unit was installed on a farm in the UK and FCSs, FCMs and FCLs were studied, in conjunction with a series of three experiments on FCSs, between 2000 and 2004. Various models were included in the test unit; some early models had design problems, most of which were overcome, although one model had to be abandoned and was removed from the market. The better designs and especially the 40 and 60 hen groups performed well on a commercial scale. The experiments included four group sizes (six, seven, eight and ten hens), four stocking densities above, at and below 762 cm^2 per hen, two minimum cage heights (38 and 45 cm) and four brown genotypes. Most flocks were beak trimmed. Stocking density of 762 cm^2 per hen seemed to produce optimum results; cage height had little effect within the range applied. Mortality was generally higher (about 5–6%) with intact beaked hens and lower (about 3–4%) with beak trimmed ones (Drakley *et al.*, 2002; Croxall and Elson, 2004). A full report is available (see Defra, 2006).

There were few behavioural differences. Hens housed at space allowances of 609 cm^2 (ten hens in full-size cages) and 762 cm^2 per hen (eight hens) had longer mean feeding bouts than hens housed at 870 cm^2 per hen (seven hens). Even at 12 cm feed trough length per hen all (brown) hens did not feed simultaneously as they grew older (Albentosa *et al.*, 2007). The same is true in NCs, of course, but in both cases there are times (e.g. during nesting) when the feeder space is less occupied and those hens feeding can do so simultaneously.

Group size in FCs was considered an important variable during the LAYWEL project (Blokhuis *et al.*, 2007). There are several reasons for this. The main one from a welfare point of view is the consideration that the potential for cannibalism and feather pecking may increase in larger groups and that the combination of use of different genotypes and especially the ban of beak trimming in some countries should be considered. As mentioned later there are countries which have limited the group size in FCs. On the other hand, a larger cage provides a greater total surface for birds to move around and share, and possibly increase their bone strength. As with floor housing of hens, the fact that physically more of the birds in a larger group could be attacked by individual 'peckers' and cause cannibalism is a clear possibility. Hence, in theory, the use of brown non-beak trimmed genotypes in larger groups would be considered as a potential risk. Consequently in countries where beak trimming is banned (Finland, Norway and Sweden) the proportion of brown genotypes is very low.

In a three batch (2006–2010) study at Funbo-Lövsta Research Centre in Sweden, Wall and Tauson (2010b), Bolander (2011) and Wall (2011) studied group sizes of eight, ten (maximum allowed in Sweden is 16) and 20, 40 and 60 birds of white and brown intact beaked cage-reared genotypes. Results varied between batches of birds, genotypes and group sizes. In general, production and mortality were quite similar and acceptable, egg quality and hygienic conditions (plumage in white birds) worse in groups of 20 and 40, and use of nests considerably poorer in groups with >20 birds (especially in brown birds). The tonic immobility test – often used as a measure for fear levels – showed a significant increase as group size increased in the brown genotypes but not the white. Also, brown genotypes used the litter areas more than white. Catching birds e.g. at depopulation was probably less stressful to them in the FCSs than the larger groups (20–60), mainly because birds were easier to reach and catch. Unexpectedly, bone strength was not increased in larger group sizes. This may suggest that the necessary vertical movements performed in FCSs in order to perch and to reach the raised litter box compensate for the greater amount of horizontal movement to easily use the litter mats or walk along the cage length in between parallel perches generally installed in the larger areas of FCLs. Barnett *et al.* (2009) reported that welfare traits at increased group size were inferior to those in smaller groups and suggested that group size may even be more important than the inclusion of furniture per se. Larger colony sizes (20, 40 and 80 hens) of intact beaked brown and white genotypes were also studied by Sandilands *et al.* (2009). Production and mortality (5–6%, brown hens being slightly higher than white) were acceptable at all colony sizes especially in view of the fact that the birds were not beak trimmed; white hens gave the best overall performance. A full report is available (see Defra, 2008).

A wide spectrum of aspects of FCs has featured in more recent studies. These include: perch design, width and position (Valkonen *et al.*, 2005, 2009; Cox *et al.*, 2009a); a comparison of seven different perch widths (Struelens *et al.*, 2009); nest floor linings and nest location (Wall and Tauson, 2010a,b); design of litter area, provision of litter material and effect of group size (Guinebretiere *et al.*, 2009; Wall and Tauson, 2009a,b, 2010b); litter facilities – boxes or mats (Cox *et al.*, 2009b); soiling of litter mats (Cox *et al.*, 2009c); resource provision (Shimmura *et al.*, 2009); individually examined items of furniture – perch, dust bath, nest box (Barnett *et al.*, 2009); and finally rearing methods for FCs (e.g. Moe *et al.*, 2010). The latter reported that stress levels in birds reared in cages and laying in FCSs were lower than if birds were reared on a litter floor, which partly also agrees with Weitzenbürger *et al.* (2005) and Wall and Tauson (2005).

A useful development in FCLs has been to place the nest box well away from the litter scratching mats. This is possible where there is sufficient cage length to fit the nest box at one end of the cage and the litter mat towards the other end, so that they are 2 to 3 m apart. Lighting in the aisles is then positioned so that the nest boxes are in the area of lowest light intensity and the litter mats have the highest intensity. Nest boxes in each adjoining pair of cages are adjacent to each other, well away from a light source, and litter mats are similarly adjacent at the opposite end of one cage and the near end of the next, but close to a light source. This arrangement favours good secluded nest usage and increased activity at the litter mats; it tends to keep mats cleaner and minimize the number of eggs laid on them.

Several devices have proved efficient in reducing egg shell damage: an 'egg saver' wire mounted under the feed trough near the cage front which holds eggs immediately behind it for a short period before releasing them to roll slowly on to the egg collection belt, a low hanging flap at the front of the nest outlet to reduce the speed of eggs before they roll out on to the egg belt (Wall and Tauson, 2002) and intermittent movement of that belt to avoid any major build-up of eggs in front of the nest boxes.

PERFORMANCE ASSESSMENTS

In several northern European countries some sort of performance assessment was applied to FCs in respect of production, health, mortality and welfare before they came into widespread commercial use.

The most thorough of these was in Sweden where CCs were phased out between 1999 and 2002 and replaced by either FCSs or NCs. The assessments were carried out without beak trimming (not allowed) under the 7th paragraph of the compulsory Swedish Animal Welfare Ordinance and according to the New Technique Evaluation Programme of the Swedish Board of Agriculture (Tauson and Holm, 2002). In general, production figures and plumage were acceptable; both production and mortality were very similar to those found in CCs. After these evaluation tests in commercial units, FCs were finally approved for full commercial use in Sweden.

Also in Sweden, studies were conducted to compare various aspects of FCSs and aviaries (e.g. Tauson, 2002; Wall and Tauson, 2007; Wall *et al.*, 2008) and also FCSs and CCs (e.g. Tauson and Abrahamsson, 1994; Abrahamsson *et al.*, 1996). Most of them dealt with improvements in e.g. perch design for foot condition, location of perches to enhance egg quality and findings on improved bone strength by 10 to 15%. Furthermore, the use of nest opening/closing mechanisms and studies on individual patterns of hens' daily use of litter during full laying cycles were described and recorded.

In France, Michel and Huonnic (2003) compared performance and welfare in FCs, CCs and NCs (aviaries) and later Huneau-Salaun *et al.* (2010) compared the effects of these three systems on eggshell contamination. In Germany, Leyendecker *et al.* (2005) compared bone strength in CCs, FCs and NCs (aviaries). In Japan, Shimmura *et al.* (2010) carried out a multifactorial investigation of CCs, FCs and an aviary.

As mentioned immediately above, a few studies have compared results from two or three housing systems for laying hens but none until recently have included a whole range of systems. Two large-scale studies involving most known systems housing many large flocks of laying hens were carried out in the UK between 2005 and 2009.

The first was conducted by ADAS Gleadthorpe on behalf of a British Egg Industry Council consortium. This study included 39 beak trimmed large laying flocks on commercial poultry farms in five systems: CCs (eight flocks), FCs (16), single-level barns (three), multi-tier aviaries (three) and conventional + organic free-range systems (nine). Performance and welfare were assessed according to the Swedish standards (Tauson and Holm, 2002). The cooperating egg producers were selected because they were known to practise high management standards and keep good records. Performance was generally good: one important welfare indicator was mortality, for which large differences were seen between systems. The highest mean cumulative mortality during the laying year occurred in free range flocks (14%) and the lowest in FCs (3%), most of which were FCLs (Elson and Croxall, 2006; Croxall and Elson, 2007).

The second large-scale UK study in which systems of egg production were compared was carried out by the University of Bristol and funded by the Department for Environment, Food and Rural Affairs (Defra). The study involved 26 beak trimmed brown hybrid laying flocks in four different systems: CCs (six), FCs (six), single-level barns (seven) and free range (seven). As in the ADAS study, mortality was lowest in FCs and in a range of welfare measures FCs came out best. The full report is available (Defra, 2009), but one conclusion was:

> This study did not include a detailed analysis of all hen behaviours but, considering the indicators of physical wellbeing and stress response that were measured, the welfare of hens in the FC system appeared to be better than that of hens in other systems. (Nicol *et al.*, 2009.)

Thus Sherwin *et al.* (2010) stated 'The lowest prevalence of [welfare] problems occurred in hens in furnished cages'. Similar conclusions were reported by

Wall and Tauson (2005) who compared, on a research station, FCSs, litter floor systems and multi-tier aviaries with brown and white intact beaked laying hens focusing on plumage condition, mortality, health and stress measurements.

LEGISLATIVE REQUIREMENTS

Since Sweden was the first Member State to adopt legislative requirements for FCs in 1998, well in advance of others, their statute, which required 600 cm^2 of usable cage area per hen plus a further 150 cm^2 for nest and dust bath area, formed the basis of FC requirements in CD 99/74/EC. Additional requirements in Sweden, not required by CD 99/74/EC, were a maximum group size of 16 birds per cage, minimum opening times of the litter area (opening not more than 8 h hours after lights-on and minimum opening time of 5 h continuously), a maximum of three tiers of cages and rearing to be carried out in a 'similar way', i.e. in cages. As in Finland and Norway, beak trimming has been prohibited in Sweden since the early 1970s. In Denmark the maximum number of hens allowed in FCs is ten per cage and the maximum number of tiers is three.

CD 99/74/EC required that, over a period of 12 years leading up to 1 January 2012, all hens in all Member States should be housed in FCs or NCs and no CCs should remain in use after that date. The majority are then expected to be in FCs. Unfortunately, the review of CD 99/74/EC and resulting report due in 2005 were delayed until 2008. In several countries it was expected that this report might alter the requirements for FCs; this meant that in most Member States few, if any, FCs were installed until then. In 2008 about 260 million (63%) hens in the EU were still in CCs (Elson, 2010), the majority of which will have to be housed in FCs within 3 years; this is unlikely to be achieved (Elson, 2009a; EUWEP, 2010). Casey (2010) reported that 'EUWEP believes over 100m hens [in the EU] will still be in conventional cages after the 1 January 2012 deadline'.

The minimum requirements of CD 99/74/EC (see additional national requirements in some Member States mentioned earlier) have been implemented by national laws in all Member States (e.g. WOFAR, 2007). They include, for FCs, the provision of:

- at least 750 cm^2 of cage area per hen, of which 600 cm^2 (80%) shall be usable (at least 45 cm high with a maximum floor slope of 14%) and the remaining 150 cm^2 (20%) shall be at least 20 cm high;
- a nest (not regarded as usable area);
- litter, i.e. friable material enabling hens to satisfy their ethological needs, such that pecking and scratching are possible;
- a feed trough allowing at least 12 cm length per hen;
- a drinking system appropriate to the size of the group – if nipple or cup drinkers are used, at least two of them must be within reach of each hen;
- a minimum aisle width between cage blocks of 90 cm;

- at least 35 cm between the floor of the building and the bottom tier of cages; and
- suitable claw shorteners.

In each Member State the Central Competent Authority (CCA; e.g. Defra in the case of the UK) must ensure that a specified number of farm inspections for animal welfare are carried out each year, including the requirements of CD 99/74/EC for FCs. Every few years for the past 10 years (e.g. 2004, 2006 and 2009 in the case of the UK) the EU SANCO Food and Veterinary Office (FVO) has made inspection missions on animal welfare, including that of laying hens, to Member States to ensure that CCAs are checking and enforcing e.g. the requirements of CD 99/74/EC. Its mission reports make recommendations to CCAs to remedy any deficiencies (FVO, 2010). Some of these FVO missions have revealed differences in interpretation of CD 99/74/EC between and within Member States. Some of these may affect the design of FCs and stocking density within them.

Some Member States (e.g. Austria) have decided to phase out FCs, and others (e.g. Germany and the Netherlands) to require what they regard as higher standards such as greater space and height; but FCs meeting the requirements of CD 99/74/EC will be allowed to be used for the foreseeable future in most Member States under EU legislation.

POSSIBLE FUTURE DEVELOPMENTS

A general conclusion of the European Food Safety Authority (EFSA, 2005) report and opinion on welfare aspects of various systems of keeping laying hens was that 'recent research and development and commercial experience have led to considerable improvements in design of systems, particularly FCs, and management of birds in FCs and NCs'. However, although as indicated above FCs were studied for several years before coming into commercial use, and have subsequently been in use on a considerable scale for the past 10 years or so, the system is still relatively new and improvements are still appearing. Some aspects are better developed than others and some design features still need attention (Elson, 2009b). Areas in which research and development are still required include the number of birds in a group (colony), physical size of the cage, the provision of litter and the best form of litter, lighting and light intensity, and catching and handling hens during depopulation.

Group (colony) size

It was thought for some years that small group sizes in FCSs would give the best results and, in certain circumstances, this may still be the case; e.g. with certain brown genotypes that have intact beaks. However, in recent years larger groups of up to 90 hens per FCL have been installed and results have generally been as good as those from FCSs, especially with beak trimmed hens.

The largest group size used in a published study was 80 hens (Sandilands *et al.*, 2007) where 20, 40 and 80 intact beaked brown and white genotype hen groups were compared; the full report is available (Defra, 2008) and stated that 'no single colony size was found to be superior to others in the factors assessed'. In large commercial installations management of FCLs has been claimed to be easier than that of FCSs. At present the optimum group size is not known: only that performance, mortality rate and welfare seem to be good up to at least 60 (beak trimmed) laying hens. A study of group sizes from ten through several stages up to 120 or even 150 laying hens of various genotypes with intact and trimmed beaks would be required to establish the optimum. The technique studied by Wall *et al.* (2004), using pop holes in the rear partition of larger cages to enable birds of lower rank to escape into the adjacent group of the same cage, worked well and might be a way to reduce feather pecking in larger cages.

Size of cage

The depth of FCLs in commercial use is gradually increasing. About 120 to 130 cm was typical until recently but several models are now 150 cm deep and one FCL manufacturer offers models up to 220 cm deep. It is unlikely that this could be increased much more without running into practical problems e.g. during depopulation. However, the length of FCLs has also increased, so far up to about 5 m, without any obvious drawbacks. Increasing the size of FCLs does offer more flexibility, e.g. to increase the distance between nest boxes and litter areas, to introduce different light intensities in certain areas (possibly requiring internal illumination) and to allow more space for the exercise of some behaviours. Cage size is therefore a likely area of further development.

Litter provision and litter material

The requirements of CD 99/74/EC for litter are minimal and cage manufacturers often only install a small mat designed to receive a scattering of (feed) litter occasionally. While this does stimulate foraging (pecking, scratching, feed particle selection and consumption) it seems only to allow partial, often interrupted, dust bathing. The limited size of the mats and amount of litter may increase the risk of feather pecking, especially in hens with intact beaks (Weitzenbürger *et al.*, 2005). Studies are required to establish the best litter type, depth and frequency of provision to enable hens to 'satisfy their ethological needs' and minimize feather pecking. Two approaches to litter provision have emerged: (i) on mats within the usable cage area, laid over the cage floor; and (ii) in separate compartments of the cage, usually above the nest boxes. Studies are required to establish which approach best satisfies the hens' preferences, without welfare drawbacks, in order to enable optimum cage design.

Lighting and light intensity

These are important especially in terms of minimizing feather and injurious pecking in non-beak trimmed birds, ensuring optimum egg output and meeting hens' requirements for various behaviours. Thus a higher light intensity may be better for foraging and dust bathing in the litter area and a lower one better in nest boxes. In some FCLs, especially deep ones where some of the feed troughs are inside the cages, internal lighting may be required as well as that provided in the aisles. Applied research and development is required in this area.

Techniques to prevent feather pecking

A technique to interest and occupy hens and, at the same time, blunt their beaks and redirect pecking from their feathers to a suitable object would be a great innovation. Some kind of attractive, nutritional, abrasive block fixed in appropriate places, e.g. on to cage partitions, might be envisaged; if a promising material object could be found it should be tested. The provision of greater amounts of suitable litter in which hens can forage freely may partially serve this purpose.

Catching and handling hens during depopulation

Catching and handling hens during depopulation can have a serious detrimental effect on bird welfare. Sandilands et al. (2005) studied the impact of production system on the welfare of laying hens at depopulation. They found high levels of catching damage, as indicated by new bone fractures, in CCs, much lower ones in FCs and intermediate ones in NCs. They also observed that cage gate openings, through which hens have to be drawn on removal, are about twice the size in FCs compared with CCs. Training catching teams to handle hens gently, i.e. singly by both legs while supporting the breast, is important to minimize damage (Defra, 2002).

As mentioned earlier, Tauson and Wall (2010b) found that handling hens at depopulation was easier in FCSs than in FCLs because they were easier to reach and catch. However, recent experience with large commercial FCL flocks in the UK suggests that damage can be minimized by using well thought out techniques. These might include: (i) working from both sides of deep cages; (ii) removing hens in dim light during the night directly from the perches before they have left them; and (iii) herding the remainder either to one side or one end of the FCL before catching them, possibly using nest boxes as collection areas.

Having protected hen welfare to high standards throughout the production period it is important to preserve it during depopulation. A project to achieve this to an even greater degree in FCs would be well justified in view if their increasingly widespread use. One possible technique in FCLs would be to remove, rather than just slide open, several gates at once to create much larger openings.

In contrast to some of the above suggestions, increasing the space per hen and providing greater height plus perches at different levels, as in the German Kleingruppenhaltungs (KGH) design of FCL, although giving good results (Flock, 2009), does not seem to produce any real benefits over standard FCLs and certainly increases production costs (Elson, 2009c). Perches at more than one level have been studied in cages at several centres in Europe in the form of the get-away cage (e.g. Elson, 1976) and the high cage with elevated perches (Moinard *et al.*, 1998), with the detrimental effects mentioned earlier in this chapter. Hence, it remains to be seen whether similar problems with poor hygienic conditions from falling droppings and/or vent pecking of hens on perches by others below also occur in KGH FCLs.

CONCLUSIONS

FCs were conceived over 30 years ago when the welfare deficiencies of barren CCs were realized. Their use was intended to provide more space and furniture to enhance the hens' behavioural repertoire without introducing the disadvantages of loose and/or extensive housing; this has largely been achieved. Intensive study of design features has enabled improved performance and hen welfare. Group size has been an important consideration and has revealed much variation in damaging pecking in different genotypes with or without beak treatment. Regulations on the latter vary from country to country and affect design, group size and management.

The trend has been to move from FCSs, used mainly in Scandinavia so far, where beak trimming has long been banned in most conditions, to FCMs and FCLs, which were subsequently developed and used in certain other countries. These have generally performed well when carefully managed with certain interventions e.g. beak trimming at a very young age and well-controlled light intensity.

FCs provide birds with less space than floor systems. However, large-scale studies in which the performance and welfare have been compared across all currently available systems lead us to conclude that they are as good as, or better, in FCs than in any other system. CD 99/74/EC, which requires the demise of all CCs in the EU by 2012, accelerated the move into FCs and it seems likely that the majority of laying hens in Europe will be housed in this system for the foreseeable future. This should enhance the welfare of European laying hens. However, FCs have potential for further improvement especially in terms of cage and group size, litter and lighting provision, the development of a technique to blunt beaks and thereby redirect pecking away from feathers, and catching and handling during depopulation.

REFERENCES

Abrahamsson, P., Tauson, R. and Appleby, M.C. (1995) Performance of four hybrids of laying hens in modified and conventional cages. *Acta Agricultura Scandinavica* 45, 286–296.

Abrahamsson, P., Tauson, R. and Appleby, M.C. (1996) Behaviour, health and integument of four hybrids of laying hens in modified and conventional cages. *British Poultry Science* 37, 521–540.

Albentosa, M.J., Cooper, J.J., Luddem, T., Redgate, S.E., Elson, H.A. and Walker, A.W. (2007) Evaluation of the effects of cage height and stocking density on the behaviour of laying hens in furnished cages. *British Poultry Science* 48, 1–11.

Appleby, M.C. (1998) The Edinburgh Modified Cage: group size and space allowance on brown laying hens. *Journal of Applied Poultry Research* 7, 152–161.

Appleby, M.C. and Hughes, B.O. (1995) The Edinburgh Modified Cage for laying hens. *British Poultry Science* 36, 707–718.

Appleby, M.C., Walker, A.W., Nicol, C.J., Lindberg, A.C., Freire, R., Hughes, B.O. and Elson, H.A. (2002) Development of furnished cages for laying hens. *British Poultry Science* 43, 489–500.

Bareham, J.R. (1976) A comparison of the behaviour and production of laying hens in experimental and conventional battery cages. *Applied Animal Behaviour Science* 2, 291–303.

Barnett, J.L., Cronin, G.M., Tauson, R., Downing, J.A., Janardha, V., Lowental, J.W. and Butler, K.L. (2009) The effects of a perch, dust bath, and nest box, either alone or in combination as used in furnished cages, on the welfare of laying hens. *Poultry Science* 88, 456–470.

Bell, D.D. (1995) A case study with laying hens. In: *Proceedings of Animal Behavior and the Design of Livestock and Poultry Systems International Conference*, Indianapolis, Indiana, 19–21 April 1995. National Resource, Agriculture, and Engineering Service, Ithaca, New York, pp. 307–319.

Blokhuis, H.J., Fiks-van Niekerk, T., Bessei, W., Elson, A., Guemene, D., Kjaer, J.B., Levrino, G.A., Nicol, C.J., Tauson, R., Weeks, C.A. and van de Weerd, H.A. (2007) The LAYWEL project: welfare implications of changes in production systems for laying hens. *World's Poultry Science Journal* 63, 101–114.

Bolander, C. (2011) Effects of group size and genotype in furnished cages on plumage condition, H:L and TI. Masters thesis. Swedish University of Agricultural Sciences, Uppsala, Sweden.

Casey, S. (2010) Enriched cages: France ill-prepared for 2012 cage ban. *Poultry World* October issue, 4.

Cox, M., De Baere, K., Vervaet, E., Zoons, J. and Fiks-van Niekerk, T. (2009a) Effect of perch material and profile on the use of perches. *World's Poultry Science Journal, Abstracts of 8th European Symposium on Poultry Welfare* p. 19.

Cox, M., De Baere, K., Vervaet, E., Zoons, J. and Fiks-van Niekerk, T. (2009b) Facilities for providing litter: litter boxes and scratching mats. *World's Poultry Science Journal, Abstracts of 8th European Symposium on Poultry Welfare* p. 22.

Cox, M., De Baere, K., Vervaet, E., Zoons, J. and Fiks-van Niekerk, T. (2009c) Method for scoring dirtiness of scratching materials in enriched cages. *World's Poultry Science Journal, Abstracts of 8th European Symposium on Poultry Welfare*, p. 20.

Croxall, R. and Elson, A. (2004) Production of eggs in furnished cages. *ADAS Science Review 2003–2004*, p. 49; available at: http://www.adas.co.uk/LinkClick.aspx?fileticket=adapA OSsgP4%3d&tabid=211&mid=664 (accessed 19 October 2011).

Croxall, R.A. and Elson, H.A. (2007) Comparative welfare of laying hens in a wide range of egg production systems assessed by criteria in Swedish animal welfare standards. *British Poultry Abstracts* 3, 15–16.

Defra (2002) Catching and transport. In: *Code of Recommendations for the Welfare of Livestock: Laying Hens*. Department for Environment, Food and Rural Affairs, London, pp. 23–24.

Defra (2006) Effects of stocking density, cage height on behaviour, physiology and production of laying hens in enriched cages – AW0226. http://randd.defra.gov.uk/Default.aspx?Menu=Menu&Module=More&Location=None&ProjectID=9859&FromSearch=Y&Publisher=1&SearchText=AW0226&SortString=ProjectCode&SortOrder=Asc&Paging=10#Description (accessed 10 October 2011).

Defra (2008) A study to compare the health and welfare of laying hens in different types of enriched cage – AW0235. Project AW0235. http://randd.defra.gov.uk/Default.aspx?Menu=Menu&Module=More&Location=None&ProjectID=13598&FromSearch=Y&Publisher=1&SearchText=AW0235&SortString=ProjectCode&SortOrder=Asc&Paging=10#Description (accessed 10 October 2011).

Defra (2009) A comparative study to assess the welfare of laying hens in current housing systems – AW1132. http://randd.defra.gov.uk/Default.aspx?Menu=Menu&Module=Moreocation=None&ProjectID=12672&FromSearch=Y&Publisher=1&SearchText=AW1132&SortString=ProjectCode&SortOrder=Asc&Paging=10#Description (accessed 10 October 2011).

Drakley, C., Elson, H.A. and Walker, A.W. (2002) Production efficiency of laying hens at four stocking densities in furnished cages of two heights. In: *Proceedings of the 11th European Poultry Conference*, Bremen, Germany, 6–10 September 2002. *European Poultry Science* 66 (Special Issue), p. 37. Eugen Ulmer GmbH, Stuttgart-Hohenheim, Germany.

EFSA (2005) The welfare aspects of various systems of keeping laying hens. *The EFSA Journal* 197, 1–23; available at: http://www.efsa.europa.eu/en/scdocs/scdoc/197.htm (accessed 3 October 2010).

Elson, A. (2004) The laying hen: systems of egg production. In: Perry, G.C. (ed.) *Welfare of the Laying Hen*. CABI Publishing, Wallingford, UK, pp. 67–80.

Elson, H.A. (1976) New ideas on laying cage design – the 'Get-away' cage. In: *Proceedings of the 5th European Poultry Conference*, Malta, 5–11 September 1976. World's Poultry Science Association (Malta Branch), pp. 1030–1041.

Elson, H.A. (1990) Recent developments in laying cages designed to improve bird welfare. *World's Poultry Science Journal* 46, 34–37.

Elson, H.A. (2009a) Housing systems for laying hens in Europe; current developments and technical results. In: Cepero, R. (ed.) *Proceedings of the 46th WPSA Symposium Cientifico de Avicultura*, Zaragoza, Spain, 30 September–2 October 2009. World's Poultry Science Association (Spanish Branch), pp. 57–68.

Elson, H.A. (2009b) Welfare friendly poultry systems. *ISA Focus* 2.

Elson, H.A. (2009c) Effects of EU Directive 99/74/EC on egg production systems, economics and business in Europe and beyond. In: *VIII Konference Ceske narodni vetve svetore drubezarske organizace (WPSA)*, Brno, Czech Republic, 15–16 October 2009. World's Poultry Science Association (Czech Republic Branch), p. 2.

Elson, H.A. (2010) Housing systems for laying hens in Europe: current and likely future developments in terms of production, economics and welfare. In: *Proceedings of the WPSA 45th Annual Convention*, Jerusalem, 8–10 March 2010. World's Poultry Science Association (Israel Branch), pp. 18–19.

Elson, H.A. and Croxall, R.A. (2006) European study on the comparative welfare of laying hens in cage and non-cage systems. *European Poultry Science* 70, pp. 194–198. Eugen Ulmer GmbH, Stuttgart-Hohenheim, Germany

EU Council Directive 99/74/EC (1999) Laying down minimum standards for the protection of laying hens. *Official Journal of the European Communities* L 203, 03/08/1999, 53–57.

EUWEP (2010) EU egg processors discuss trade concerns with parliament's AG chief. *Poultry World* July issue, 4.

Fiks-van Niekerk, Th.G.C.M. and Elson, H.A. (2005) Categories of housing systems for laying hens. *Polish Animal Science Papers and Reports* 23, 283–284.

Fiks-van Niekerk, Th.G.C.M., Reuvekamp, B.F.J. and van Emous, R.A. (2001) Furnished cages for larger groups of laying hens. In: Oester, H. and Wyss, C. (eds) *Proceedings of the 6th WPSA European Symposium on Poultry Welfare*, Zollikofen, Switzerland, 1–4 September 2001. World's Poultry Science Association (Swiss Branch), pp. 20–22.

Flock, D.K. (2009) WPSA support for German exit from conventional cages. *Lohmann Information* 44, 61–64.

FVO (2010) EU SANCO Food and Veterinary Office – Inspection Report Database. http://ec.europa.eu/food/fvo/ir_search_en.cfm (accessed 3 October 2010).

Guinebretiere, M., Huonnic, D., de Treglode, M., Huneau-Salaun, A. and Michel, V. (2009) Furnished cages for laying hens: effect of group size and litter provision on laying, feeding, perching and dust bathing behaviours. *World's Poultry Science Journal, Abstracts of 8th European Symposium on Poultry Welfare*, p. 17.

Harrison, R. (1964) *Animal Machines: The New Factory Farming Industry*. Vincent Stuart Ltd, London.

HSUS (2010) California: all shell eggs sold state-wide to be cage-free by 2015 – State Bill AB 1437. http://www.humanesociety.org/news/press_releases/2010/07/ab1437_passage_070610.html (accessed 10 October 2011).

Hughes, B.O. and Black, A.J. (1976) Battery cage shape: its effect on diurnal feeding pattern, egg shell cracking and feather pecking. *British Poultry Science* 17, 327–336.

Huneau-Salaun, A., Michel, V., Huonnic, D., Balaine, L. and le Bouquin, S. (2010) Factors influencing bacterial eggshell contamination in conventional cages, furnished cages and free-range systems for laying hens under commercial conditions. *British Poultry Science* 51, 163–169.

Jendral, M.J. and Rathgeber, B.M. (2009) Productivity, health and welfare of 3 strains of laying hens housed in conventional battery cages and furnished colony cages. *Poultry Science* 88 (Suppl. 1), 75.

Lay, D.C. Jr, Fulton, R.M., Hester, P.Y., Karcher, D.M., Kjaer, J.B., Mench, J.A., Mullens, B.A., Newberry, R.C., Nicol, C.J., O'Sullivan, N.P. and Porter, R.E. (2011) Hen welfare in different housing systems. *Poultry Science* 90, 278–294.

Leyendecker, M., Hamann, H., Hartung, J., Kamphues, J., Neumann, U., Surie, C.S. and Distl, O. (2005) Keeping laying hens in furnished cages and an aviary housing system enhances their bone stability. *British Poultry Science* 46, 536–544.

Mench, J.A., Sumner, D.A. and Rosen-Molina, J.T. (2011) Sustainability of egg production in the United States – the policy and market context. *Poultry Science* 90, 229–240.

Michel, V. and Huonnic, D. (2003) A comparison of welfare, health and production performance of laying hens reared in cages or in aviaries. *British Poultry Science* 44, 775–776.

Moe, R.O., Guémené, D., Bakken, M., Larsen, H.J.S., Shini, S., Lervik, S., Skjerve, E., Michel, V. and Tauson, R. (2010) Effects of housing conditions during the rearing and laying period on adrenal reactivity, immune response and heterophil to lymphocyte (H/L) ratios in laying hens. *Animal* 4, 1709–1715.

Moinard, C., Morisse, J.P. and Faure, J.M. (1998) Effect of cage area, cage height and perches on feather condition, bone breakage and mortality of laying hens. *British Poultry Science* 39, 198–202.

Nicol, C.J., Brown, S.N., Haslam, S.M., Hothersall, B., Melotti, L., Richards, G.J. and Sherwin, C.M. (2009) The welfare of laying hens in four different housing systems in the UK. *World's Poultry Science Journal, Abstracts of 8th European Symposium on Poultry Welfare*, p. 12.

Pohle, K. and Cheng, H.W. (2009) Comparative effects of furnished and battery cages on egg production and physiological parameters in White Leghorn hens. *Poultry Science* 88, 2042–2051.

Prip, M. (1976). Hysteria in laying hens. In: *Proceedings of the 5th European Poultry Conference*, Malta, 5–11 September 1976. World's Poultry Science Association (Malta Branch), pp. 1062–1075.

Sandilands, V., Sparks, N., Wilson, S. and Nevison, I. (2005) Laying hens at depopulation: the impact of the production system on bird welfare. *British Poultry Abstracts* 1, 23–24.

Sandilands, V., McDevitt, R.M. and Sparks, N.H.C. (2007) Effects of enriched cage design and colony size on production, health and welfare in two strains of laying hens. *British Poultry Abstracts* 3, 16–17.

Sandilands, V., Baker, L. and Brocklehurst, S. (2009) The reaction of brown and white strains of hens to enriched cages. *British Poultry Abstracts* 5, 31–32.

Sherwin, C.M., Richards, G.J. and Nicol, C.J. (2010) Comparison of the welfare of layer hens in four housing systems in the UK. *British Poultry Science* 51, 488–499.

Shimmura, T., Azuma, T., Eguchi, K., Uetake, K. and Tanaka, T. (2009) Effects of separation of resources on behaviour, physical condition and production of laying hens in furnished cages. *British Poultry Science* 50, 39–46.

Shimmura, T., Hirahara, S., Azuma, T., Suzuki, T., Eguchi, K., Uetake, K. and Tanaka, T. (2010) Multi-factorial investigation of various housing systems for laying hens. *British Poultry Science* 51, 31–42.

Struelens, E., Tuyttens, F.A.M., Ampe, B., Odberg, F., Sonck, B. and Duchateau, L. (2009) Perch width preferences of laying hens. *British Poultry Science* 50, 418–423.

Swanson, J.C., Mench, J.A. and Thompson, P.B. (2011) Introduction – the socially sustainable egg production project. *Poultry Science* 90, 227–228.

Tauson, R. (1985) Mortality in laying hens caused by differences in cage design. *Acta Agricultura Scandinavica* 35, 165–174.

Tauson, R. (2002) Furnished cages and aviaries: production and health. *World's Poultry Science Journal* 58, 49–63.

Tauson, R. and Abrahamsson, P. (1994) Foot- and skeletal disorders in laying hens. Effects of perch design, hybrid, housing system and stocking density. *Acta Agriultura Scandinavica Section A* 44, 110–119.

Tauson, R. and Holm, K.E. (2002) *Evaluation of Victorsson Furnished Cage for 8 Laying Hens According to the 7th Swedish Animal Welfare Ordinance and the New-Technique Evaluation Program at the Swedish Board of Agriculture*. Report No. 251. Swedish University of Agricultural Sciences, Uppsala, Sweden.

Tauson, R. and Holm, K.E. (2005) Mortality, production and use of facilities in furnished small group cages for layers in commercial egg production in Sweden 1998 – 2003. *Polish Animal Science Papers and Reports* 23, 95–102.

Valkonen, E., Valaja, J. and Venalainen, E. (2005) The effects of dietary energy and perch design on the performance and condition of laying hens kept in furnished cages. *Polish Animal Science Papers and Reports* 23, 103–110.

Valkonen, E., Rinne, R. and Valaja, J. (2009) The effects of perches in furnished cages. *World's Poultry Science Journal, Abstracts of 8th European Symposium on Poultry Welfare*, p. 18.

Wall, H. (2011) Production performance and proportion of nest eggs in layer hybrids housed in different designs of furnished cages. *Poultry Science* 90, 2153–2161.

Wall, H. and Tauson, R. (2002) Egg quality in furnished cages for laying hens – effects of crack reduction measures and hybrids. *Poultry Science* 81, 340–349.

Wall, H. and Tauson, R. (2005) Produktion, befjädring och stress hos två hybrider i olika inhysningssystem. *Fjäderfä* 2005(10), 58-60.

Wall, H. and Tauson, R. (2007) Redesutnyttjande och äggkvalitet i inredda burar. *Fjäderfä* 2007(9), 88–91.

Wall, H. and Tauson, R. (2009a) Ökad gruppstorlek i inredda burar – hur påverkas dödlighet och äggkvalitet? *Fjäderfä* 2009(6), 20–24.

Wall, H. and Tauson, R. (2009b) Är det positivt för hönan med större gruppstorlek i inredda burar? *Fjäderfä* 2009(8), 14–17.

Wall, H. and Tauson, R. (2010a) Kan plastnät ersätta turf i inredda burar? *Fjäderfä* 2010(3), 34–37.

Wall, H. and Tauson, R. (2010b) *Utformning av inredda burar med fokus på alternativa gruppstorlekar, reden och ströbad.* Report of research contracts Dnr. 31-570/08. Swedish Board of Agriculture, Jonkoping, Sweden.

Wall, H., Tauson, R. and Elwinger, K. (2004) Pophole passages and welfare in furnished cages for laying hens. *British Poultry Science* 45, 20-27.

Wall, H., Tauson, R. and Elwinger, K. (2008) Effects of litter substrate and genotype on layers' use of litter, exterior appearance, and heterophil:lymphocyte ratios in furnished cages. *Poultry Science* 87, 2458–2465.

Weeks, C.A. and Nicol, C.J. (2006) Behavioural needs, priorities and preferences of laying hens. *World's Poultry Science Journal* 62, 297–308.

Weitzenbürger, D., Vits, A., Hamann, H. and Distl, O. (2005) Effect of furnished small group housing systems and furnished cages on mortality and causes of death in two layer strains. *British Poultry Science* 46, 553–559.

WOFAR (2007) The Welfare of Farmed Animals (England) Regulations 2007. SI 2007 No. 2078. http://www.opsi.gov.uk/si/si2007/uksi_20072078_en_3 (accessed 10 October 2011).

World Poultry (2010) American Humane Certified endorses colonies [FCLs] for hens. http://worldpoultry.net/news/american-humane-certified-endorses-colonies-for-hens-7602.html (accessed 10 October 2011).

CHAPTER *12*

Performance, Welfare, Health and Hygiene of Laying Hens in Non-Cage Systems in Comparison with Cage Systems

T.B. Rodenburg, K. De Reu and F.A.M. Tuyttens

ABSTRACT

This chapter compares the performance, welfare, health and hygiene of laying hens in different types of non-cage systems, focusing on barn, free range and organic systems. These non-cage systems are compared with each other and with cage systems. This comparison shows that both between barn, free range and organic systems and between non-cage and cage systems large differences can be identified. Moving from conventional cages to furnished cages, barn, free range and finally organic systems results in increasing environmental complexity, which is positive for some aspects of hen welfare, but also increasing risks for performance, health and hygiene, which is negative for other aspects of hen welfare. For the improvement of the welfare of laying hens in non-cage systems and furnished cages, we recommend that the focus should be on creating a better match between the animals and their husbandry environment. Good examples are the development of new housing designs that combine the benefits of non-cage systems with improved performance, health and hygienic status. Further, promising approaches in animal breeding and optimizing rearing environments are expected to yield major improvements in the welfare of laying hens in non-cage systems and furnished cages.

INTRODUCTION

Following Council Directive 1999/74/EC, from 2012 conventional cages are prohibited in the European Union (EU) and laying hen housing systems had to change from (mainly) conventional cages to furnished cages and non-cage systems (Rodenburg *et al.*, 2005; Tauson, 2005). In this chapter we describe the performance, welfare, health and hygiene of laying hens in non-cage systems, focusing on barn, free range and organic systems. These non-cage systems are compared with each other and with cage systems. The pros and cons of each system are discussed. Further, we also describe possible improvements that can lead to a better match between the animals and their husbandry environment. For these improvements we focus both on adaptations in housing and husbandry and on possibilities for animal breeding.

DESCRIPTION OF NON-CAGE SYSTEMS

Broadly, two types of non-cage systems can be distinguished: (i) floor housing systems; and (ii) aviary systems. The term 'non-cage systems' is currently preferred over the term 'alternative systems', that was used previously in the EU, because it is more specific (Fiks-van Niekerk and Elson, 2005). In both types of non-cage systems, laying hens are kept in large flocks. Flock size can vary from approximately 5000 birds to 30,000 birds (Rodenburg et al., 2008b). Birds have access to nests, perches and a large pecking and scratching area with litter (one-third of the floor area), following the requirements laid down in EU Council Directive 1999/74/EC. In floor housing systems, all birds are kept on a single level, usually with a central slatted area which includes perches, feeders, drinkers and nests and a pecking and scratching area with litter on both sides of the slatted area. In aviary systems birds have access to different levels or tiers, with a maximum of four tiers. Tiers can contain perches, nests, drinkers and feeders. The scratching area is located on the floor of the house, similar to floor housing systems.

Different types of non-cage systems can also be defined by the presence or absence of an outdoor run. With respect to outdoor run access, three types of systems can be distinguished: (i) barn systems; (ii) free range systems; and (iii) organic systems. These three types of non-cage systems correspond to the way table eggs are marketed in the EU: barn eggs (code 2), free range eggs (code 1) and organic eggs (code 0). Cage eggs form a separate category (code 3). The differences in requirements for the three types of non-cage systems were laid down in Commission Regulations No 889/2008 (organic) and No 557/2007 (barn and free range). The most important differences are summarized in Table 12.1.

In barn and free range systems, a maximum of 9 birds m^{-2} can be kept, equalling 1111 cm^2 of floor space per bird. In organic systems, birds have more space with a maximum of 6 birds m^{-2}, equalling 1667 cm^2 of floor space per bird. In free range and organic systems, birds additionally have access to an

Table 12.1. Minimum requirements for barn, free range and organic systems.

	Barn	Free range	Organic
Space per bird (cm^2)	1111	1111	1667
Outdoor space per bird (m^2)	–	4	4
Outdoor access (h day^{-1})	–	8	8
Group size	Unlimited	Unlimited	Max. 3000
Beak trimming	Yes	Yes	No

outdoor run, allowing at least 4 m^2 per bird for at least 8 h day^{-1}. On many free range and organic farms, covered outdoor runs, or winter gardens, are used as transitions from the barn to the outdoor run (Häne et al., 2000). If this area is continuously available to the hens, it can be added to the available indoor floor space. These areas are mainly used as pecking and scratching areas and are normally open sided, providing daylight to the birds. In some cases, winter gardens are also combined with barn systems without outdoor access. Further differences between organic systems and the other two systems are that in organic systems, group size is restricted to 3000 birds and beak trimming is normally not allowed (it can be authorized by the competent authority if intended to improve health, welfare or hygiene on a case-by-case basis). Also, the outdoor run for organic hens has to be managed according to organic regulations (no artificial fertilizer, no pesticides) and the diet has to be at least of 95% organic origin, due to rise to 100% in 2012 (Acamovic et al., 2008).

Based on egg production figures in 2008, the main egg producers in the EU are France, Germany, Italy, the Netherlands, Spain and the UK (Viaene and Verheecke, 2009). Together they produced around 75% of the total EU egg production. Large differences can be observed in the uptake of the different non-cage systems throughout the EU. In France, Germany, Italy and Spain, the vast majority of eggs (70 to 95%) still came from conventional cage systems (van Horne and Achterbosch, 2008). This may, however, have changed in the last 4 years, due to the conventional cage ban. The number of furnished cage farms is still limited in most parts of Europe, except in Sweden where up to 40% of hens are already housed in furnished cages. There, conventional cages were prohibited in 1998 (van Horne and Achterbosch, 2008). Barn systems are particularly popular in Sweden, the Netherlands, Switzerland and Austria, accounting for 30 to 60% of egg production. Free range systems can mainly be found in the UK, Switzerland and Austria. Switzerland is the only country with a full cage ban (including furnished cages), which came into force in 1992. Organic egg production systems form a niche market throughout the EU (Windhorst, 2005). Denmark has the largest share of organic egg production (15%) when compared with Danish conventional egg production, followed by Austria (4%), France (2%), Sweden (2%) and the UK (2%).

COMPARISON OF BARN, FREE RANGE AND ORGANIC SYSTEMS

The main difference between barn systems on the one hand and free range and organic systems on the other is of course the absence or presence of the outdoor run. As stated by Knierim (2006), access to an outdoor run increases the space per bird, increases the number of environmental stimuli and allows the hens more freedom to choose between different environments. Patzke et al. (2009) studied brain morphology in hens from conventional cages, barn systems and free range systems and detected morphological differences in the hippocampus between hens from cages and hens from free range systems, probably related to the increased spatial complexity of free range systems compared with cages. No differences were found between hens from barn systems and hens from the other two systems. In agreement with these results, Krause et al. (2006) showed that free range access can positively influence learning performance in chicks, at least in the short term. Access to an outdoor run gives the hens access to natural light and fresh air with increased opportunities for foraging behaviour, for which laying hens are strongly motivated (Dixon, 2008). Daylight may be very important for behavioural development. Indeed, results from Bestman et al. (2009) indicate that the absence of daylight during rearing can increase the risk of feather pecking during the laying period, which can result in increased mortality rates. Shimmura et al. (2008) compared the behaviour of hens in barn systems and free range systems and found that free range hens showed more foraging behaviour than birds in a barn system. In the barn system, more aggressive pecking and feather pecking were observed. It has been shown that foraging motivation plays an important role in the development of feather pecking (Blokhuis, 1986; Newberry et al., 2007). In line with this observation, access to and proper use of an outdoor run have been shown to reduce the risk of feather pecking and cannibalism (Green et al., 2000; Bestman and Wagenaar, 2003; Lambton et al., 2010). Therefore, access to an outdoor run can improve performance by reducing mortality due to feather pecking and cannibalism. This is an important benefit, as beak trimming is likely to be prohibited in many European countries in the near future (van Horne and Achterbosch, 2008). At present, beak trimming is prohibited in Scandinavia and Austria and regulated in Germany, Switzerland, the Benelux countries and the UK. In these latter countries, beak trimming is likely to be prohibited in the near future. The other EU Member States follow the European legislation regarding beak trimming, applying mild beak trimming before 10 days of age. For the transition period from beak trimmed birds to non-beak trimmed birds, new methods that are less painful for the birds such as infrared beak trimming (Gentle and McKeegan, 2007) may offer opportunities.

Although an outdoor run has positive effects on behaviour and may reduce feather pecking, it also presents increased risks of predation and health problems (Miao et al., 2005; van de Weerd et al., 2009) (see also chapter 4, this volume). Health problems reported in free range hens include bacterial infections, such as Pasteurella infections (Hegelund et al., 2006), and gastrointestinal helminth infections (Permin et al., 1999). For instance in the case of Pasteurella infections, high mortality levels have been reported of about 55% (Hegelund et al., 2006). Gastrointestinal helminths are found more frequently in droppings of free range flocks than in droppings of indoor flocks (Permin et al., 1999). Endoparasitic infestation in itself is not necessarily a threat to laying hen health and welfare, but the parasites can transmit other infections such as Salmonella or Histomonas (Knierim, 2006). Red mite (Dermanyssus galinae) infestations are found in all poultry production systems, but they have been reported to be more severe in free range systems (Guy et al., 2004). Red mites feed on the hens' blood during the night and can cause considerable stress to the flock, anaemia and increased mortality rates (Kilpinen *et al.*, 2005). In organic flocks these health issues present a greater challenge than in free range flocks, because in organic flocks use of antibiotics and anthelmintics is restricted. To reduce the risk of infection, it is important to keep the parasitic pressure and the presence of bacteria to a minimum in all parts of the outdoor run. This can be done by encouraging hens to make proper use of the entire range area (to draw away the birds from the area near the house) or by rotating access to various parts of the range or by using mobile housing systems that can be frequently moved to a clean area (Knierim, 2006).

Predation is another risk related to free range access. Hegelund *et al.* (2006) recorded mortality levels and causes of mortality in 18 flocks of organic laying hens and reported an average of 6.4% 'lost' or taken by predators. Reducing the risk of predation can be done by improved fencing (against e.g. foxes) or improved range cover (against aerial predators). Hens also have a preference for an outdoor run that provides cover compared with an open outdoor run, so providing cover will also help to increase range use (Hegelund *et al.*, 2005; Zeltner and Hirt, 2008).

Considerable differences in mortality have been reported in barn, free range and organic flocks, although no comparative studies are available. In barn flocks, Rodenburg *et al.* (2008b) reported mortality levels of about 8%. Whay *et al.* (2007) recorded mortality in 25 free range flocks in the UK and reported a median of 7% mortality (varying between 2 and 21%). In organic flocks, Hegelund *et al.* (2006) reported on average 23% mortality. However, as described previously, in their study some problem flocks were included with very high mortality levels (54 and 62%). The fact that in organic flocks beak trimming is already prohibited, and in free range and barn flocks it is not, will also contribute to increased risks of high mortality levels in organic systems. Another problem in the study by Hegelund *et al.* (2006) was mortality due to piling, which causes suffocation of the birds near the bottom. This problem and its causes are still poorly understood, although it has been reported to contribute up to 25% to mortality in organic flocks in a UK survey (Sparks *et al.*, 2008). Another risk factor in organic flocks is the organic diet. There have been suggestions that, since methionine levels are often lower in organic diets than in conventional diets, this could pose an additional risk for feather pecking in organic laying hens (van Krimpen *et al.*, 2005). To date, however, few results have been found that point in this direction (Kjaer and Sørensen, 2002; Acamovic *et al.*, 2008). It has been suggested that the lack of methionine may be compensated by foraging in the outdoor run.

It seems that the risk of high mortality rates is greater in organic flocks than in flocks from barn and free range systems. This can negatively affect egg production in organic flocks, because increased mortality will of course result in a reduced number of eggs per hen housed.

COMPARISON WITH CAGE SYSTEMS

Apart from non-cage systems, furnished cages are also allowed in most European countries since the ban on conventional cages in 2012. Exceptions are Switzerland, where a full cage ban is in place, Germany and the Netherlands, where only large furnished cages are allowed. In furnished cages, hens have more space than in conventional cages (750 cm^2 versus 550 cm^2 per bird). Due to the increased group size, hens also have more 'shared space' in furnished cages than in conventional cages. In addition, they have access to a nest, a perch and an area with some litter for pecking, scratching and dust bathing. Depending on the system, birds are kept in relatively small groups, ranging from five to 100 birds (Rodenburg *et al.*, 2005). Fiks-van Niekerk and Elson (2005) suggested a distinction between three categories of furnished cages: small (up to 15 hens), medium (15–30 hens) and large (more than 30) (see also chapter 11, this volume).

De Mol *et al.* (2006) compared welfare of laying hens in various housing systems, using a computer model (FOWEL) based on statements from literature. They found that, based on their model, feeding level (*ad libitum* or restricted), space per hen (450 to 2000 cm^2), water availability (*ad libitum* or restricted) and the presence or absence of perches and nests had the strongest effect on the overall welfare score. Interestingly, access to a free range resulted in only a small improvement in the welfare score in their model. The welfare score of cage systems was low, of barn systems intermediate, and of organic systems high.

Rodenburg *et al.* (2008c) developed a welfare assessment protocol that could be used to compare welfare in furnished cages and non-cage systems. This protocol includes an integrated welfare score, in which each welfare indicator was assigned a certain weight by a group of independent experts. This protocol was then used to assess welfare in six furnished cage flocks and seven flocks in non-cage systems without access to an outdoor run (Rodenburg *et al.*, 2008b). In non-cage systems, birds were found to be more active and made greater use of resources (scratching area, perches) than in furnished cages. These birds also had stronger bones and were less fearful than birds in furnished cages, as measured in a tonic immobility test. On the other hand, birds in furnished cages had lower mortality rates, a lower incidence of keel bone fractures and lower air-borne dust concentrations in their atmosphere (Rodenburg *et al.*, 2008b). When all the welfare indicators were integrated into an overall welfare score, there were no significant differences between systems. These results indicate that furnished cages and non-cage systems have both strong and weak points in terms of their impact on animal welfare (Rodenburg *et al.*, 2008b). Clearly, in both systems there is room for improvement and both systems are still being developed further. However, the room for improvement of animal welfare may be larger for non-cage systems than for furnished cages. In furnished cages, the cage environment restricts the further improvements possible, whereas in non-cage systems improvements in housing design and management could still result in major improvements in animal welfare. Comparable results were found in a similar multifactorial investigation by Shimmura *et al.* (2010). They compared conventional cages, furnished cages and non-cage systems. They found that, although birds in non-cage systems have more freedom to express normal behaviour and an immune status similar to cage systems, non-cage systems received a lower score for providing an environment in which hens are free from pain, injury and disease compared with cage systems. The reverse was found in conventional cages. The furnished cages were in the middle between conventional cages and non-cage systems.

De Reu et al. (2009) studied the effects of housing system on bacterial contamination of the eggs in the same flocks as studied by Rodenburg et al. (2008b). They found lower levels of contamination in terms of total count of aerobic bacteria (4.75 versus 4.98 log cfu per eggshell) on eggshells from furnished cages compared with non-cage systems (excluding floor eggs). Similar results were found by Huneau-Salaün et al. (2010) in a comparison between conventional cages and non-cage systems in France. This is probably related to the higher air-borne dust levels in non-cage systems (Rodenburg et al., 2008b), resulting in a higher bacterial load (total count of aerobic bacteria) of the air (De Reu et al., 2005). Concerning Gram-negative bacteria and Enterobacteriaceae, no significant difference in average eggshell contamination between both systems could be shown (De Reu et al., 2005, 2009). Although counts of aerobic bacteria on eggshells were higher for nest eggs from non-cage systems, the observed differences in contamination were not considered to be biologically significant. Therefore, it could be concluded that housing system had no serious adverse effects on bacterial eggshell contamination (ignoring outside nest eggs and floor eggs) (De Reu et al., 2009). Limited information is available on the influence of housing system on egg content contamination. Recent research does not indicate large differences in egg content contamination between eggs from cage and non-cage systems (De Reu et al., 2008b). In a preliminary study by De Reu et al. (2008a), egg content contamination of nest eggs was 1.9% for furnished cages compared with 2.3% for non-cage systems.

An overview by Duwulf et al. (2009), on published observational studies evaluating the effect of housing system on the prevalence of *Salmonella Enteritidis* infections, indicated that cage systems had an increased risk of being *Salmonella* positive compared with non-cage systems. However, the authors indicated that there was not necessarily a causal relationship between the type of housing and the *Salmonella* infection. It is more likely that the housing system was a proxy of other farm characteristics. The authors summarized a number of important farm characteristics that may be related to the probability of a *Salmonella* infection: herd and flock size, stocking density, stress, age of the building, carry-over of infections from previous flocks, pests and vaccinations. Some of these risk factors are also of special importance to free range and organic systems: as contact between the birds and the environment is made in the outdoor run, extra attention should be given to pest control and minimizing contact with other species (Meerburg et al., 2004) (see chapter 5, this volume).

Mortality rates are generally higher in non-cage systems than in cage systems. In a comparison of mortality in conventional cages, furnished cages and non-cage systems (with and without free range access), the lowest mortality rates were found in conventional and furnished cage systems and the highest in free range flocks (Elson and Croxall, 2006). Fossum et al. (2009) also compared mortality in furnished cage systems, barn systems and free range systems and found higher mortality levels in barn and free range systems than in furnished cages. Similarly, Rodenburg et al. (2008b) reported mortality rates of 3% in furnished cages, compared with 8% in barn systems. Michel and Huonnic (2004) also found that mortality was higher in aviaries than in conventional cages but remained within acceptable limits (up to 5%). On the other hand, Aerni *et al.* (2005) performed a systematic review on the comparison of mortality in conventional cages and aviary systems and did not find a difference in mortality between the systems. It seems that at least the risk of high mortality rates is greater in non-cage systems than in cage systems. This is probably related to the fact that flock management is much more demanding in non-cage systems. Further, the large flock size in non-cage systems poses a double threat: (i) feather peckers and cannibals can target many more victims in a large flock than in a cage environment; and (ii) it is much more difficult to identify and remove the birds or groups that cause problems with feather pecking and cannibalism.

Elson and Croxall (2006) also compared performance between conventional cages, furnished cages and non-cage systems and found that feed usage and feed conversion ratios were higher in non-cage systems than in both cage systems. These results were confirmed in the systematic review by Aerni *et al.* (2005). This difference in feed efficiency needs to be recovered from the higher egg prices for eggs from non-cage systems compared with cage systems. In general, eggs from non-cage systems are sold at higher prices than eggs from cage systems. This is a considerable disadvantage for furnished cages: here also the housing costs are higher compared with conventional cages, but the egg can still only be marketed as a cage egg.

Eggshell dirt and cracks have been reported as negative implications of mainly furnished cages and to a lesser extent of non-cage systems (especially floor eggs) compared with conventional cage eggs (Mertens *et al.*, 2006; De Reu *et al.*, 2010). However, the research available also indicates that the problem represented by dirty and cracked eggs can be overcome with the correct design of furnished cages and nest boxes and good egg collection procedures (Guesdon *et al.*, 2006; De Reu *et al.*, 2009).

ADAPTATIONS FOR THE FUTURE

From this review it becomes clear that large differences exist between the three different types of non-cage systems, but also between non-cage systems and cage systems. Let us start with focusing on the comparison between barn, free range and organic system: should we provide laying hens with outdoor access or not? There are clear advantages (foraging opportunities, daylight, space, environmental variation) but also clear disadvantages (health and hygiene risks, predation). One interesting approach has been taken by Groot Koerkamp and Bos (2008), who developed new housing designs for laying hens based on the needs of the hens, the farmer and society. This approach was interesting because it resulted in a discussion on the actual needs of laying hens in relation to outdoor access: do hens really need to go outside or can we also provide the key stimuli in an indoor system? This resulted in the development of the 'Plantage' system and the 'Rondeel®' system. The Rondeel system provides the hens with a large, indoor pecking and scratching area on artificial grass with ample daylight and a smaller, forest-like outdoor run which can be closed in case of health or food safety risks (Fig. 12.1).

The Rondeel system was developed further by a commercial company. The first Rondeel in the Netherlands was opened in 2010 and received the same label from the Dutch Society for Protection of Animals as organic eggs. Similarly, the Plantage system was also developed further into a novel type of production system for organic laying hens: the 'De Lankerenhof' farm. Here, adaptations include novel rearing methods and an improved layout of the house that helps to reduce problems with feather pecking and cannibalism in organic laying hens. Innovations like this may enable us to combine good behavioural opportunities of the hens with high health and food safety standards. The development of the indoor pecking and scratching area in the Rondeel system is also in line with the results from De Mol *et al.* (2006), who showed that outdoor access in itself may not be the most important factor in the welfare of laying hens. Whether the Rondeel system also has lower mortality levels than free range and organic systems will become clearer in the near future, when more experience has been obtained with this new production system (Fig. 12.2).

When comparing non-cage systems with cage systems, it becomes clear that behavioural opportunities are much better in non-cage systems. Birds make better use of litter and of raised perches in non-cage systems and are more active. This also results in birds with stronger bones (probably related to the amount of exercise) that are less fearful (Rodenburg *et al.*, 2008b). However, the higher mortality rates in non-cage systems are a cause for concern. In the case of free range and organic systems, mortality could be even

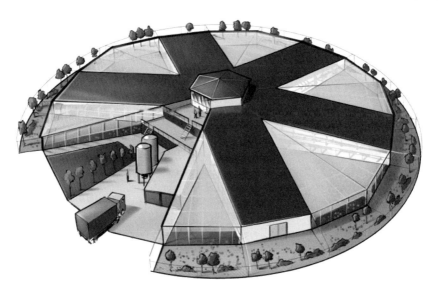

Fig. 12.1. Overview of the Rondeel® system, with five arms that each contain an aviary system and between the arms the covered pecking and scratching areas. The outer circle contains the outdoor run which is covered in mesh. (Image courtesy of Rondeel BV.)

Fig. 12.2. Laying hens in the Rondeel; with the outdoor run in the foreground and the pecking and scratching area in the background. (Image courtesy of Rondeel BV.)

higher due to predation. Causes for high mortality in the study by Rodenburg *et al.* (2008b) included: feather pecking and cannibalism, health problems, infections with red mites and smothering. The presence of two behavioural problems in this list of causes for high mortality, namely feather pecking and smothering, confirms that behaviour of laying hens in large flocks is much more difficult to manage than behaviour of laying hens in cage systems. Therefore, solutions for these behavioural problems in non-cage systems should be sought in improving breeding and rearing (including housing and management) methods for these birds. The aim here should be to create an optimal match between the bird and its social and physical environment.

Novel genetic selection methods developed by Bijma *et al.* (2007a,b), based on the group selection method described by Muir (1996), are expected to improve the behaviour of laying hens in large flocks. Ellen *et al.* (2007) started a selection experiment based on these methods, selecting for low mortality in group-housed hens. This selection experiment was set up in collaboration between ISA BV, the layer breeding division of Hendrix Genetics, and Wageningen University. In the first generation of selection a 10% difference in mortality was already found between the low mortality line and the control line, although this difference remains to be confirmed in the next generations (Rodenburg *et al.*, 2010). When behaviour of these lines was investigated, it was found that selection for low mortality leads to animals that show less cannibalism (Rodenburg *et al.*, 2009b) and are less fearful and less sensitive to stressors in a range of behavioural tests. Consistent results have been found both in young chicks (Rodenburg *et al.*, 2009b) and in adult birds (Bolhuis *et al.*, 2009; Rodenburg *et al.*, 2009a). As this selection method seems to affect both damaging behaviour and fearfulness, it may offer perspectives for reducing mortality due to feather pecking and smothering in commercial flocks.

Improving rearing methods also holds great promise for improving the performance and welfare of laying hens in non-cage systems. Among both scientists and practitioners, there is increasing agreement that the rearing environment should match the laying environment as closely as possible (van de Weerd and Elson, 2006). There is evidence that early access to litter and perches (Gunnarsson *et al.*, 1999; Huber-Eicher and Sebo, 2001; Heikkila *et al.*, 2006) improves performance and welfare of hens housed in extensive systems during the laying period. Further, rearing factors that mimic maternal care seem to hold great promise for commercial flocks (Rodenburg *et al.*, 2008a). Jensen *et al.* (2006) reported that providing chicks with dark brooders, that mimic the hen's body, strongly reduced feather pecking. This is probably because they help to separate active and resting chicks. If maternal care is provided, chicks have been shown to be less fearful (Rodenburg *et al.*, 2009b), show more foraging behaviour and to have lower mortality levels due to feather pecking (Riber *et al.*, 2007). Translating the beneficial effects of maternal care to application in commercial systems offers perspective for further reducing behavioural problems in non-cage systems.

CONCLUSIONS

Clearly, large differences exist in performance, welfare, health and hygiene of laying hens in non-cage systems and furnished cages. Both between barn, free range and organic systems and between non-cage and cage systems large differences were identified. Moving from conventional cages to furnished cages, barn, free range and finally organic systems, there is a general trend of increasing environmental complexity, which is positive for hen welfare, but also of increasing risks for performance, health and hygiene, which is negative for hen welfare. For the improvement of furnished cages and non-cage systems, we should focus on creating a better match between the animals and their husbandry environment. Good examples are the development of new housing designs that combine the benefits of non-cage systems with improved performance, health and hygienic status, such as the Rondeel system. Further, promising approaches in animal breeding and optimizing rearing environments are expected to yield major improvements in the welfare of laying hens in non-cage systems.

ACKNOWLEDGEMENTS

We would like to thank ISA BV, the layer breeding division of Hendrix Genetics, for their involvement in the genetic research described in this chapter. Thanks to Rondeel BV for supplying the figures on the Rondeel system. T.B. Rodenburg was supported by the Dutch Technology Foundation STW, Applied Science Division of NWO and the Technology Program of the Ministry of Economic Affairs.

REFERENCES

Acamovic, T., Sandilands, V., Kyriazakis, I. and Sparks, N. (2008) The effect of organic diets on the performance of pullets maintained under semi-organic conditions. *Animal* 2, 117–124.

Aerni, V., Brinkhof, M.W.G., Wechsler, B., Oester, H. and Fröhlich, E. (2005) Productivity and mortality of laying hens in aviaries: a systematic review. *World's Poultry Science Journal* 61, 130–142.

Bestman, M.W.P. and Wagenaar, J.P. (2003) Farm level factors associated with feather pecking in organic laying hens. *Livestock Production Science* 80, 133–140.

Bestman, M., Koene, P. and Wagenaar, J.-P. (2009) Influence of farm factors on the occurrence of feather pecking in organic reared hens and their predictability for feather pecking in the laying period. *Applied Animal Behaviour Science* 121, 120–125.

Bijma, P., Muir, W.M., Ellen, E.D., Wolf, J.B. and van Arendonk, J.A.M. (2007a) Multilevel selection 2: estimating the genetic parameters determining inheritance and response to selection. *Genetics* 175, 289–299.

Bijma, P., Muir, W.M. and van Arendonk, J.A.M. (2007b) Multilevel selection 1: quantitative genetics of inheritance and response to selection. *Genetics* 175, 277–288.

Blokhuis, H.J. (1986) Feather-pecking in poultry: its relation with ground-pecking. *Applied Animal Behaviour Science* 16, 63–67.

Bolhuis, J.E., Ellen, E.D., Van Reenen, C.G., De Groot, J., Ten Napel, J., Koopmanschap, R., De Wries Reilingh, G., Uitdhaag, K.A., Kemp, B. and Rodenburg, T.B. (2009) Effects of genetic group selection against mortality on behaviour and peripheral serotonin in domestic laying hens with trimmed and intact beaks. *Physiology & Behavior* 97, 470–475.

De Mol, R.M., Schouten, W.G.P., Evers, E., Houwers, H.W.J. and Smits, A.C. (2006) A computer model for welfare assessment of poultry production systems for laying hens. *NJAS – Wageningen Journal of Animal Sciences* 54, 157–168.

De Reu, K., Grijspeerdt, K., Heyndrickx, M., Zoons, J., De Baere, K., Uyttendaele, M., Debevere, J. and Herman, L. (2005) Bacterial eggshell contamination in conventional cages, furnished cages and aviary housing systems for laying hens. *British Poultry Science* 46, 149–155.

De Reu, K., Heyndrickx, M., Grijspeerdt, K., Rodenburg, B., Tuyttens, F., Uyttendaele, M. and Herman, L. (2008a) Estimation of the vertical and horizontal bacterial infection of hen's table eggs. *World's Poultry Science Journal* 64(Suppl. 2), 142.

De Reu, K., Messens, W., Heyndrickx, M., Rodenburg, B., Uyttendaele, M. and Herman, L. (2008b) Bacterial contamination of table eggs and the influence of housing systems. *World's Poultry Science Journal* 64, 5–19.

De Reu, K., Rodenburg, T.B., Grijspeerdt, K., Messens, W., Heyndrickx, M., Tuyttens, F.A.M., Sonck, B., Zoons, J. and Herman, L. (2009) Bacteriological contamination, dirt, and cracks of eggshells in furnished cages and noncage systems for laying hens: an international on-farm comparison. *Poultry Science* 88, 2442–2448.

De Reu, K., Messens, W., Grijspeerdt, K., Heyndrickx, M., Rodenburg, B., Uyttendaele, M. and Herman, L. (2010) Influence of housing systems on the bacteriological quality and safety of table eggs. In: *Proceeding of the 21st Annual Australian Poultry Science*

Symposium, Sydney, Australia, 1–3 February 2010. The Poultry Research Foundation (University of Sydney) and The World's Poultry Science Association (Australian Branch), Sydney Australia, pp. 74–81.

Dewulf, J., Van Hoorebeeke, S. and van Immerseel, F. (2009) Epidemiology of *Salmonella* infection in laying hens with special emphasis on the influence of the housing system. In: *Proceedings of the XIIIth Symposium on the Quality of Eggs and Egg Products*, Turku, Finland, 21–25 June 2009. The World's Poultry Science Association (WPSA), Turku, Finland, p. 12.

Dixon, L.M. (2008) Feather pecking behaviour and associated welfare issues in laying hens. *Avian Biology Research* 1, 73–87.

Ellen, E.D., Muir, W.M. and Bijma, P. (2007) Genetic improvement of traits affected by interactions among individuals: sib selection schemes. *Genetics* 176, 489–499.

Elson, H.A. and Croxall, R. (2006) European study on the comparative welfare of laying hens in cage and non-cage systems. *Archiv für Geflugelkunde* 70, 194–198.

Fiks-van Niekerk, T.G.C.M. and Elson, H.A. (2005) Categories of housing systems for laying hens. *Animal Science Papers and Reports* 23, 283–284.

Fossum, O., Jansson, D.S., Etterlin, P.E. and Vagsholm, I. (2009) Causes of mortality in laying hens in different housing systems on 2001 to 2004. *Acta Veterinaria Scandinavica* 51, 9.

Gentle, M.J. and McKeegan, D.E.F. (2007) Evaluation of the effects of infrared beak trimming in broiler breeder chicks. *The Veterinary Record* 160, 145–148.

Green, L.E., Lewis, K., Kimpton, A. and Nicol, C.J. (2000) Cross-sectional study of the prevalence of feather pecking in laying hens in alternative systems and its associations with management and disease. *The Veterinary Record* 147, 233–238.

Groot Koerkamp, P.W.G. and Bos, A.P. (2008) Designing complex and sustainable agricultural production systems: an integrated and reflexive approach for the case of table egg production in the Netherlands. *NJAS – Wageningen Journal of Life Sciences* 55, 113–138.

Guesdon, V., Ahmed, A.M.H., Mallet, S., Faure, J.M. and Nys, Y. (2006) Effects of beak trimming and cage design on laying hen performance and egg quality. *British Poultry Science* 47, 1–12.

Gunnarsson, S., Keeling, L.J. and Svedberg, J. (1999) Effect of rearing factors on the prevalence of floor eggs, cloacal cannibalism and feather pecking in commercial flocks of loose housed laying hens. *British Poultry Science* 40, 12–18.

Guy, J.H., Khahavi, M., Hlalel, M.M. and Sparagano, O. (2004) Red mite (*Dermanyssus gallinae*) prevalence in laying units in Northern England. *British Poultry Science* 45, S15–S16.

Häne, M., Huber-Eicher, B. and Frohlich, E. (2000) Survey of laying hen husbandry in Switzerland. *World's Poultry Science Journal* 56, 21–31.

Hegelund, L., Sørensen, J.T., Kjaer, J.B. and Kristensen, I.S. (2005) Use of the range area in organic egg production systems: effect of climatic factors, flock size, age and artificial cover. *British Poultry Science* 46, 1–8.

Hegelund, L., Sorensen, J.T. and Hermansen, J.E. (2006) Welfare and productivity of laying hens in commercial organic egg production systems in Denmark. *NJAS – Wageningen Journal of Animal Sciences* 54, 147–156.

Heikkila, M., Wichman, A., Gunnarsson, S. and Valros, A. (2006) Development of perching behaviour in chicks reared in enriched environment. *Applied Animal Behaviour Science* 99, 145–156.

Huber-Eicher, B. and Sebo, F. (2001) Reducing feather pecking when raising laying hen chicks in aviary systems. *Applied Animal Behaviour Science* 73, 59–68.

Huneau-Salaün, A., Michel, V., Huonnic, D., Balaine, L. and Le Bouquin, S. (2010) Factors influencing bacterial eggshell contamination in conventional cages, furnished cages and free-range systems for laying hens under commercial conditions. *British Poultry Science* 51, 163–169.

Jensen, A.B., Palme, R. and Forkman, B. (2006) Effect of brooders on feather pecking and cannibalism in domestic fowl (*Gallus gallus domesticus*). *Applied Animal Behaviour Science* 99, 287–300.

Kilpinen, O., Roepstorff, A., Permin, A., Nørgaard-Nielsen, G., Lawson, L.G. and Simonsen, H.B. (2005) Influence of *Dermanyssus gallinae* and *Ascaridia galli* infections on behaviour and health of laying hens (*Gallus gallus domesticus*). *British Poultry Science* 46, 26–34.

Kjaer, J.B. and Sørensen, P. (2002) Feather pecking and cannibalism in free-range laying hens as affected by genotype, dietary level of methionine + cystine, light intensity during rearing and age at first access to the range area. *Applied Animal Behaviour Science* 76, 21–39.

Knierim, U. (2006) Animal welfare aspects of outdoor runs for laying hens: a review. *NJAS – Wageningen Journal of Life Sciences* 54, 133–146.

Krause, E.T., Naguib, M., Trillmich, F. and Schrader, L. (2006) The effects of short term enrichment on learning in chickens from a laying strain (*Gallus gallus domesticus*). *Applied Animal Behaviour Science* 101, 318–327.

Lambton, S.L., Knowles, T.G., Yorke, C. and Nicol, C.J. (2010) The risk factors affecting the development of gentle and severe feather pecking in loose housed laying hens. *Applied Animal Behaviour Science* 123, 32–42.

Meerburg, B.G., Bonde, M., Brom, F.W.A., Endepols, A.N., Jensen, A.N., Leirs, H., Lodal, J., Singleton, G.R., Pelz, H.J., Rodenburg, T.B. and Kijlstra, A. (2004) Towards sustainable management of rodents in organic animal husbandry. *NJAS – Wageningen Journal of Life Sciences* 52, 195–205.

Mertens, K., Bamelis, F., Kemps, B., Kamers, B., Verhoelst, E., De Ketelaere, B., Bain, M., Decuypere, E. and De Baerdemaeker, J. (2006) Monitoring of eggshell breakage and eggshell strength in different production chains of consumption eggs, *Poultry Science* 85, 1670–1677.

Miao, Z.H., Glatz, P.C. and Ru, Y.J. (2005) Free-range poultry production – a review. *Asian-Australian Journal of Animal Science* 18, 113–132.

Michel, V. and Huonnic, D. (2004) A comparison of welfare, health and production performance of laying hens reared in cages or aviaries. *British Poultry Science* 44, 775–776.

Muir, W.M. (1996) Group selection for adaptation to multiple-hen cages: selection program and direct responses. *Poultry Science* 75, 447–458.

Newberry, R.C., Keeling, L.J., Estevez, I. and Bilcik, B. (2007) Behaviour when young as a predictor of severe feather pecking in adult laying hens: the redirected foraging hypothesis revisited. *Applied Animal Behaviour Science* 107, 262–274.

Patzke, N., Ocklenburg, S., Van der Staay, F.J., Gunturkun, O. and Manns, M. (2009) Consequences of different housing conditions on brain morphology in laying hens. *Journal of Chemical Neuroanatomy* 37, 141–148.

Permin, A., Bisgaard, M., Frandsen, F., Pearman, M., Kold, J. and Nansen, P. (1999) Prevalence of gastrointestinal helminths in different poultry production systems. *British Poultry Science* 40, 439–443.

Riber, A.B., Wichman, A., Braastad, B.O. and Forkman, B. (2007) Effects of broody hens on perch use, ground pecking, feather pecking and cannibalism in domestic fowl (*Gallus gallus domesticus*). *Applied Animal Behaviour Science* 106, 39–51.

Rodenburg, T.B., Tuyttens, F.A.M., De Reu, K., Herman, L., Zoons, J. and Sonck, B. (2005) Welfare, health and hygiene of laying hens housed in furnished cages and in alternative housing systems. *Journal of Applied Animal Welfare Science* 8, 211–226.

Rodenburg, T.B., Komen, H., Ellen, E.D., Uitdehaag, K.A. and van Arendonk, J.A.M. (2008a) Selection method and early-life history affect behavioural development, feather pecking and cannibalism in laying hens: a review. *Applied Animal Behaviour Science* 110, 217–228.

Rodenburg, T.B., Tuyttens, F.A.M., De Reu, K., Herman, L., Zoons, J. and Sonck, B. (2008b) Welfare assessment of laying hens in furnished cages and non-cage systems: an on-farm comparison. *Animal Welfare* 17, 363–373.

Rodenburg, T.B., Tuyttens, F.A.M., De Reu, K., Herman, L., Zoons, J. and Sonck, B. (2008c) Welfare assessment of laying hens in furnished cages and non-cage systems: assimilating expert opinion. *Animal Welfare* 17, 355–361.

Rodenburg, T.B., Bolhuis, J.E., Koopmanschap, R.E., Ellen, E.D. and Decuypere, E. (2009a) Maternal care and selection for low mortality affect post-stress corticosterone and peripheral serotonin in laying hens. *Physiology & Behavior* 98, 519–523.

Rodenburg, T.B., Uitdehaag, K.A., Ellen, E.D. and Komen, J. (2009b) The effects of selection on low mortality and brooding by a mother hen on open-field response, feather pecking and cannibalism in laying hens. *Animal Welfare* 18, 427–432.

Rodenburg, T.B., Bijma, P., Ellen, E.D., Bergsma, R., de Vries, S., Bolhuis, J.E., Kemp, B. and van Arendonk, J.A.M. (2010) Breeding amiable animals? Improving farm animal welfare by including social effects in breeding programmes. *Animal Welfare* 19, S77–S82.

Shimmura, T., Suzuki, T., Hirahara, S., Eguchi, Y., Uetake, K. and Tanaka, T. (2008) Pecking behaviour of laying hens in single-tiered aviaries with and without outdoor area. *British Poultry Science* 49, 396–401.

Shimmura, T., Hirahara, S., Azuma, T., Suzuki, T., Eguchi, Y., Uetake, K. and Tanaka, T. (2010) Multi-factorial investigation of various housing systems for laying hens. *British Poultry Science* 51, 31–42.

Sparks, N.H.C., Conroy, M.A. and Sandilands, V. (2008) Socio-economic drivers for UK organic pullet rearers and the implications for poultry health. *British Poultry Science* 49, 525–532.

Tauson, R. (2005) Management and housing systems for layers – effects on welfare and production. *World's Poultry Science Journal* 61, 477–490.

van de Weerd, H.A. and Elson, A. (2006) Rearing factors that influence the propensity for injurious feather pecking in laying hens. *World's Poultry Science Journal* 62, 654–664.

van de Weerd, H.A., Keatinge, R. and Roderick, S. (2009) A review of key health-related welfare issues in organic poultry production. *World's Poultry Science Journal* 65, 649–684.

van Horne, P.L.M. and Achterbosch, T.J. (2008) Animal welfare in poultry production systems: impact of EU standards on world trade. *World's Poultry Science Journal* 64, 40–52.

van Krimpen, M.M., Kwakkel, R.P., Reuvekamp, B.F.J., van der Peet-Schwering, C.M.C., Den Hartog, L.A. and Verstegen, M.W.A. (2005) Impact of feeding management on feather pecking in laying hens. *World's Poultry Science Journal* 61, 663–685.

Viaene, J. and Verheecke, W. (2009) *Overzicht van de Belgische pluimvee- en konijnenhouderij in 2008* (in Dutch). VEPEK, Ghent, Belgium, p. 104; available at: http://www.agecon.ugent.be/vepek/overzicht2008_okt09.pdf (accessed 10 October 2011).

Whay, H.R., Main, D.C.J., Green, L.E., Heaven, G., Howell, H., Morgan, M., Pearson, A. and Webster, A.J.F. (2007) Assessment of the behaviour and welfare of laying hens on free-range units. *The Veterinary Record* 161, 119–128.

Windhorst, H.W. (2005) Development of organic egg production and marketing in the EU. *World's Poultry Science Journal* 61, 451–462.

Zeltner, E. and Hirt, H. (2008) Factors involved in the improvement of the use of hen runs. *Applied Animal Behaviour Science* 114, 395–408.

CHAPTER 13

Housing and Management of Broiler Breeders and Turkey Breeders

I.C. de Jong and M. Swalander

ABSTRACT

This chapter describes the housing and management of broiler breeders and turkey breeders in Europe. The majority of broiler breeders in Europe are the standard, fast growing genotype, but 18–20% of the broiler breeders are dwarf parental females that produce standard and alternative (medium or slow growing) broilers. Broiler breeder housing systems are very similar in rearing and production; a low percentage of birds are in cages, and alternative systems are not used. Broiler breeders are generally housed in climate-controlled houses with litter floors during the rearing period and partially slatted floors during the production period. Males and females are reared separately until 18–21 weeks of age and then transferred to the production farm where they are housed together until 60–65 weeks of age. The restricted feeding regime during rearing is generally seen as one of the major welfare issues in broiler breeders as it leads to chronic hunger and frustration of the feeding motivation. The majority (>95%) of turkey breeders in Europe are of either heavy or heavy medium genotype with white plumage. The remainder of the turkey market consists of small strain white or coloured birds for whole bird seasonal production. Both conventional large strain turkeys and small strain traditional turkeys are used for outdoor/alternative production systems. Rearing of breeding turkeys is floor based on deep litter, and predominantly in environmentally controlled housing. Males and females are reared separately until 29 weeks of age and then transferred to laying facilities. Male parent stock is selected at 16–18 weeks paying attention to health, fitness, plumage and conformation. Laying facilities are either open-sided houses or controlled environment for breeder females, and typically controlled environment housing for breeder males. Breeding turkeys are kept in production until 56–60 weeks of age (i.e. 24–28 weeks lay). Quantitative feed restriction is applied in breeder males from selection age (16–18 weeks) to end of production, to maximize fitness and reproduction. Breeder females are fed unrestricted throughout rearing but on a lower-protein diet to avoid fatness of the hens. Injurious pecking is generally seen as the most important welfare issue in flocks where beak trimming is not applied.

BROILER BREEDERS

This section describes housing and management of broiler breeders in Europe. The housing and management of standard broiler breeders (fast growing genotype) is described, including welfare, health and productivity. In general this also applies to alternative breeds (dwarf parental females) producing standard, slow or medium growing broilers. Where there are differences between standard broiler breeders and alternative breeds this is indicated in the text.

Broiler breeder housing systems are very similar for both rearing stock and adults: a very low percentage of birds are housed in cages in Europe, and alternative systems are not used. Outdoor systems are not used because of the large impact that an infection would have on the offspring. Aviary systems are not commonly used when birds are mature because the risk of bone fractures is too high due to the large weight of the birds. As the prevention of floor eggs is important, housing systems must not have many corners or dark places. In addition, the interaction between males and females should be promoted, e.g. by preventing females from 'hiding' from the males (J.H. van Middelkoop 2010, personal communication). All of these arguments have contributed to the development of a very uniform housing system for broiler breeders that is described below.

Parent stock management manuals provided by the breeding companies are generally used as a guideline to construct houses or establish management practices (Laughlin, 2009), although aspects such as legislation, local climate and local traditions may lead to differences between countries. The description of housing systems and management of broiler breeders is based on these management guides (Cobb, 2008; Aviagen, 2009; Hubbard, 2009a), field experiences and discussions with experts or breeder representatives and gives a general picture of the situation in Europe. Additional references are provided where available.

The estimated number of broiler parent stock (broiler breeders) in Europe is about 75 million birds (O. van Tuijl, Aviagen, 2010, personal communication). The majority of birds are used to produce standard broiler chickens that grow in about 42 days of age to a live weight of about 2.5 kg. France, the UK, Poland, Spain, Italy, Germany and the Netherlands are the largest producers in the European Union (EU) (FAO, 2008). Dwarf parental females are used to produce either slow growing (e.g. organic, Label Rouge) or medium growing broilers (by breeding coloured dwarf hens with slow growing or fast growing males, respectively) or standard normal size broilers (heavier white dwarf hens breeding with standard fast growing males), and represent the majority of broiler breeder hens in France and about 18–20% of the European population of broiler breeders (EFSA, 2010). Medium growing broilers reach a live weight of 2.2 kg in 56–63 days, while slow growing broilers reach a live weight of 2.2 kg in 84 days.

Selection of broiler breeders

The supply of broiler breeding stock to a wide variety of markets and environmental conditions has been a driver for the evolution of balanced breeding goals. Selection is consequently applied on a broad range of traits including: (i) fitness and welfare traits such as skeletal integrity, cardiovascular fitness and liveability; (ii) efficiency (e.g. breast meat yield and feed conversion); (iii) reproductive performance (e.g. egg production, fertility, hatchability); and (iv) quality traits (e.g. meat quality, feathering). The primary poultry breeders' use of advanced statistical techniques (e.g. Best Linear Unbiased Prediction, BLUP) and maintaining large breeding populations ensure that both commercial fitness and reproductive traits improve simultaneously and that the genetic resources are sustained (Avendano *et al.*, 2010). Alternative strains of broilers are selected to maintain a slower growth rate achieving a higher age at maturity, robustness (leg health and liveability) as well as qualitative characteristics such as feather colour and plumage quality. Thus a combination of BLUP and phenotypic selection is typically used.

Housing and management

Rearing period

The main goal of the rearing period is to provide birds of ideal weight, condition and stage of sexual maturity as they enter the production house. Major factors influencing the development of the breeder birds are nutrition and feeding management, environmental control and health status (Leeson and Summers, 2000).

Male and female broiler breeders are reared in separate groups in houses with concrete floors covered with litter (Hocking, 2004). Standard broiler breeder units are mechanically ventilated and windowless, although in some countries (e.g. Sweden) houses have windows or open-sided houses are used (where the sides of the house are open or covered with mesh curtains, e.g. Eastern Europe). On average the standard group size during rearing is approximately 3000 birds with 10,000 to 25,000 birds per farm (EFSA, 2010). Cage rearing is done at only a very small number of European farms, mainly in Poland (EFSA, 2010; O. van Tuijl, Aviagen, 2010, personal communication).

Mutilations to control the damaging effect of undesirable behaviours are often standard practice although there are differences between countries depending on national legislation or hybrid of bird (EFSA, 2010). Beak trimming to reduce pecking damage is usually carried out in males and females either at the hatchery or at the farm. Exact figures are not available, but in Europe beak trimming is not allowed in Sweden, Norway and Finland and commonly carried out in other countries (Fiks *et al.*, 2006). In addition, in most male parent stock the inner and/or hind toes are clipped and the males of some lines or crosses are despurred at the hatchery to prevent damage to the hens during mating. Detoeing is carried out at the hatchery using a hot blade or hot wire. It is allowed and commonly carried out in European countries (Fiks *et al.*, 2006). Despurring is done at the hatchery by pressing the spurs briefly against a hot wire or blade (Fiks *et al.*, 2006). It is not allowed in Denmark, Norway and Sweden. As not all lines or crosses require despurring it is not as commonly done as detoeing (Fiks *et al.*, 2006). Comb dubbing is an uncommon procedure, carried out in less than 10% of the males upon customer request (EFSA, 2010). It used to be done to reduce the size of large male combs, in order to prevent comb damage and inactivity of the males (from combs covering their eyes and thus preventing mating). Nowadays broiler breeder males' combs are relatively smaller (EFSA, 2010).

In most countries the stocking density during rearing is not limited by legislation. Manuals of the breeding companies are used as guidelines and common stocking densities are 4–8 birds m^{-2} for males and 7–10 birds m^{-2} for females. Lower stocking densities are usually applied in open-sided houses. The target weight of a hen is approximately 40 g at day 1 and increases to 1800–1900 g at 18 weeks of age. Target weight for males is approximately 2600 g by 18 weeks of age. For alternative breeds (dwarf parental females) recommended stocking densities are 9–10 hens m^{-2} but stocking densities may be higher (15 birds m^{-2}) in practice (EFSA, 2010). During the first few days chicks are housed under continuous light. 1 h light per day is common after 2 weeks of age. Light intensity during rearing is in general 10–20 lux, although lower intensities may be applied to prevent feather pecking in non-beak trimmed flocks.

There is no manure removal during the lifetime of a rearing or breeding flock, thus the manure becomes an integrated part of the litter (Hocking, 2004). Wood shavings, peat and straw are often used as litter materials. Breeding companies recommend providing the rearing birds with perches or raised platforms to help them learn to jump in order to facilitate nesting behaviour during the production period. Whole room heating is generally used, with an environmental temperature of 30°C at day 1 which is then gradually decreased until it reaches about 20°C.

Feed can be crumbs, mash or pellets and provided in feeder pans or feeder tracks. Pelleted feed can also be scattered on the floor using so-called 'spin-feeders'. Feed restriction is applied to male and female broiler breeders of fast growing lines, to achieve set target body weights at a particular age (Renema *et al.*, 2007). If broiler breeders were fed unrestricted during their entire life, they would grow too rapidly and become far too heavy to maintain good health before reaching the age of sexual maturity. In addition their fertility would be negatively affected (e.g. Decuypere *et al.*, 2006; Renema *et al.*, 2007). Males and females follow separate feeding programmes which is the main reason for housing them separately during rearing. Feed restriction is in general applied from about 2 to 3 weeks of age onwards. The restriction level varies between one-quarter to one-third of the intake of an unrestricted fed bird and is most severe between 8 and 15 weeks of age (De Jong *et al.*, 2002). After 15 weeks of age the amount of feed is increased to support the onset of egg production. Figure 13.1 shows a typical growth profile of broiler breeders of a fast growing strain. In Europe, 6/1, 5/2 or 4/3 feeding programmes are often applied which means that birds receive feed on six, five or four consecutive days followed by one, two or three days without feed (EFSA, 2010). The reason for

(a)

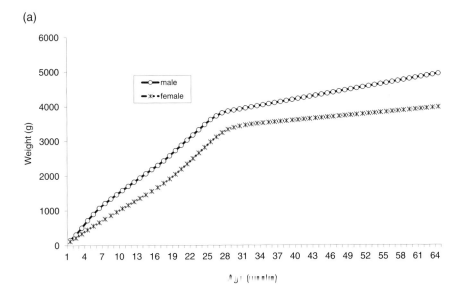

(b)

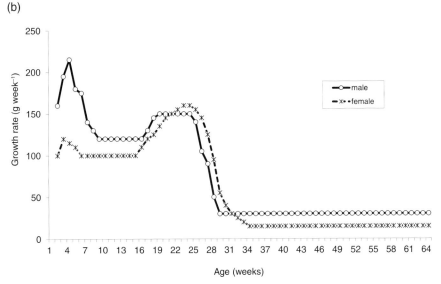

Fig. 13.1. Growth profiles of Aviagen Ross 308 male and female breeders, showing (a) cumulative body weight and (b) weekly growth rate.

using these feeding programmes is that the amount of feed given on a daily basis is very small which, if fed daily, may mean feed does not reach all birds equally (Laughlin, 2009). However, in some countries legislation requires daily feeding. Sufficient feeder space and fast feed distribution is important to ensure bird weight uniformity and to minimize aggression around feeding. Alternative breeds do not have to be feed restricted or are much less restricted compared with standard breeds (Decuypere *et al.*, 2006).

Water is supplied automatically by bell drinkers, nipples or cups. Water is usually available *ad libitum* during the first weeks of life. Thereafter water access may be limited to a couple of hours around feeding time or on other occasions during the day. This is done to prevent spilling and over-drinking resulting in wet litter (Leeson and Summers, 2000).

Production period

The production period usually starts between 18 and 23 weeks of age and lasts until 60–65 weeks of age. During the production period, the main goal of management of adult breeders is to maximize the production of fertile eggs. It is important to maintain the health status of the flock while allowing for a continued, but slow increase in body weight in order to keep production at a high level. Major criteria for monitoring birds for management purposes include body weight, body condition (conformation and fat deposition), egg production, hatchability, egg weight and egg mass (Leeson and Summers, 2000).

Natural mating is the norm in Europe as opposed to artificial insemination of the hens. Males and females are transported from the rearing farm to the production farm and housed together. Male birds are often placed a few days earlier than the females. Males are selected at the end of the rearing period, before transfer to the production house, on the basis of their body weight, feather cover, body, leg and toe condition. Males and females become sexually mature between 18 and 23 weeks of age. It is important that males and females are equally mature to prevent problems with over-mating. Over-mating and aggressive behaviour of males leads to fearful females hiding in the nests (Millman *et al.*, 2000). Immature males should not be transferred to the production house as they will not become reproductive. The percentage of males placed in the production house varies between 8 and 11% of the total flock, decreasing to 7–9.5% of males due to selection at 23 weeks of age when egg production starts. There is variation dependent on country and individual farm management. Male selection continues during the production period. Important selection criteria are reproductive activity (inactive males have a full coat of new feathers and the fluff around the vent area is intact), extreme body weight and leg condition. At 60 weeks of age the proportion of males has decreased to 6% of the flock due to selection (Hocking, 2004). During the production period, 15–25% of the placed males are culled for reasons of selection (EFSA, 2010). In some countries 'spiking' is a common procedure. Inactive males in bad condition are removed and replaced by younger mature males, with the objective of maintaining fertility to the end of the breeding period. Spiking introduces a risk of introduction of pathogens because birds of another farm are introduced into the flock (Leeson and Summers, 2000).

Common group size is 3000 to 8000 birds per flock with approximately 10,000 to 30,000 birds per farm. Standard production houses in Northern Europe are windowless and mechanically ventilated (Hocking, 2004). However, in Sweden and France windows are often present in the houses, although these may be covered with curtains in case they conflict with the lighting schedule or cause increased feather pecking (EFSA, 2010). Open-sided houses can be found in Italy, France, Spain and Eastern Europe (EFSA, 2010). Whole room heating is used in many countries and temperature is adjusted to 20°C. A few farms in the Netherlands and Germany have multi-tier colony cages with natural mating, automatic nests and perches but without litter. They are decreasing in number in the Netherlands (R.A. van Emous, Lelystad, 2010, personal communication). A small number of farms, mainly in southern Europe and Poland, house their birds in single sex groups in conventional cages and use artificial insemination (about 1–2% of the European parent stock) (EFSA, 2010; O. van Tuijl, Aviagen, 2010, personal communication). Cages are used for standard breeds as well as for alternative breeds (EFSA, 2010).

Similar to the rearing period, in the production period stocking density is often not limited by legislation (although Sweden, Norway and the Netherlands do have legislation for the production period limiting stocking density). Stocking density in Europe varies between 5 and 8.5 birds m^{-2} with the lowest densities in open-sided houses. Stocking density for alternative breeds is about 9–10 birds m^{-2}. Female weights increase from 1.8–1.9 kg at 18 weeks of age to 3.5–4.0 kg at 60 weeks of age. Body weights of female dwarf hens vary between 2.1–2.6 kg at the peak of lay or around to 3.3–3.4 kg for standard females at the peak of lay (EFSA, 2010). Male weights increase from about 2.6 kg at 18 weeks of age up to 4.8–5.5 kg at 60 weeks of age.

Feed can be provided in feeder pans or tracks. Males and females have separate feeding systems, where males are typically fed using pans hanging high enough to prevent the females eating from them. On the female pans or tracks, grills are placed to prevent the males eating from them. Males have wider heads than females so that they cannot put their heads through the grills. Weight control is important during the production period and separate feeding is applied for males and females. Feed is less severely restricted compared with the rearing period, but feeding is carefully controlled during the production period for males and females. Females are restricted to about 45–80% of *ad libitum* intake until the peak of lay (Bruggeman *et al.*, 1999) and to about 80% of *ad libitum* intake after the peak of lay (Hocking, 2009). These figures are not provided for males. However, males should not lose weight but not become too heavy as it has adverse effects on fertility. Especially after 30 weeks of age small but weekly body weight increases are necessary to maintain fertility in males (Hocking, 2009) (see also Fig. 13.1). For the females, the aim is to start egg production at 23–25 weeks of age. Egg production and body condition determine the amount of feed provided. When the flock reaches 5–10% production a larger increase in feed is advised until peak production (around 30 weeks of age). After peak production feed intake is decreased slowly to prevent fat deposition and too sharp a decrease in egg production (Hocking, 2009) (see also Fig. 13.1). Feed is provided daily, either in the morning half an hour after lights-on or about 5–8 h after lights-on. This makes it more likely that hens will lay their eggs in the nests, rather than being attracted by running feed hoppers.

Water is usually provided using bell drinkers but nipples are becoming more popular. Also similar to the rearing period, water is often restricted during the laying period to prevent over-drinking and spilling (Leeson and Summers, 2000). In general, water is supplied during feeding until at least 2 h after feeding and during 1 h in the afternoon before turning the lights off. Water restriction is not applied in all countries.

Most commonly, production houses have a litter area and a raised slatted area on which the nests are positioned. One-third of the floor area as raised slats is common in Europe. The height of the slats is no more than 60 cm above the litter area. Straw or wood shavings are often used as litter. Usually the manure is removed after a production round (Hocking, 2004). Automated collection nests are most common in European countries but individual nests with litter and manual egg collection are also used. Water is provided on the slatted area and feed is usually provided in the litter area. Perches are not very common in the EU (EFSA, 2010).

According to the recommendations of the breeding companies, at the age of transfer to the production house the light period increases from 8 h to 15–16 h lights on at 28 weeks of age. Light intensity increases to 40–60 lux between 19 and 21 weeks of age. Lower light intensities may be applied for non-beak trimmed birds. For open-sided houses or houses with daylight a different approach may be necessary.

Productivity

The ultimate output of a broiler breeding operation is a high proportion of viable 1-day-old commercial chicks to grow for meat production. The measure of output starts with the total egg production, reduced to the number of hatching eggs by factors which determine an egg unsuitable for incubation, such as faecal contamination, egg handling at breeder farm and during transport, cracks, storage time and storage conditions at hatchery, presence of floor eggs or dirty nest eggs, and finally by the hatching success. Hatching success is determined by both the fertility of the eggs and the hatchability of fertile eggs (Laughlin, 2009). Figures on productivity in European countries are scarce and therefore we give here the expected performance taken from breeder guides, which are summarized in Table 13.1, and a few data from field studies. In 2005, average hatchability at UK farms was about 82% of eggs set (Laughlin, 2009). At Dutch farms, average hatchability between the years 2004 and 2007 was 80–81% of eggs set with the number of hatching eggs per hen housed varying between 147 and 157 (R.A. van Emous, Lelystad, 2010, personal communication). According to the breeder manuals, egg production in alternative strains producing slow or medium growing broilers is higher, resulting in a higher number of day-old chicks per hen housed.

Table 13.1. Overview of reproductive performance of parent stock broiler hens for different market segments (Aviagen, 2007a,b; Hubbard, 2009b).

Strain type	Standard	High yield	Slow growing
No. of eggs at 64 weeks age	175	166	211
Hatch of eggs set (%)	84.8	86.6	85.0
Day-old chicks per hen housed	148	144	179

Welfare

Feed restriction

With respect to broiler breeder welfare, much research attention has been paid to the consequences of severe feed restriction during rearing. Due to the genetic selection for fast growth and high breast muscle yields broiler breeders have a high capacity for feed intake when fed unrestricted. Feed restriction in females controls multiple ovulations and prevents poor fertility during the production period. In males, feed restriction maximizes fertility and the ability to mate (Hocking, 2004, 2009). In addition, it prevents birds of both sexes from becoming overweight and developing pathological conditions such as lameness and premature death that have a negative effect on broiler breeder welfare (Mench, 2002). However, there is substantial evidence that feed restriction also has negative effects on broiler breeder welfare, especially during the rearing period (e.g. Mench, 2002; D'Eath et al., 2009). Feed restriction in general starts around 2–3 weeks of age and food allocations during rearing are about one-quarter to one-third of the intake of unrestricted fed birds. During rearing, feed restricted broiler breeders consume their daily ration in 15 min (De Jong et al., 2002). Feed restriction is most severe between 8 and 16 weeks of age (De Jong and Jones, 2006). The consequences of feed restriction include chronic hunger and the performance of abnormal behaviours such as stereotyped pecking at non-food objects, pacing and over-drinking. These behaviours are characteristic of frustration due to unfulfilled feeding motivation (e.g. Hocking et al., 1993; Savory et al., 1993; De Jong et al., 2003; D'Eath et al., 2009). Aggression due to competition around feeding time may lead to injuries in the birds (Hocking and Jones, 2006). In addition, some studies indicate that there are also physiological indicators of stress like increased corticosterone concentrations or heterophil/lymphocyte ratios (e.g. Hocking et al., 1993; Savory et al., 1996; De Jong et al., 2002, 2003). However, there are criticisms regarding the validity of these measures as hunger indicators. The metabolic function of corticosterone makes stress-based measures problematic for use as measures of hunger. In addition, it is difficult to relate peripheral measures to the animal's subjective experience of hunger (e.g. D'Eath et al., 2009).

Compared with the rearing period, feed restriction is less severe during the production period, where it varies from 45–80% of the intake of unrestricted fed birds until peak lay (Bruggeman et al., 1999) to 80% of the intake of unrestricted fed birds after the peak lay (Hocking et al., 2002). Stereotyped pecking behaviour can be observed during the first weeks of the production period (Zuidhof *et al.*, 1995; De Jong *et al.*, 2005a) but also after peak lay (Hocking *et al.*, 2002), although the time spent stereotypic pecking is much less than in the rearing period. Due to the increase in growth potential of the birds over the past 30 years, body weight targets have remained more or less the same; however, the degree of feed restriction in broiler breeders has increased and this trend is likely to continue (Renema *et al.*, 2007).

Some research has focused on the development of alternative feeding strategies to reduce the negative effects of feed restriction on bird welfare while maintaining the desired growth rate. Diet dilution using only increased fibre content turned out not to be a viable alternative, as the results of a number of different studies were contradictory or the effects found on behaviour were very small (Hocking *et al.*, 2004; Jones *et al.*, 2004; De Jong *et al.*, 2005a; Hocking, 2006). Although these diets may result in more normal feeding behaviour and could potentially improve welfare through increased satiety, it is possible that metabolic hunger still remains (D'Eath *et al.*, 2009). A combination of a chemical appetite suppressant and oat hulls seemed to be more promising as it had larger effects on the behaviour of the birds. Stereotypic pecking was absent in this treatment group and time spent sitting increased significantly (Sandilands *et al.*, 2005, 2006). However, it is no alternative for commercial practice as this will not be acceptable for consumers and farmers (for ethical reasons and with respect to food safety) (Hocking and Bernard, 1993). Feeding broiler breeders with spin-feeders promoted foraging behaviour but did not reduce behavioural and physiological indicators of hunger or frustration, such as feed intake motivation, oral behaviours and plasma corticosterone concentrations (De Jong *et al.*, 2005b).

Thus far, the use of alternative genotypes (dwarf females) that do not need to be feed restricted is the only solution for the welfare problem of feed restriction (Decuypere *et al.*, 2006), but this will be economically unacceptable to many breeding companies and farmers without compensatory inducements such as subsidies and/or premium prices as the progeny needs more time to grow to slaughter weight (De Jong and Jones, 2006).

Other welfare issues

Many broiler breeder flocks undergo mutilations such as beak trimming in males and females, and toe trimming and spur trimming in males. Beak trimming is carried out to prevent the birds from performing damaging feather pecking behaviour (Gentle and McKeegan, 2007). In addition, beak trimming, toe trimming and spur trimming in males are carried out to prevent injuries to hens due to mating (Henderson *et al.*, 2009). When the infrared method is used for beak trimming (Henderson *et al.*, 2009) it is done at the hatchery (e.g. UK, Germany). When the hot or cold blade method is used it may be done at the farm before 10 days of age (EFSA, 2010). Mutilations and the handling of the birds involved can be stressful and cause acute and/or chronic pain in the birds (e.g. Cheng, 2006); however, not mutilating the birds may also have negative consequences for welfare, especially for the females.

Broiler breeder males may display aggressive behaviours towards females, especially during the performance of sexual behaviour (Millman *et al.*, 2000; De Jong *et al.*, 2009). This rough sexual behaviour of the males leads to females having wounds on the back of their heads where males have pecked and grabbed them with their beaks, and on their body and beneath the wings where males' claws have torn the skin (Duncan, 2009). Thus far, it is not clear what causes this rough male mating behaviour. It cannot be explained by a general higher level of aggression in broiler breeder males or aggression due to feed restriction (Millman and Duncan, 2000). Possibly selection on fertility as well as housing conditions play a role (De Jong *et al.*, 2009) but more research is necessary to find causes and develop solutions to the problem. Apart from rough mating behaviour, forced copulations leading to injuries and fear in females can be caused by over-mating (Leone and Estevez, 2008). Over-mating occurs when males reach sexual maturity earlier than females. It is therefore important to carefully control sexual development of males and females. Inadequate management, i.e. large variation in body weight, may lead to males reaching sexual maturity earlier than females. Over-mating can be prevented by postponing mixing of males and females and/or adjusting the mating ratio to a lower number of males.

Environmental enrichment is not very common in broiler breeder houses, in rear or in lay. Often perches are regarded as enrichment (Estevez, 2009), but they can be regarded more as an essential element in broiler breeder houses (EFSA, 2010). Providing perches or raised platforms at an early age improves the skills of the birds to jump, to enter nest boxes and to find resources. In addition, the development of good navigation skills in females may be relevant so that they can move quickly to prevent overactive males during the early production period (Estevez, 2009).

Hocking and Jones (2006) studied the provision of bunches of string and bales of wood shavings during the rearing period. Although the bales of wood shavings were attractive for the birds, there was no evidence of reduced aggression or feather damage. Estevez (1999) showed that vertically placed cover panels in the production house were effective to control excessive mating problems in commercial farms. Cover panels provided females with shelter and attracted the birds to the litter area, thereby increasing the males' mating opportunities. In a later study it was shown that cover panels also improved reproductive performance in broiler breeder flocks, by not only attracting females to the litter area but also reducing male–male competition for females and over-mating (Leone and Estevez, 2008).

Health

Biosecurity, disease control and hygiene are essential criteria in broiler breeder houses and management, as the health of the breeders has the potential to affect the health of large numbers of commercial broilers. Therefore, it is crucial to manage the environment as disease-free as possible and vaccinations are essential (Hocking, 2004; Collett, 2009; Cserep, 2009). Vaccinations are applied at regular intervals, starting at the hatchery with vaccinations against Marek's disease and often also against infectious bronchitis and Newcastle disease (EFSA, 2010; I.C. de Jong, Lelystad, 2010, personal communication). The vaccination programme during rearing and production varies between countries. During the rearing period broiler breeders are vaccinated against a number of infectious diseases. Usually during the production period they only receive vaccinations against infectious bronchitis. In addition, broiler breeders are blood sampled at regular intervals to check for infections such as Newcastle disease and avian influenza (EFSA, 2010).

Collett (2009) summarizes the scarce data on prevalence of disease in broiler breeders. The most common causes of death in females are reproductive disorders. During rearing mortality ranges between 3 and 7% (Collett, 2009; EFSA, 2010). Female mortality during the production period ranges between 10 and 13% (Hocking and McCorquodale, 2008) with 30% of the mortality between the 25–35 weeks period predominantly due to metabolic-induced diseases (Collett, 2009). Male mortality during the production period is about 10–13%, mainly due to synovitis, tenosynovitis and acute heart failure (Collett, 2009; EFSA, 2010). Overweight breeders are more susceptible to injuries and diseases like tendon rupture, prolapse and sudden death syndrome (Collett, 2009). Mortality in alternative, dwarf females is said to be lower compared with standard intensive breeds (Decuypere et al., 2006). During rearing mortality in alternative females is about 5% and during the production period mortality is estimated at between 6 and 7%, probably due to their slower growth making them less susceptible to metabolic diseases (EFSA, 2010).

Broiler breeders are housed in the same accommodation for a long period of time (i.e. 40–45 weeks during production) and litter management is therefore important (Hocking, 2004), to prevent foot pad dermatitis and hock burns.

TURKEY BREEDERS

This section aims to describe the commonly used management systems in Europe for turkey breeders from day-old poults through rearing and the entire production cycle. As with broiler breeders, the various turkey breeders issue a range of management guides and technical advice sheets (Aviagen Turkeys, 2005, 2007, 2008) which are used by existing customers to optimize their management systems and for new customers to serve as guidelines when constructing new turkey breeder houses, commercial houses and hatcheries.

The European turkey market is segmented into three distinct categories (Table 13.2): (i) heavy strains; (ii) heavy medium strains; and (iii) specialist/niche strains used for whole birds or Christmas turkey production. The management principles described in this chapter are focused primarily towards the heavy and heavy medium turkey sector; however management practices are largely applicable across all segments.

The total turkey breeder parent stock market in Europe is approximately 3.0–3.5 million parent stock female breeders per annum (R.A. Hutchinson, Aviagen, 2010, personal communication), thus about 4–5% of the chicken

Table 13.2. Typical live weights and processing ages for the three main turkey market strains.

Type of strain	Male live weight and age	Female live weight and age
Heavy	20.5–21.5 kg at 20–21 weeks	10.5–11.0 kg at 16 weeks
Heavy medium	14.0–15.5 kg at 16–18 weeks	6.5–8.0 kg at 12–14 weeks
Specialist/niche	12.0–14.0 kg at 20–22 weeks	5.5–6.0 kg at 20–22 weeks

parent stock market. The largest turkey producers in the EU are France, Germany, Italy, Poland, the UK and Hungary. These top six producing countries taken together are responsible for approximately 90% of the EU turkey meat production (USDA, 2007).

The vast majority of turkeys processed today have white plumage, with some specialist strains offering different colour variants (e.g. bronze, black, auburn). The move of the turkey industry towards white plumage (which is a colour recessive) was largely due to the fact that white feathered turkeys, with their lack of pigmentation, give a carcass that is not discoloured by the pigmentation in feather follicles. Specialist and niche strains are mainly used in the UK and France for either Christmas turkey production or Label type production. The EU market share for specialist or niche strains is less than 5% of total turkey meat production.

Selection of turkeys

As for broiler breeders, the breeder and commercial turkeys are supplied by the breeders to customers worldwide, and thus the turkeys are selected to perform to a high level in a variety of environmental conditions. BLUP techniques are used to achieve a balanced selection between: (i) fitness including liveability, skeletal development and leg health; (ii) growth and efficiency characteristics such as daily gain and feed conversion rate; (iii) reproductive performance of the parent male (fertility) and females (egg production, fertility, hatchability); and (iv) qualitative traits such as meat quality and plumage.

Heavy strain turkeys are predominantly used for further processed meat production. It is worth noting, however, that heavy strain turkeys are used successfully in Europe for commercial outdoor production as well as alternative small and coloured strains. The alternative small strain turkeys are typically developed for whole bird seasonal production. Selection is applied on phenotype and focused predominantly for conformation, and to achieve this with a slow growth pattern. Mainly hens are grown as commercial birds, due to their smaller size, slower growth and better conformation than males. Processing is done at 20–22 weeks achieving a growth rate of 35–40 g day^{-1} (Hockenhull Turkeys, 2010). This can be compared with growth of 95 g day^{-1} in heavy strains. The older age at processing achieves a higher degree of subcutaneous skin coverage, improving the appearance and cooking quality of the whole bird.

Housing and management

Rearing period

The goal during the rearing period is to prepare the breeder male and female as best as possible for the production period, for the breeder male and female to reach ideal condition with regard to weight, plumage, conformation and health. Factors affecting the ability to reach this goal are nutrition, stocking density, environmental control, quality of management/stockmanship and health.

Male and female turkey breeders are housed separately from day-old. The birds are placed in houses which are environmentally controlled (mechanical ventilation) and highly insulated. Buildings are typically windowless to allow for light control during rearing; however some countries have windows or open-sided (curtain lined) houses allowing for limited natural light. In all systems, standard or alternative, light control is applied to females in rearing, in order to enable appropriate light stimulation and onset of lay. Turkeys are kept on floor-based systems, no cages are used. The buildings have concrete floors and the litter material is wood shavings, straw or peat.

Guidelines recommend stocking density during rearing to reach a maximum of 36–38 kg m^{-2} at the end of rear (at approximately 28–30 weeks of age) for environmentally controlled houses and 25 kg m^{-2} for open-sided, naturally ventilated houses (FAWC, 1995). This is equivalent to a maximum of 3.0–3.5 females m^{-2} depending on the type of strain used (in an environmentally controlled building). No selection aside from culling for defects is normally applied to females. Similar to broiler breeders, males are supplied as a proportion of females, typically around 10% of the female number. This allows for phenotypic selection of the parent stock males for primarily fitness and conformation but also weight. Males are recommended to reach a stocking density of 2 males m^{-2} prior to the selection stage at 14–16 weeks of age, and 1 male m^{-2} at end of rearing.

Two different types of rearing schemes are used in Europe. Most common is for the turkeys to be reared in one house up to age of transfer to the laying farm, at 28–30 weeks of age. Less common for breeders (but more common for commercial birds) are 'brood-and-move' systems which use specialized farms to brood the birds up to an age of 5–6 weeks and then transfer the birds to a second farm for continued rearing. Typical flock sizes for turkey breeders are 4000–8000 breeder females, which are housed in groups of 2500–3000 birds. Typical breeder male number is 500–800 males at day-old reared together as one group. Either spot brooding or whole house brooding is used. Recommended spot temperature directly under brooder at day-old is 38°C, with an ambient temperature of 28°C. Temperature is gradually reduced over time, to reach 20°C in week 7. Water is supplied by bell drinkers; normally two different sizes are used during the turkeys' growth cycle, to suit best the turkeys' body size and allow for easy access to the water. Nipple drinkers are less common in turkeys as they tend to offer restricted water supply and thus slow growth. The day-old turkeys are fed a starter crumb, followed by pelleted feed.

In some instances for parent stock, mash feed is used throughout, but is not recommended by breeders. Mash feed can limit the ability of the breeder hen in hotter climates to feed sufficiently to sustain egg production.

Weight profiles as recommended by the breeding companies vary according to the strain type; as an example, for a heavy strain the target weight of females at point of light stimulation (29 weeks) is 12.7 kg, equivalent to a growth rate of 63 g day^{-1}. Contrary to broiler breeders, turkey breeding females do not routinely undergo quantitative feed restriction. The feeding schedule is *ad libitum*, but the diet has a lower nutrient concentration than diets used in earlier life, thus achieving a controlled level of growth. Crouch *et al.* (2002), however, showed some advantages for egg production by restricting feed early in the growth phase, but advantages were offset by lower poult quality and higher sensitivity to broodiness by environmental temperature fluctuations. The topic of quantitative feed restriction of breeder females is currently of great interest to breeding companies, given its impact on cost per poult produced; however more research is needed in this area.

For males the target weight at transfer (29 weeks) is 25.4 kg or an average body weight gain of 154 g day^{-1}. The peak growth (Fig. 13.2) occurs at 14–16 weeks of age for the male, at time of selection. Post-selection, the males are put into a phase of restricted feeding to control the growth. Controlling the growth has large benefits on the fitness of the male, reduces fatness and improves in particular persistency of fertility (Hulet and Brody, 1986). There are two different methods applied to control growth of breeder males: (i) qualitative restriction; and (ii) quantitative restriction.

Qualitative restriction uses a special low protein/high energy diet (i.e. 10.1% protein, 13.39 MJ kg^{-1}, 3200 kcal kg^{-1}) fed *ad libitum*, known as male holding diet. By widening the protein to energy ratio the birds eat primarily to satisfy their energy demands, and thus eat less. This ration is used throughout the production period. The overall effectiveness of qualitative weight control is dependent on the level of protein intake, which can be difficult to calculate.

Quantitative restriction has the advantage that the procedure is controlled by the stockman/manager and gives most reliable results if performed correctly. The target growth rate should be 400–500 g week^{-1}, and this means approximately 400–500 g feed day^{-1}. On a breeder rearing farm, round feeders on a winch system are most suitable, allowing sufficient space per bird (25–35 cm) around the feeders. Allowing this space and free access to water are crucial for successful feeding and healthy weight control. Water for both males and females is administered *ad libitum* via bell drinkers.

Beak trimming is applied in breeding turkeys in order to control feather and vent pecking behaviour (Grigor *et al.*, 1995). Beak trimming is mainly performed at day-old using infrared techniques, but also hot or cold blade cutting up to 21 days of age is an acceptable method (FAWC, 1995). Toe trimming by infrared technique is infrequent. For the first 36 h the turkey poults are recommended to have continuous light at an intensity of 100 lux, with 1 h of conditioning darkness (Aviagen Turkeys, 2005). At 36 h up to 14 weeks of age a day length of 14 h is typically used, with approximately 50 lux

(a)

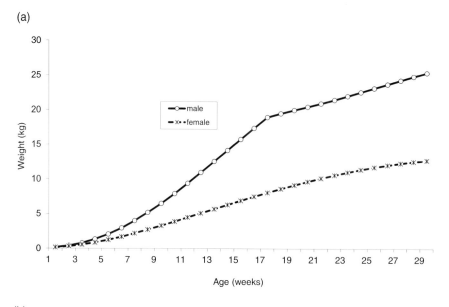

(b)

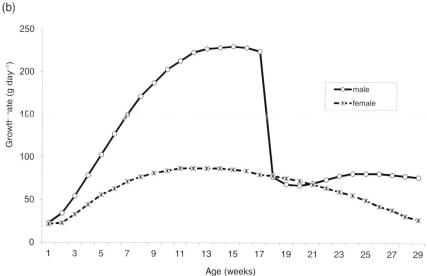

Fig. 13.2. Growth profiles for BUT Big6 male and female breeders, showing (a) cumulative body weight and (b) daily growth rate.

intensity. In order to facilitate photostimulation, males are reduced in day length to 10 h (25 lux) at 14 weeks, females reduced down to 8 h (50 lux) at 18 weeks. Light stimulation is commenced at 24 weeks for males to stimulate growth of the testicles and onset of semen production. Females are light stimulated at transfer at 28–30 weeks of age. The photostimulation triggers responses in prolactin which in turn triggers onset of lay (Proudman and Siopes, 2005).

Production period

The main goal of management during the production period is to maximize the number of fertile hatching eggs, through carefully controlled management, optimizing the environment and maintaining a high health status. The parameters that are monitored are growth pattern (weight loss/gain), lay pattern (peak, persistency of lay), total hen housed production, mortality, incidence of broodiness, fertility and hatchability.

Males will already be selected at 16–18 weeks based on leg health, conformation and weight. Typically there will be 10% of males supplied with the female breeders at day-old, at onset of lay approximately 6–7% of males will be used (taking into account liveability and selection of the parent stock males) which would be equivalent to 1:14–1:16 mating ratio. Artificial insemination is used, and males are kept in separate housing allowing for specialized management with regard to lighting and feed programmes.

Turkey breeders are normally transferred from the rearing farm to the laying farm at 28–30 weeks of age, depending on production system. The time to onset of lay is normally 2 weeks, giving age at first egg of 30–32 weeks of age. Stocking density in lay is generally not limited by legislation, and is more governed by aspects of nest box ratio than kg m^{-2}. A typical nest box ratio is 5.5–6.5 birds per nest or, for heavy strains, typically 1.8–2.0 birds m^{-2} floor space (or 22.8–25.4 kg m^{-2}). The lighting recommendation in lay is to achieve at least 100 lux for 14 h day^{-1} for females. For males they should be kept at minimum 25 lux for 14 h day^{-1}.

Standard lay length for turkeys ranges between 24 and 28 weeks, thus giving a range in age at depletion of 54–60 weeks of age. The females are light stimulated at transfer whereas the males are light stimulated at 24 weeks of age (i.e. on the rearing farm). Standard flock size varies between 4000 and 8000 breeder females, with approximately 1500–2000 females per house. Open-sided breeder houses are widely used (e.g. France, Italy, UK), with environmentally controlled houses used in some countries (e.g. Germany, Scandinavia). In order to enable better control of house temperature and air quality, mechanical ventilation and also tunnel ventilation are becoming more popular particularly in locations with higher summer temperatures. Fogging or misting systems are also common in environmentally controlled and open-sided houses, in order to control both temperature (through convection) and dust levels. Typically turkey breeder females are not adversely affected by low temperatures; hence it is not common to use extra heating sources in layer buildings. Both manual and automatic nests are used successfully for breeding turkeys. Litter material on the floor of the turkey breeder houses is straw or wood shavings, whereas wood shavings are routinely used in the nest box areas in order to ensure cleanliness of the eggs laid.

Following the rearing period males continue to be feed restricted during the production period, with a controlled growth rate from 30 weeks to end of production of 300–400 g week^{-1}, which represents approximately 70% of *ad libitum* feed intake. Females increase by 100–150 g week^{-1} in growth in the three weeks leading into onset of lay, and thereafter lose weight reaching an approximate 8% weight loss from lighting to week 8 after light stimulation, 37 weeks of age. Thereafter there is a slow weight gain in the hens by 60–100 g week^{-1} coming to a plateau at 16 weeks of production, i.e. 48 weeks of age. Hens are normally provided with fixed feeding programmes in lay, fed *ad libitum* at a rate of approximately 2.1 kg week^{-1} (heavy strains) on a low energy/high fibre diet (Aviagen Turkeys, 2005).

Artificial insemination is used throughout the turkey industry to maximize fertility and minimize mating damage to female turkeys. Care is taken in training of the male turkeys in order that the semen collected is of good quality and that there is minimum stress to the male. If males are not trained properly the volume of semen will be reduced or they will produce poor quality semen. It is good practice to collect semen from males at least twice prior to use on females, in order to ensure good semen quality. First insemination takes place 14–16 days after light stimulation. At this time 85% of the females or more should be showing crouching behaviour. In the first week inseminations should be done three times, to achieve high early fertility. After the first week insemination is carried out once weekly. At the beginning and at the end of production there is a requirement to inseminate with a higher concentration of sperm cells in order to maintain high levels of fertility. In the first three inseminations 250 million sperm cells are required, between 1 and 8 weeks 200 million sperm cells required, to be increased towards end of production to a total of 300 million sperm cells per inseminated hen. The quality of semen and sperm concentration can be measured using methods such as SQA (Sperm Quality Analyser) (Neuman *et al.*, 2002), mobility or packed cell volume (King *et al.*, 2000).

Productivity

The total output of breeding poults varies depending on the type of breeding strain, and is a combination of total eggs produced and the success in establishing high fertility and hatchability levels. The use of artificial insemination gives a greater chance of achieving high levels of fertility. Typical fertility levels for turkeys range between 87 and 92% live embryos depending on stage of production. Total hatchability of eggs set averages 80–84% depending on strain (Aviagen Turkeys, 2007, 2008). A comparison of productivity of strains from different turkey segments is described in Table 13.3. Specialist or niche strains typically use either a heavy medium hen paired with a specialist male, or use both specialist female and male. Reproductive performance for specialist/niche strains is thus equivalent to (or in some cases lower than) heavy medium breeders.

Welfare

Primary breeders operate balanced selection programmes, taking into account both classic production traits (weight, feed conversion rate, reproduction) and health/welfare traits. Significant progress has been made with regard to leg

Table 13.3. Overview of reproductive performance of parent stock turkey hens for different market segments (Aviagen Turkeys, 2007, 2008).

Strain type	Heavy	Heavy medium	Specialist/niche
No. of eggs at 24 weeks production	104.3	123.1	110–123
Hatch of eggs set (%)	83.2	85.7	82–85
Day-old poults per hen housed	86.8	105.5	90–105

health of turkeys, utilizing extensive walking ability and X-ray testing techniques. As an example, the prevalence of tibial dyschondroplasia in male line turkeys has been reduced from 30–40% to 5% in the last 5 years (M. Swalander, Aviagen, 2010, unpublished data). Improving leg strength has a positive impact both on liveability and welfare in the form of a lower prevalence of injurious pecking, as any birds displaying leg weakness are of higher risk of being pecked. Havenstein *et al.* (2007a) reported higher levels of liveability in modern type turkeys, compared with a 1966 random-bred control line. Furthermore, evidence of improved immune systems with regard to phagocytic response was found in the modern turkey compared with the control line, indicating balanced selection between health and commercial traits (Havenstein *et al.*, 2007b).

Beak trimming, either through the use of infrared treatment or hot or cold blade techniques, is routinely used in the turkey industry in order to control incidence of feather pecking. Grigor *et al.* (1995) studied the impact of beak trimming on mortality and pecking. They found that none of the beak trimming methods used (Bio-beaking, hot and cold blade at 6 or 21 days of age) significantly affected bird behaviour or production characteristics. The impact on mortality and injurious pecking was significant however, indicating an overall benefit in animal welfare. These results were later confirmed by the Farm Animal Welfare Council's report on turkey welfare (FAWC, 1995), stating that beak trimming only influenced behaviour to a minor extent and yet had beneficial effects in reducing feather damage and mortality. Beak trimming using the above methods did not result in neuroma formation; all treatments resulted in an area at the tip of the beak which lacked sensory afferent nerve fibres and sense organs. In some countries toe trimming is practised on breeder hens by use of infrared techniques.

Feather pecking leading to injured birds is a major problem in non-beak trimmed flocks. Feather pecking can be observed in turkeys already at 4 days of age (Veldkamp, 2010). Although Sherwin *et al.* (1999a, 1999b) showed that environmental enrichment (wheat straw, visual barriers, chains, ropes) reduced injurious pecking, this could not be confirmed by others (Veldkamp, 2010). It has been reported that feather pecking is also common in systems with covered outdoor ranges ('winter garden') (Veldkamp, 2010). To reduce injurious pecking behaviour reducing light intensity (under 5 lux) is the most commonly applied management method (Martrenchar, 1999).

Hocking *et al.* (1999) compared the behavioural and hormonal responses to feed restriction versus *ad libitum* fed male and female turkey breeders. Albeit some behavioural differences were seen in the direction of behaviour, in particular wall pecking, it was concluded based on the hormonal response that turkeys may be better able to adjust physiologically to the demands of food restriction than broiler breeders and that there were few deleterious consequences of restricting male turkeys after 18 weeks of age, although difficult to quantify through objective measures as discussed by D'Eath *et al.* (2009).

Health

As for broiler breeders, the health and biosecurity of breeder turkey flocks is important given the impact of the breeder animals on the commercial generations. In Europe, both rearing and laying breeder farms normally operate on strict all in/all out systems, allowing time between flocks to perform cleaning and disinfection. Biosecurity procedures to restrict infection by organisms from the outside environment in combination with vaccination schemes generally ensure good health status of the European turkey industry. Breeder companies have invested heavily in eradication programmes to enable provision of breeder flocks free from mycoplasmas and Salmonellae. The principal aim of disease control programmes in turkey breeders is to protect them from infection with pathogens which can result in disease and/or loss of production in the breeder hens themselves, or can be vertically transmitted to their progeny, and to protect the consumer from zoonotic pathogens such as Salmonellae. As a result vaccination programmes for breeder turkeys in Europe would include vaccination against Newcastle disease, pasteurella and turkey rhinotracheitis, and may be more elaborate dependent on local disease risk. For example, it may be appropriate to include vaccinations against avian encephalomyelitis, haemorrhagic enteritis virus, pox, ornithobacterium rhinotracheitis and paramyxovirus-3 depending on the disease history in the area. The biosecurity measures will be designed to prevent flocks from being exposed to these pathogens and others for which vaccination is generally not practised, such as avian influenza, mycoplasmas and Salmonellae, although this is again dependent on local challenges. Breeder hens in alternative outdoor systems would tend to be at higher risk of a number of classical poultry diseases such as histomoniasis (black head), erypsilis, fowl cholera and of infection with endoparasites; this is evident in turkeys as with other poultry species (Vits *et al.*, 2005). The risk of disease from many of these infectious organisms increases in an area as the size and density of poultry units increase, the longer outdoor sites have been in use and in particular if they are multi-age. Outdoor systems also tend to be at higher risk of contracting those infections that may be present in the wild bird population including *Mycoplasma gallisepticum*, Newcastle disease and avian influenza. At the hatchery and during both rearing and production, blood tests, swabs from the birds and environmental swabs are taken in order to detect any infections in the breeding stock. Legislation governing trade in poultry in the EU requires breeding stock and therefore progeny to be free from *Salmonella* Enteritidis (SE), *Salmonella* Typhimurium (ST), *Salmonella pullorum/gallinarum*, *Salmonella arizona*, *Mycoplasma meleagridis* and

M. gallisepticum. Testing every 3 weeks for five salmonellas of public health significance which include SE and ST is a requirement for breeders. It is mandatory to test for all these diseases at prescribed intervals but breeders commonly test much more frequently as part of their biosecurity programmes. Bacteriological and serological testing methods are conducted on the samples in either Government or Government-approved laboratories, which in the UK have to be independently accredited to the laboratory quality testing standard ISO 17025.

CONCLUSIONS

The majority of broiler breeders in Europe are the standard, fast growing genotype and the majority of turkey breeders in Europe are the heavy and heavy medium strains. Management of both broiler and turkey breeders is focused on maximizing the production of day-old chickens and poults, respectively, to grow for meat production. Weight management and a high health status are important factors to successful production results for both species but management of broiler and turkey breeders obviously differs in various aspects that are described above. Broiler breeders are housed in floor systems during rearing, and partially slatted floors during production. Except for a small percentage of birds housed in cages in Europe, no other alternative housing systems (like outdoor systems) are used. The main reason for the lack of outdoor systems in broiler breeders is the importance of biosecurity and disease control. The most important welfare issue in broiler breeders is the feed restriction applied during rearing, leading to chronic hunger and frustration of the feeding motivation. Turkeys are always housed on the floor through both rearing and production periods. The majority of breeder turkeys are housed in open-sided buildings, with outdoor systems more frequent in commercial turkeys. In turkeys the most important welfare challenge is feather pecking in systems where beak treatment is not permitted.

REFERENCES

Avendano, S., Watson, K. and Kranis, A. (2010) Genomics in poultry breeding – from utopias to deliverables. In: *Proceedings of the 9th World Congress on Genetics Applied to Livestock Production*, Leipzig, Germany; available at: http://www.kongressband.de/wcgalp2010/assets/pdf/0049.pdf (accessed 17 October 2011).

Aviagen (2007a) Ross 308 Parent Stock: Performance Objectives. http://en.aviagen.com/assets/Tech_Center/Ross_PS/Ross-308-PS-PO-2011.pdf (accessed 20 June 2011).

Aviagen (2007b) Ross 708 Parent Stock: Performance Objectives. http://en.aviagen.com/assets/Tech_Center/Ross_PS/Ross-708-PS-PO-2011.pdf (accessed 20 June 2011).

Aviagen (2009) Ross 308 Parent Stock Management Manual. http://en.aviagen.com/assets/Tech_Center/Ross_PS/ROSS_308_Manual.pdf (accessed 20 June 2011).

Aviagen Turkeys (2005) Management Essentials for Breeder Turkeys. http://en.aviagen.com/assets/Tech_Center/Turkeys_breeders/ATI/mgt_guides/Breeder_MgtEss.pdf (accessed 20 June 2011).

Aviagen Turkeys (2007) BUT Big6 Commercial Performance Goals. http://en.aviagen.com/ assets/Tech_Center/Turkeys_commercial/ATL/Mgt_Guide/ATL_Big6_Commercial_ 0050.pdf (accessed 20 June 2011).

Aviagen Turkeys (2008) BUT10 Commercial Performance Goals. http://en.aviagen.com/ assets/Tech_Center/Turkeys_commercial/ATL/Mgt_Guide/ATL_BUT10_ Commercial_0054.pdf (accessed 20 June 2011).

Bruggeman, V., Onagbesan, O., D'Hondt, E., Buys, N., Safi, M., Vanmontfort, D., Berghman, L., Vandesande, F. and Decuypere, E. (1999) Effects of timing and duration of feed restriction during rearing on reproductive characteristics in broiler breeder females. *Poultry Science* 78, 1424–1434.

Cheng, H. (2006) Morphopathological changes and pain in beak trimmed laying hens. *World's Poultry Science Journal* 62, 41–52.

Cobb (2008) *Cobb Breeder Management Guide.* Cobb-Vantress, Siloam Springs, Arkansas; available at: http://www.cobb-vantress.com (accessed 20 June 2011).

Collett, S.R. (2009) Managing current disease challenges in breeders. In: Hocking, P.M. (ed.) *Biology of Breeding Poultry.* CABI Publishing Wallingford, UK, pp. 414–434.

Crouch, A.N., Grimes, J.L., Christensen, V.L. and Krueger, K.K. (2002) Effect of feed restriction during rearing on large white turkey breeder hens. 2. Reproductive performance. *Poultry Science* 81, 16–22.

Cserep, T. (2009) Vaccination: theory and practice. In: Hocking, P.M. (ed.) *Biology of Breeding Poultry.* CABI Publishing, Wallingford, UK, pp. 377–390.

D'Eath, R.B., Tolkamp, B.J., Kyriazakis, I. and Lawrence, A.B. (2009) 'Freedom from hunger' and preventing obesity: the animal welfare implications of reducing food quantity or quality. *Animal Behaviour* 77, 275–288.

Decuypere, E., Hocking, P.M., Tona, K., Onagbesan, O., Bruggeman, V., Jones, E.K.M., Cassy, S., Rideau, N., Metayer, S., Jego, Y., Putterflam, J., Tesseraud, S., Collin, A., Duclos, M., Trevidy, J.J. and Williams, J. (2006) Broiler breeder paradox: a project report. *World's Poultry Science Journal* 62, 443–453.

De Jong, I.C. and Jones, B. (2006) Feed restriction and welfare in domestic birds. In: Bels, V. (ed.) *Feeding in Domestic Vertebrates.* CABI Publishing, Wallingford, UK, pp. 120–135.

De Jong, I.C., Van Voorst, S., Ehlhardt, D.A. and Blokhuis, H.J. (2002) Effects of restricted feeding on physiological stress parameters in growing broiler breeders. *British Poultry Science* 43, 157–168.

De Jong, I.C., Van Voorst, S. and Blokhuis, H.J. (2003) Parameters for quantification of hunger in broiler breeders. *Physiology & Behavior* 78, 773–783.

De Jong, I.C., Enting, H., Van Voorst, S., Ruesink, E.W. and Blokhuis, H.J. (2005a) Do low density diets improve broiler breeder welfare during rearing and laying? *Poultry Science* 84, 194–203.

De Jong, I.C., Fillerup, M. and Blokhuis, H.J. (2005b) Effect of scattered feeding and feeding twice a day during rearing on parameters of hunger and frustration in broiler breeders. *Applied Animal Behaviour Science* 92, 61–76.

De Jong, I.C., Wolthuis-Fillerup, M. and Van Emous, R.A. (2009) Development of sexual behaviour in commercially-housed broiler breeders after mixing. *British Poultry Science* 50, 151–160.

Duncan, I.J.H. (2009) Mating behaviour and fertility. In: Hocking, P.M. (ed.) *Biology of Breeding Poultry.* CABI Publishing, Wallingford, UK, pp. 111–132.

EFSA (European Food Safety Authority) (2010) Scientific opinion on welfare aspects of the management and housing of the grand-parent and parent stock raised and kept for breeding purposes. *The EFSA Journal* 8: 1667; doi:10.2903/j.efsa.2010.1667.

Estevez, I. (1999) Cover panels for chickens: a cheap tool that can help you. *Poultry Perspectives* 1, 4–6.

Estevez, I. (2009) Behaviour and environmental enrichment in broiler breeders. In: Hocking, P.M. (ed.) *Biology of Breeding Poultry*. CABI Publishing, Wallingford, UK, pp. 261–283.

FAWC (1995) *Report on the Welfare of Turkeys*. Farm Animal Welfare Council, London; available at: http://www.fawc.org.uk/reports/turkeys/turkrtoc.htm (accessed 20 June 2011).

FAO (2008) Food and Agriculture Organization of the United Nations FAOSTAT database. http://faostat.fao.org (accessed 20 June 2011).

Fiks, T.G.C.M., De Jong, I.C., Veldkamp, T., Van Emous, R.A. and Van Middelkoop. J.H. (2006) *Literature study mutilations poultry*. PraktijkRapport Pluimvee 19. Animal Sciences Group, Lelystad, the Netherlands.

Gentle, M.J. and McKeegan, D.E.F. (2007) Evaluation of the effects of infrared beak trimming in broiler breeder chicks. *The Veterinary Record* 160, 145–148.

Grigor, P.N., Hughes, B.O. and Gentle, M.J. (1995) An experimental investigation of the costs and benefits of beak trimming in turkeys. *The Veterinary Record* 136, 257–265.

Havenstein, G.B., Ferket, P.R., Grimes, J.L., Quereshi, M.A. and Nestor, K.E. (2007a) Comparison of the performance of 1966- versus 2003-type turkeys when fed representative 1966 and 2003 turkey diets: growth rate, livability, and feed conversion. *Poultry Science* 86, 232–240.

Havenstein, G.B., Ferket, P.R., Grimes, J.L., Quereshi, M.A. and Nestor, K.E. (2007b) A comparison of the immune response of 2003 commercial turkeys and a 1966 random-bred strain when fed representative 2003 and 1966 turkey diets. *Poultry Science* 86, 241–248.

Henderson, S.N., Barton, J.T., Wolfenden, A.D., Higgins, S.E., Higgins, J.P., Kuenzel, W.J., Lester, C.A., Tellez, G. and Hargis, B.M. (2009) Comparison of beak-trimming methods on early broiler breeder performance. *Poultry Science* 88, 57–60.

Hockenhull Turkeys (2010) Performance Goals. http://www.hockenhullturkeys.co.uk/productsandprices.php (accessed 20 June 2011).

Hocking, P.M. (2004) Measuring and auditing the welfare of broiler breeders. In: Weeks, C.A. and Butterworth, A. (eds) *Measuring and Auditing Broiler Welfare*. CABI Publishing, Wallingford, UK, pp. 19–35.

Hocking, P.M. (2006) High-fibre pelleted rations decrease water intake but do not improve physiological indexes of welfare in food-restricted female broiler breeders. *British Poultry Science* 47, 19–23.

Hocking, P.M. (2009) Feed restriction. In: Hocking, P.M. (ed.) *Biology of Breeding Poultry*. CABI Publishing, Wallingford, UK, pp. 307–330.

Hocking, P.M. and Bernard, R. (1993) Evaluation of putative appetite suppressants in the domestic fowl. *British Poultry Science* 34, 393–404.

Hocking, P.M. and Jones, E.K.M. (2006) On-farm assessment of environmental enrichment for broiler breeders. *British Poultry Science* 47, 418–425.

Hocking, P.M. and McCorquodale, C.C. (2008) Similar improvements in reproductive performance of male line, female line and parent stock broiler breeders genetically selected in the UK and in South America. *British Poultry Science* 49, 282–289.

Hocking, P.M., Maxwell, M.H. and Mitchell, M.A. (1993) Welfare assessment of broiler breeder and layer females subjected to food restriction and limited access to water during rearing. *British Poultry Science* 34, 443–458.

Hocking, P.M., Maxwell, M.H. and Mitchell, M.A. (1999) Welfare of food restricted male and female turkeys. *British Poultry Science* 40, 19–29.

Hocking, P.M., Maxwell, M.H., Robertson, G.W. and Mitchell, M.A. (2002) Welfare assessment of broiler breeders that are food restricted after peak of lay. *British Poultry Science* 43, 5–15.

Hocking, P.M., Zaczek, V., Jones, E.K.M. and McLeod, M.G. (2004) Different concentrations and sources of dietary fibre may improve the welfare of female broiler breeders. *British Poultry Science* 45, 9–19.

Hubbard (2009a) *Hubbard Classic Management Guide Parent Stock*. Hubbard, Walpole, New Hampshire; available at: http://www.hubbardbreeders.com/managementguides/index.php?id=12 (accessed 20 June 2011).

Hubbard (2009b) *Hubbard JA57 Performance Summary*. Hubbard, Walpole, New Hampshire; available at: http://www.hubbardbreeders.com/managementguides/index.php?id=33 (accessed 20 June 2011).

Hulet, R.M. and Brody, T.B. (1986) Semen quality and fat accumulation in prepuberal and postpuberal male turkeys as affected by restricted feeding. *Poultry Science* 65, 1972–1976.

Jones, E.K.M., Zaczek, V., McLeod, M.G. and Hocking, P.M. (2004) Genotype, dietary manipulation and food allocation affect indices of welfare in broiler breeders. *British Poultry Science* 45, 725–737.

King, L.M., Kirby, J.D., Froman, F.P., Sonstegard, T.S., Harry, D.E., Darden, J.R., Marini, P.J., Walker, R.M., Rhoads, M.L. and Donoghue, A.M. (2000) Efficacy of sperm mobility assessment in commercial flocks and the relationships of sperm mobility and insemination dose and fertility in turkeys. *Poultry Science* 79, 1797–1802.

Laughlin, K.F. (2009) Breeder management: how did we get here? In: Hocking, P.M. (ed.) *Biology of Breeding Poultry*. CABI Publishing, Wallingford, UK, pp. 9–25.

Leeson, S. and Summers, J.D. (2000) *Broiler Breeder Production*. University Books, Guelph, Canada.

Leone, E.H. and Estevez, I. (2008) Economic and welfare benefits of environmental enrichment for broiler breeders. *Poultry Science* 87, 14–21.

Martrenchar, R. (1999) Animal welfare and production of turkey broilers. *World's Poultry Science Journal* 55, 143–152.

Mench, J.A. (2002) Broiler breeders: feed restriction and welfare. *World's Poultry Science Journal* 58, 23–30.

Millman, S.T. and Duncan, I.J.H. (2000) Effect of male-to-male aggressiveness and feed-restriction during rearing on sexual behaviour and aggressiveness towards females by male domestic fowl. *Applied Animal Behaviour Science* 70, 63–82.

Millman, S.T., Duncan, I.J.H. and Widowski, T.M. (2000) Male broiler breeder fowl display high levels of aggression toward females. *Poultry Science* 79, 1233–1241.

Neuman, S.L., McDaniel, C.D., Frank, L., Radu, J., Einstein, M.E. and Hester P.Y. (2002) Utilization of a sperm quality analyser to evaluate sperm quality and quantity of turkey breeders. *British Poultry Science* 43, 457–464.

Proudman, J.A. and Siopes, T.D. (2005) Thyroid hormone and prolactin profiles in male and female turkeys following photostimulation. *Poultry Science* 84, 942–946.

Renema, R., Rustad, M.E. and Robinson, F.E. (2007) Implications of changes to commercial broiler and broiler breeder body weight targets over the past 30 years. *World's Poultry Science Journal* 63, 457–472.

Sandilands, V., Tolkamp, B.J. and Kyriazakis, I. (2005) Behaviour of food restricted broilers during rearing and lay – effects of an alternative feeding method. *Physiology & Behavior* 85, 115–123.

Sandilands, V., Tolkamp, B., Savory, C.J. and Kyriazakis, I. (2006) Behaviour and welfare of broiler breeders fed qualitatively restricted diets during rearing: are these viable alternatives to quantitative restriction? *Applied Animal Behaviour Science* 96, 53–67.

Savory, C.J., Maros, K. and Rutter, S.M. (1993) Assessment of hunger in growing broiler breeders in relation to a commercial restricted feeding programme. *Animal Welfare* 2, 131–152.

Savory, C.J., Hocking, P.M., Mann, J.S. and Maxwell, M.H. (1996) Is broiler breeder welfare improved by using qualitative rather than quantitative food restriction to limit growth rate? *Animal Welfare* 5, 105–127.

Sherwin, C.M., Lewis, P.D. and Perry, G.C. (1999a) The effects of environmental enrichment and intermittent lighting on the behaviour and welfare of male domestic turkeys. *Applied Animal Behaviour Science* 62, 319–333.

Sherwin, C.M., Lewis, P.D. and Perry, G.C. (1999b) Effects of environmental enrichment, fluorescent and intermittent lighting on injurious pecking amongst male turkey poults. *British Poultry Science* 40, 592–598.

USDA (2007) US Department of Agriculture Foreign Agricultural Service, Global Agriculture Information Network. GAIN Report E47061. http://www.fas.usda.gov/gainfiles/200707/146291790.pdf (accessed 20 June 2011).

Vits, A., Weitzenburger, D. and Distl, O. (2005) Comparison of different housing systems for laying hens in respect to economic, health and welfare parameters. *Deutsche Tierärztliches Wochenschrift* 112, 332–342.

Veldkamp, T. (2010) Beak trimming in turkey production – alternatives by means of breeding and farm management. Wageningen UR Livestock Research, Report 197. http://edepot.wur.nl/145552 (accessed 20 June 2011).

Zuidhof, M.J., Robinson, F.E., Feddes, J.J.R., Hardin, R.T., Wilson, J.L., Mckay, R.I. and Newcombe, M. (1995) The effects of nutrient dilution on the well-being and performance of female broiler breeders. *Poultry Science* 74, 441–456.

CHAPTER *14*

Alternative Systems for Meat Chickens and Turkeys: Production, Health and Welfare

T.A. Jones and J. Berk

ABSTRACT

Legislative and assurance scheme requirements for standard and alternative indoor and outdoor broiler and turkey production systems are described. Generally, health and welfare are protected to various extents by a series of input requirements, largely related to stocking density, light, environmental control parameters, environmental enrichment, mutilations and growth rate. Outcome measures (usually related to physical well-being) highlight flocks that perform poorly, and success depends on the effectiveness of the input and output measures, the reporting structure and any remedial action taken. Alternative systems represent a low market share of broiler and turkey production in the European Union (approximately 10 and 30%, respectively) and generally production costs more. Free range and organic systems are largely considered to have the potential to provide good living conditions and reduce environmental pollution. However, concerns have been raised over bird health (*Campylobacter* infection), welfare (higher foot lesions and breast blisters, lack of outdoor ranging), product quality and consumers' willingness to pay. Research shows that breed suitability is one of the largest factors in determining welfare in alternative systems, particularly for broiler chickens. More robust, hardy breeds with lower growth rates should be used; these birds are better suited to a wide range of environmental parameters and diets with lower energy density. The quality of the diet (particularly in relation to indispensible amino acids and protein balance) and the free range environment (particularly in relation to the provision of natural cover outdoors) are also highly important for both broilers and turkeys. The meat from slow growth broiler breeds is more suited to the whole bird market (as opposed to portioned or further processed) and generally contains less fat and more protein than from conventional breeds. Consumers tend to be unable to differentiate chicken products from alternative systems by odour and taste, but can differentiate by appearance and texture.

INTRODUCTION

Meat chicken (broiler) and turkey production in the EU-27 for 2009 was 8.8 and 1.8 million tonnes, respectively, equating to a respective consumption of 17.1 and 3.4 kg per capita (AVEC, 2010). The top six producing countries of the EU-27 are shown in Fig. 14.1, and account for 66.7% of the total broiler and 90.8% of the total turkey production. The UK is the largest broiler producer, while France and Germany are the largest turkey producers. Turkey production has declined in the European Union over the last decade (Proplanta, 2010), particularly in France and the UK (by 43% and 21%, respectively), whereas Germany has seen a small increase (3.6%).

(a)

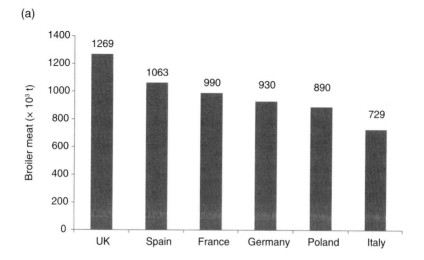

(b)

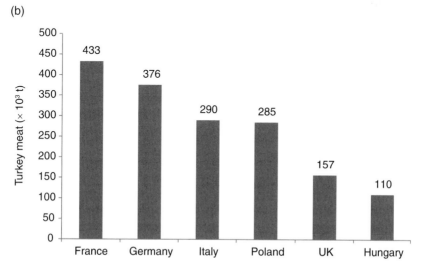

Fig. 14.1. (a) Broiler meat production and (b) turkey meat production in the top six producing countries of the EU-27 in 2009 (from AVEC, 2010).

Most broilers in the EU (approximately 90%) are produced under standard intensive systems, in flocks of up to 50,000 birds or more, in houses with controlled temperature, light and ventilation; there is litter on the floor and often the houses are windowless. The broilers are of fast growth breeds and are fed nutrient-dense diets usually inclusive of coccidiostats, synthetic amino acids and genetically modified (GM) ingredients. They are grown indoors and are slaughtered from as early as 35 days. Minimum conditions for the protection of meat chickens are set out in the revised EU directive (Council Directive, 2007) which came into force in June 2010. Turkeys are reared in a greater variety of housing. Standard housing is similar to that for chickens, whereas pole barns are less-intensive indoor systems with side curtains and natural light and ventilation, deep litter (often straw) and reduced stocking densities. Seasonal producers often have smaller sheds with rudimentary shelter. There is no Council Directive outlining minimum conditions for the protection of meat turkeys.

MEAT CHICKENS (BROILERS)

Production systems, legislation and assurance schemes

Alternative systems exist for both indoor broilers and where outdoor access (free range and organic) is given. These systems are defined in EU marketing terms (Commission Regulation, 2008), organic legislation (Council Regulation, 1999) and various accredited assurance schemes, which often improve upon the legislative standard. Legislative and assurance scheme requirements are given in Table 14.1 for indoor systems. Assured Chicken Production (ACP, 2010a) is the industry Red Tractor Farm Assurance Scheme adopted for standard production in the UK. It is aligned with the EU broiler directive, except for the permissible maximum stocking density (limited to 38 kg m^{-2}). Forthcoming British (GB) legislation (Defra, 2010) limits maximum stocking density to 39 kg m^{-2} and includes trigger levels for a series of nine post-mortem conditions including foot pad dermatitis (FPD). Once triggered, alerts will be sent to Animal Health and an action plan developed to resolve problems.

RSPCA Freedom Food (FF) Indoor, Certified (France) and Barn chicken further reduce the permissible stocking density to 30 and 25 kg m^{-2} (for RSPCA and Barn, respectively), and increase minimum slaughter age to 56 days (Barn and Certified (France)). RSPCA FF and Certified (France) limit the genetic growth rate (GR) of the strain; the former specifying less than 45 g day^{-1} averaged over the production cycle. Minimum requirements for light intensity and the provision of a dark period are now set at 20 lux and 6 h, respectively (Council Directive, 2007), increasing to 100 lux and 6–12 h under RSPCA FF standards. The latter also incorporates the requirements for natural light via windows, environmental enrichment and is the only scheme to include trigger levels for leg abnormalities and a requirement to cull birds with a gait score (GS) of 3. Gait is scored from 0 (normal) to 5 (unable to stand) in line with Kestin *et al.* (1992); birds with GS=3 walk with an obvious defect which impairs their function.

Table 14.1. European Union legislative and assurance scheme (UK and France) requirements for standard and alternative indoor broiler production systems.

Criteria	Legislation		Assured Chicken Production (Red Tractor) UK (ACP, 2010a)	Assurance Scheme	
	Council Directive (2007)	Barn (Commission Regulation, 2008)		RSPCA Freedom Food Indoor UK (RSPCA, 2008)	Certified France (CCP, 2010)
Min. age (days)	–	56	–	–	56
Max. area (birds m^{-2})	–	15	–	19	18
Max. SD (kg m^{-2})	33 D1: 39[a] D2: 42[b]	25	Planned 38	30	–
Breed	–	–	–	Slow growth <45 g day^{-1}	Intermediate growth
Feed	–	Min 65% cereal	–	No in-feed antibiotics	100% vegetable Min 65% cereal No growth factors
Light	Min. 20 lux over 80% floor area Min. 6 h D (4 h uninterrupted)	As Council Directive (2007)	As Council Directive (2007)	L: min. 8 h constant D: min. 6 h – max. 12 h Light level: av. min. 100 lux over 75% area, no area <20 lux Dim/raise light over 15 min period Natural light must be provided via openings ≥3% total floor area Shutters must be provided, opening min 0.56 m^2	As Council Directive (2007)
Ventilation	NH_3 <20 ppm CO_2 <3000 ppm Cope with a 3°C temperature lift, when 30°C or more in the shade RH <70% when outside <10°C	As Council Directive (2007)	As Council Directive (2007)	Dust <10 mg m^{-2} CO <50 ppm NH_3 <15 ppm CO_2 <5000 ppm Cope with a 3°C temperature lift RH maintained at 50–70%	As Council Directive (2007)

Table 14.1. – Continued

Criteria	Legislation		Assurance Scheme		
	Council Directive (2007)	Barn (Commission Regulation, 2008)	Assured Chicken Production (Red Tractor) UK (ACP, 2010)	RSPCA Freedom Food Indoor UK (RSPCA, 2008)	Certified France (CCP, 2010)
Enrichment	It is recommended that producers explore different types of environmental enrichment		As Council Directive (2007)	Must provide (per 1000 birds): - 1 straw bale - 2 m perch space - 1 pecking object	As Council Directive (2007)
Outcome measures	Monitor hock burn & FPD Not normally exceed: - 5.0% mortality - 1.5% PMI - 15% hock		As Council Directive (2007) Forthcoming GB legislation, levels for - daily accumulative mortality - ascites/oedema - cellulitis/dermatitis - DOA - emaciation - joint lesion - arthritis - septicaemia - respiratory total rejections - FPD	Inspection required if exceed following levels: - FPD/hock burn 4.0% - leg abnormalities 3% at 42 days - mortality 0.3% in 24 h Must also record: - breast blister - back scratch - dirty feathers	As Council Directive (2007)
Other				Birds with GS=3 or above must be culled No feed tracks permissible No bell drinkers in new systems	

Min., minimum; max., maximum; SD, stocking density; D1, Derogation 1; D2, Derogation 2; D, dark; RH, relative humidity; FPD, foot pad dermatitis; PMI, post-mortem inspection; DOA, dead on arrival; L, light; av., average; GS, gait score; –, no standard/requirement for this criterion.

[a]D1 – producers can stock up to 39 kg m^{-2} if documentation conforms (must include mortality data and hybrid/breed) and environment parameters maintained: NH$_3$ <20 ppm, CO$_2$ <3000 ppm at chicken head height; inside temperature not exceed outside temperature by more than 3°C when outside temperature in shade exceeds 30°C; average RH inside house during 48 h does not exceed 70% when outside temperature is below 10°C.

[b]D2 – producers can stock up to 42 kg m^{-2} if 2 years of monitoring indicate no deficiencies, accumulative mortality in at least seven successive flocks is below 1% plus 0.06% × slaughter age of flock in days (e.g. 3.38% at 38 days).

Legislative and assurance scheme requirements are given in Table 14.2 for outdoor systems. Free range systems increase the minimum age at slaughter to 56 days and reduce the indoor maximum stocking density to 27.5 kg m^{-2}. They also require birds to have outdoor access for half of their life at a space allowance of 1 m^2 per bird. Traditional free range and organic systems further increase slaughter age to between 70 and 81 days, reduce maximum stocking density to between 21 and 30 kg m^{-2} (if pop holes are closed at night), increase the outdoor area to between 2 and 4 m^2 per bird for one- to two-thirds of their life, and further regulate feed constituents. For organic production the diet is largely plant based of organic origin, with no synthetic amino acids or GM ingredients. Slow growing breeds are required, but not defined, by some standards but not all. Prescribed medication is permissible with increased withdrawal periods, and if antibiotics are used the birds must be sold off some schemes. Free range and organic systems often rely on natural light (post brood), although artificial light is permissible to extend day length (16 h and 100 lux is required by one scheme, i.e. RSPCA, 2008). Flock sizes are limited under organic regulations in an attempt to minimize nitrogen impact on the land to 170 kg N ha^{-1} per annum (Council Regulation, 1999).

Alternative systems accounted for 20.6% of total UK broiler production in 2009, with free range and organic systems representing only 3.1% (RSPCA, 2010a). Table 14.3 gives the approximate number and proportion of meat chickens reared under different systems in the UK in 2009; Standard Plus (includes reduced stocking density, natural light and enrichment for fast GR breeds) and RSPCA FF Indoors increased their share of production in the preceding 5 years while free range and organic remained fairly static. Alternative systems in France accounted for 27% of total production (Fig. 14.2), with free range accounting for 16% (Hubbard, 2008). Organic systems in Germany accounted for 0.63% of total broiler production in 2009 (MEG, 2011).

Safeguards for welfare

Legislation and assurance schemes aim to protect welfare by setting a series of input requirements (some more detailed and/or more stringent than others), largely related to stocking density, light, environmental control parameters, environmental enrichment, mutilations and growth rate. Various outcome measures (usually related to physical well-being) highlight flocks that performed poorly. Success depends on the effectiveness of the input and output measures, the reporting structure and any remedial action taken.

Estevez (2007) concluded that the health and welfare of broilers is compromised if stocking density exceeds 34 to 38 kg m^{-2} (depending on final body weight (BW)). The legislative second derogation to 42 kg m^{-2} in Council Directive 2007/43/EC is detrimental to welfare (Dawkins *et al.*, 2004); additionally, increasing stocking density from even relatively low levels leads to deterioration in walking ability (Dawkins *et al.*, 2004; Knowles *et al.*, 2008). While the range of monitored outcome measures is strengthened by forthcoming GB legislation, and FPD has been used to successfully monitor flock welfare at

Table 14.2. European Union legislative and assurance scheme (UE and France) requirements for standard and alternative outdoor broiler production systems.

Criteria	Legislation		Assurance Scheme			
	EU free range (Commission Regulation, 2008) ACP free range (ACP, 2010b)	EU traditional free range (Commission Regulation, 2008)	RSPCA FF Free Range UK (RSPCA, 2008)	Label Rouge France (Fanatico and Born, 2002)	Organic Soil Association, UK (Soil Association, 2010)	Organic France (PMAF, 2010)
Min. age (days)	56	81	56	81 to 110	Any age if slow GR and parent organic; 70 days if parents not organic; 81 days if fast GR	81
Max. area (birds m^{-2})	13	Fixed: 12 Mobile: 20[a]	13	Fixed: 11 Mobile: 20	Fixed: 10 Mobile: 16[a]	Fixed: 10 Mobile: 16
Max. SD (kg m^{-2})	27.5	Fixed: 25 Mobile: 40[a]	27.5		Fixed: 21 Mobile: 30[a]	
Minimum outdoor access	1/2 of life	From 6 weeks	From 28 days or less	From 6 weeks	2/3 of life	Min. 1/2 of life
Outdoor area (m^2 per bird)	1	2	1	2	2500 birds ha^{-1} (4 m^2 per bird)	4
Breed	–	Slow	Slow <45 g day^{-1}	Slow	Slow & fast	Slow/local breed preferred
Feed	Min. 70% cereal in finishing phase	Min. 70% cereal in finishing phase	No in-feed AB	100% plant based (low energy, CP); min. 75% cereal; protein supplements of peas, soybean; no feed additives or coccidiostats; synthetic AA permissible	No synthetic AA or feed additives 100% organic (5% non permissible) Soluble grit at all times	100% plant based Roughage included No synthetic AA No coccidiostats No growth factors >95% organic
Medicine	–	–		AB by prescription only	If AB used, birds are decertified & sold off scheme; increased withdrawal periods for medicines (×3)	No allopathy

Light	As Council Directive (2007)	As RSPCA FF Indoor	Natural	Can use artificial light to 16 h, must have dusk	Natural
Ventilation		As RSPCA FF Indoor	Natural		
Enrichment	For ACP, must provide (per 1000 birds): – 1 straw bale – 2 m perch space Mobile: – shelter with netting – big bales – A frames	As RSPCA FF Indoor Natural & artificial shelter: area=[(number chickens×0.3)final weight]'38 Min. overhead shade 8 m² per 1000 birds	Some producers also provide: – dividers in house – grit & whole wheat – woodland – constant ration	Perches Dust bathing areas outdoor Drinkers outdoor Soluble grit at all times	Vegetation cover
Pop holes	Min. height: 40 cm 4 m per 100 m²	Pop holes: min. height 45 cm, 50 cm wide No. of pop holes: – 1 for up to 600 birds – 1 per 700 birds & min. 2 pop holes	4 m per 100 m²	4 m per 100 m²	4 m per 100 m²
Outcome measures	ACP: as ACP Indoor	As RSPCA FF Indoor	HAACP control throughout supply chain Early organoleptic analysis by consumers & trained panel	Keep records on mortality, hock damage, % rejects & cause	
Other	Fixed: natural shelter should be encouraged	Max. 1600 m² house area per unit No more than: – 4800 birds per house – 400 m² per house	Birds with GS=3 or above must be culled No feed tracks permissible No bell drinkers in new systems	Geographic protected zones in. 21 days between flocks Max. 1600 m² house area per unit No more than: – 1100 birds per house – 400 m² per house – 4 houses per farm – 2 h or 64 miles to slaughter Product sold fresh within 9 days	Max. 1600 m² house area per unit Work towards <500 birds per house (currently max 1000)

Max. 1600 m² house area per unit
No more than:
– 200 m² per house
– 2000 birds per pen

ACP, Assured Chicken Production; FF, Freedom Food; min., minimum; max., maximum; SD, stocking density; AB, antibiotics; GS, gait score; CP, crude protein; AA, amino acids; HAACP, hazard analysis and critical control points; GR, growth rate; –, no standard/requirement for this criterion.
aFor mobile arks <150 m² maximum SD increases provided the pop holes are left open all night.

Table 14.3. The number and proportion of broilers produced under different production systems in the UK (data from RSPCA, 2010a).

Broilers	ACP	Standard plus	RSPCA FF Indoor	RSPCA FF Free Range	Free Range	Organic	Total
Number (millions)	671.1	98.3	49.6	10.4	12.9	2.6	845.6
Proportion (%)	79.4	11.7	5.9	1.3	1.5	0.3	

ACP, Assured Chicken Production; FF, Freedom Food.

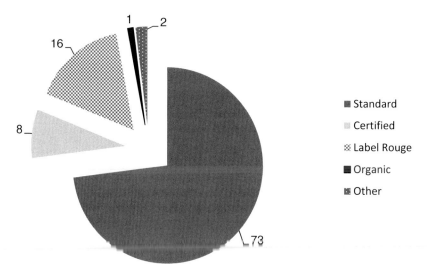

Fig. 14.2. The proportion of meat chickens reared under different standards in France (from Hubbard, 2008).

densities up to 36 kg m^{-2} in Sweden (Berg, 1998 cited by Estevez, 2007), measures directly affected by stocking density, primarily walking ability and behaviour, are not included. Currently these measures conducted on the live bird are time consuming (Kestin and Knowles, 2004), however automatic statistics are being developed (Dawkins *et al.*, 2009).

Despite genetic progress for leg health in chickens, poor walking ability is still prevalent although highly variable between flocks, with multifactorial risks (Bradshaw *et al.*, 2002). Primary risk factors in healthy flocks are those of high GR (Knowles *et al.*, 2008) and poor environmental control (Jones *et al.*, 2005). Slowing early growth and increasing activity, via introducing longer dark periods (see below), feeding less nutrient-dense diets (Leterrier *et al.*, 1998; Gordon, 2002; Welfare Quality, 2010) and feeding mash as opposed to pelleted feed (Brickett *et al.*, 2007b), can improve leg health and walking ability.

Intense genetic selection for high GR and breast meat yield with continued improvement in feed efficiency has resulted in a broiler with low activity and leg, metabolic and physiological disorders such as ascites and sudden death syndrome (SCAHAW, 2000). High GR also exacerbates the need for feed restriction and problems of chronic hunger in broiler breeder rearers. Limiting the GR potential of a breed may be the best practical solution to the welfare problems at the current time, since GR and feed conversion ratio (FCR) continue to improve year on year (McKay, 2009). Commercial trials of intermediate and fast growing breeds reared to 56 and 42 days, respectively, under the same light programme (18 h of light/6 h of dark) showed there was less mortality (1.5% versus 5.6%), FPD (12.5% versus 83.0%) and hock burn (11.5% versus 44.9%) in the intermediate than the fast birds (Cooper *et al.*, 2008).

A period of darkness is required to allow broilers to develop proper sleep patterns and diurnal behavioural rhythm (Appleby *et al.*, 1994), and sleep is required for physiological recuperation in terms of energy conservation, tissue regeneration and growth (Malleau *et al.*, 2007). Young domestic fowl rest and sleep for 12 to 16 h post hatch (Hess, 1959 cited by Malleau *et al.*, 2007), gradually reducing after the first 2 weeks (Mascetti *et al.*, 2004). Short days can reduce early growth, presumably because of reduced feeding, but do not affect BW and FCR (in fact FCR is often better) when older, due to compensatory growth (Classen *et al.*, 1991) and adaptation of feeding patterns with age (Rozenboim *et al.*, 1999; Brickett *et al.*, 2007b; Schwean-Lardner and Classen, 2010). Simulated natural brooding cycles of 40 min light/40 min dark, within a 19 h light/5 h dark programme, allowed young chicks to 14 days to show periods of synchronized high activity with lights-on and low activity with lights-off; despite having less than 10 h of light in a 24 h cycle, the broilers were able to consume enough feed to attain the same BW as those reared with 19 h of light (Malleau *et al.*, 2007). Shorter days improved welfare through fewer skeletal problems (Classen *et al.*, 1991), less mortality (Rozenboim *et al.*, 1999; Brickett *et al.*, 2007b; Schwean-Lardner and Classen, 2010), improved walking (Sanotra *et al.*, 2002; Brickett *et al.*, 2007a; Knowles *et al.*, 2008), and increased behaviour and reduced fearfulness (Sanotra *et al.*, 2002).

Few studies have examined the effects of light intensity on broiler behaviour, particularly artificial versus natural. Blatchford *et al.* (2009) showed that broilers reared under 5 lux (16 h light/8 h dark photoperiod) were less active in the photophase and exhibited less change in activity in the scotophase compared with broilers reared under 50 or 200 lux. Other studies indicate some form of spatial or temporal distribution of light intensity may benefit welfare by giving the birds areas or periods of activity and rest. Davis *et al.* (1999) showed broilers were more active (feed, drink, scratch, forage, walk) under 200 lux than under 60, 20 or 6 lux (20 h light/4 h dark), but preferred to rest and perch under 6 lux. Additionally, activity index was higher under an alternating step-up (5 to 100 lux) programme than under constant 100 or 5 lux (16 h light/8 h dark photoperiod) and was least under an alternating step-down (100 to 5 lux) programme (Kristensen *et al.*, 2006b). While physical condition, performance (Kristensen *et al*, 2006a) and immune function (Blatchford *et al.*, 2009) were not affected in these studies, 20% of broilers under 5 lux had heavy, inflamed eyes in the latter study.

Production, health and welfare – free range and organic systems

Free range and organic systems are largely considered to have the potential to provide good living conditions and reduce environmental pollution (Sundrum, 2001). There are concerns over bird health and welfare and various production issues including product quality and consumer acceptability, particularly with regard to taste and price (Hovi *et al.*, 2003; Castellini *et al.*, 2008).

Mortality

A study of 24 free range broiler farms in the UK calculated total mortality at 2.69% (range 0.01–5.96%) with total predation accounting for 0.08% (range 0–0.8%) (Moberly *et al.*, 2004). Mortality on two sites in the UK, using small mobile arks, averaged 6.0% (range 0–38.6%) in 56 flocks of Sherwood White (as hatched) and 3.2% (range 0–7.0%) in 56 flocks of Ross 308 females (Jones *et al.*, 2007). High mortality in the Sherwood White was largely due to vent pecking in hot weather and total predation accounted for 0.24% and 0.08% mortality at sites 1 and 2, respectively.

There is concern over high mortality in free range and organic systems when using fast GR breeds; this has been shown experimentally, particularly when reared to older ages (Table 14.4). Causes of mortality are linked to the excessive BW achieved (ascities, sudden death syndrome, leg problems and heat stress are cited), and indicate that fast strains may not be suited to alternative systems with longer growth cycles.

Breed suitability

It is the ambition of the organic movement to develop sustainable and environmentally friendly farming systems allowing animals a better quality of life based on more ecocentric ethics where animal welfare is supported in the underlying philosophy (Lund, 2006). The choice of a suitable breed is therefore of primary importance, as it needs to have good health and performance under a wide range of environmental conditions and with more plant-based, less energy-dense and protein-dense diets.

There are two main global breeding companies delivering highly selected lines of fast GR broilers, and one main European company delivering a wide range of alternative growth potentials. Table 14.5 shows the range of available genetic lines, the choice of which is largely determined by market weight, age and product requirements (portioned, whole carcass, etc.). In the UK, the JA 757 and Cobb-Sasso 150 are considered suitable for extensive indoor systems and the Hubbard JA757 (intermediate GR) is commonly used for organic production. Fast GR breeds, especially the females, tend to be used for free range production. Slow GR breeds tend not to be used in the UK, largely because of the small breast conformation and longer time to reach slaughter weight. In France, however, almost 50% of chicken comes from slower breeds and one-third from Label Rouge production (Quentin *et al.*, 2005).

Growth rate of the modern broiler is phenomenal, as shown in Fig. 14.3,

Table 14.4. Experimental mortality rates of fast and slow growing broiler genotypes at typical slaughter ages.

Reference	Experimental comparisons of mortality (%)				
Castellini *et al.* (2002c)	Very slow: Robusta Maculata (81 days) 4.0[b]		Slow: Kabir (81 days) 5.0[b]	Fast: Ross 208 (81 days) 10.0[a]	
Havenstein *et al.* (2003)	ACRBC, 2001 (84 days) 4.8[b]	ACRBC, 1957 (84 days) 3.2[b]		Ross 308, 2001 (84 days) 14.3[a]	Ross 308, 1957 (84 days) 9.5[a]
Quentin *et al.* (2003)	Slow: Hubbard Label (84 days) 3.3[b]		Medium: Hubbard (56 days) 2.9[b]	Fast: Hubbard (42 days) 9.1[a]	
Fanatico *et al.* (2008)	Slow: S&G poultry (91 days) Fast: Cobb female (63 days)	Slow (In) 3.0[b] Fast (In) 11.0[a]	Slow (Out) 0[b] Fast (Out) 9.0[a]		

ACRBC, Athens Canadian Random Bred Control; In, reared indoors, Out, reared with access outdoors.
Values with unlike superscript letters were significantly different (at least *P*<0.05).

Table 14.5. The genetic potential of current broiler breeds available from three major breeding companies.

Company	System	Breed	35 days			42 days		
			EW	GR	FCR	BW	GR	FCR
Ross	Standard	308/508/PM3 Arbor Acres Lohman LIR	1918–2021	54.8–57.7	1.56–1.62	2530–2652	60.2–63.1	1.73–1.77
Cobb		500/700 Cobb-Avian-48	1933–2017	57.0–57.6	1.61–1.65	2548–2626	56.6–62.5	1.76–1.77
Hubbard		Classic, Hubbard JV, Flex, F15, Yield	1830–2003	52.3–57.2	1.54–1.60	2379–2592	56.6–61.7	1.69–1.74

Company	System	Breed	49 days			56 days			70 days		
			BW	GR	FCR	BW	GR	FCR	BW	GR	FCR
Ross	Extensive	Rowan	No data available								
Cobb	Indoor	Cobb-Sasso 150	2110	43.1	1.92	2475	44.2	2.0	3135	44.8	2.23

Company	System	Breed	56 days			63 days			70/77 days		
			BW	GR	FCR	BW	GR	FCR	BW	GR	FCR
Hubbard	Differential	Various[a]	1657–2389	29.6–42.7	2.06–2.16	2296–2697	36.4–42.8	2.21–2.31	2215–2651	31.6–37.8	2.37–2.63
	Slow	I657/S757N/ S757/S666 S86							2273	29.5	2.48–2.65

BW, body weight (g); GR, growth rate ($g\ day^{-1}$); FCR, feed conversion ratio.
For Ross breeds, see www.aviagen.com/ss/broiler-breeders
For Cobb breeds, see www.cobb-vantress.com/Products/Default.aspx
For Hubbard breeds, see www.hubbardbreeders.com
[a]Progeny from various male crosses with JA57 and Redbro (male and female line) females (examples: Gris Barre (JA) Cou Nu (+/– naked neck), JA957, JA757, New Hampshire, Master Gris, Redpac).

reaching 90 g day^{-1} at 42 days (Aviagen, 2009a). The Ross bird takes a third of the rearing time (32 days versus 105 days) and three times less feed to produce a 1815 g bird with an FCR of 1.47 compared with a non-selected random-bred 1957 bird with an FCR of 4.42 (Havenstein *et al.*, 2003). Genetic selection accounts for 85 to 90% of the progress with improvements in diet accounting for the remaining 10 to 15%. This was also illustrated by Gordon (2002) who tested the different growth potentials on commercial and Label type feeds, shown in Fig. 14.4. The very slow breeds did not meet the target weight of 2.15 kg in 81 days, but their growth potential indicted they would in 96 to 105 days. Breeds such as the Light Sussex are considered better suited to community orchard and agro-forestry systems where they show potential for reducing weeds and pests (Horsted *et al.*, 2005).

Product quality and sensory acceptability

Other than BW and size, there are considerable differences between fast and slow GR broiler breeds, which can be modified by access outdoors and pasture intake. Fast GR breeds have a greater breast percentage (Brown *et al.*, 2008) and lower wing, leg and frame percentage than slow GR breeds (Fanatico *et al.*, 2005, 2008). They also have a greater abdominal fat content (Fanatico *et al.*, 2007) since rapid growth enhances late maturing tissue (fat) (Castellini *et al.*, 2002a). Breast muscle fibres in fast GR breeds have a larger diameter (Berri *et al.*, 2005), and the large breast is correlated to low activity (Gordon, 2002) and alterations in walking style (Corr *et al.*, 2003).

Slow GR strains have higher protein content in the muscle (Berri *et al.*, 2005; Fanatico *et al.*, 2007) and a greater yellow colour to the skin. They

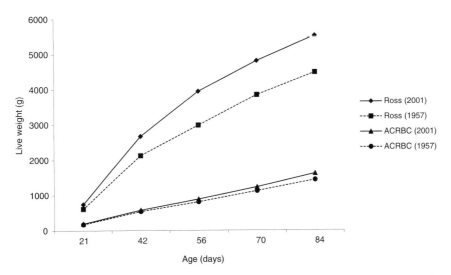

Fig. 14.3. The growth rate of the modern broiler (Ross 308) compared with a 1957 random-bred breed (Athens-Canadian Randombred Control, ACRBC) fed commercial diets that were typical in 1957 and 2001 (from Havenstein *et al.*, 2003).

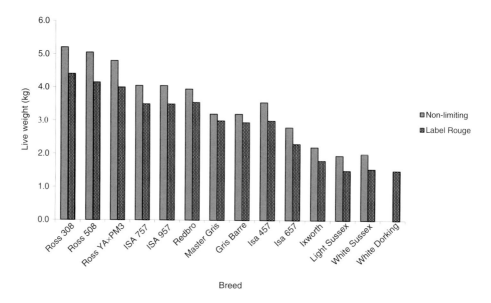

Fig. 14.4. Body weight at 81 days of various breeds of meat chickens fed commercial non-limiting and Label-type diets (Gordon, 2002).

have a faster rate of post-mortem pH decline, resulting in a lower ultimate pH (causing shrinkage of the contractile fibres so reducing water-binding ability and giving the meat a more yellow appearance), reduced water-holding capacity and greater drip loss (Castellini et al. 2002a, 2008; Quentin et al., 2003; Berri et al., 2005, Fanatico et al., 2007; Brown et al., 2008). Slow GR strains are therefore better suited to the whole carcass market rather than the portioned market (Berri et al., 2005).

Access outdoors increased the leg and reduced the wing percentage in slow GR breeds (Fanatico et al., 2008) and improved tibia strength in fast GR breeds (Fanatico et al., 2005). It increased protein and reduced fat in the breast of both GR types (Castellini et al , 2002b; Fanatico et al., 2007), and increased the yellow colour and pH of slow breeds (Fanatico et al, 2007). Access outdoors led to smaller muscle cross-sectional area and therefore lower muscle mass of the breast (to reduce the risk of insufficient diffusion of oxygen and metabolites) and larger thigh muscle cross-sectional area (fibre hypertrophy due to increased activity) in differential GR meat type strains (Polak et al., 2010). Semi-confined conditions suited two differential GR strains better than confined conditions, improving BW and FCR (Santos et al., 2005).

Pasture contains bioactive compounds (xanthophils, hypocholesterolenic, anticarcinogens) which, if ingested, are of potential benefit to chickens, but the high fibre may reduce feed efficiency and lower GR by reducing the passage time of feed through the digestive system. Ponte et al. (2008a) found that access to clover pasture increased final BW, carcass yield and breast yield of Redbro males (56 days), but did not improve FCR. There was no effect of pasture on tenderness, juiciness and flavour, but the overall acceptability (by a trained panel) of clover fed birds was raised compared with standard birds. The authors estimated the grass biomass in the crop at slaughter represented 2.5–4.5% of dry matter (18–26% fresh basis), which could be increased with feed restriction (Ponte *et al.*, 2008b). Feed restriction reduced BW and worsened FCR, but the additional clover intake had a major effect on the fatty acid profile of the meat, significantly increasing the *n*–3 polyunsaturated fatty acids, which play a large role in human and bird health.

Organic chicken is generally thought to be better flavoured; however consumer acceptance is difficult to judge. Trained panels were able to differentiate between standard, free range, corn-fed and organic chicken mostly on appearance and texture rather than odour and flavour (Lawlor *et al.*, 2003; Jahan *et al.*, 2005). Standard chicken tended to be preferred (Lawlor *et al.*, 2003, Fanatico *et al.*, 2006; Brown *et al.*, 2008), although results were variable, while chicken from slow breeds was perceived to be drier (Berri *et al.*, 2005). The yellow colour of corn-fed chicken was perceived to be bad in Northern Ireland (Kennedy *et al.*, 2005).

Protein and nutrient balance

There is a need to improve the knowledge of nutrient requirements in diverse environments and diverse breeds (Hovi *et al.*, 2003; see also MacLeod and Bentley, Chapter 15, this volume). Modern poultry nutrition is highly developed; additives give cost-effective improvements in carbohydrate and mineral digestion and synthetic amino acids give correct protein balance according to growth and maintenance demands of the breed. Synthetic amino acids and additives are not permissible in organic rations. Energy requirements are met by oil, fat (usually soybean oil) and cereal (usually wheat in the UK), and the primary source of protein is soybean (Walker and Gordon, 2003). As broilers move from starter to finisher rations (usually in several phases) there is typically an increase in energy and decrease in protein content. Table 14.6 provides a comparison of standard and alternative system feed specifications. Organic rations provide high crude protein (CP) levels in an attempt to drive protein synthesis, while low specification rations provide a balanced diet required for slow growth.

Birds do not require CP per se but require sufficient nitrogen from protein to synthesize dispensible (non-essential) amino acids. Increasing CP levels without balancing indispensible (essential) amino acids, those the body is unable to synthesize, does not lead to better utilization of the protein; this is only as good as the first limiting amino acid, usually lysine and methionine. Since plant-based ingredients are relatively low in indispensible amino acids, fish meal is included as a source in the non-organic portion of organic rations (currently 5% is permissible). Synthetic amino acids are allowed in Label diets.

The addition of methionine to a semi-organic ration improved breast yield and reduced abdominal fat in a dual-purpose breed, with concomitant improvements in BW and FCR, indicating better utilization of the protein (Koreleski and Świątkiewicz, 2008). BW also increased linearly with increasing lysine (Quentin *et al.*, 2005; Plumstead *et al.*, 2007; Mushtaq *et al.*, 2009) in

Table 14.6. Examples of different feed rations according to production system.

	Standard specification (BOCM Pauls Layer)		Organic specification (UK) (Vitrition)		Low specification (UK) (BOCM Pauls Farmgate)	
	Starter	Finisher	Starter	Finisher	Starter	Finisher
Metabolizable energy (MJ kg⁻¹)	(13.40)[a]	(12.85)	12.15	12.12	(12.30)	(12.50)
Crude protein (%)	20.5	19.5	25.0	18.0	18.0	14.5
Oil (%)	5.0	5.0	3.54	3.22	3.5	3.75
Fibre (%)	3.5	3.5	4.12	4.06	5.0	3.5
Lysine (%)	na	na	1.24	0.89	na	na
Methionine (%)	0.55	0.56	na	na	0.35	0.25
Methionine + cystine (%)			0.93	0.67	na	na
Wheat (%)	40	40	na	(70)	40	40

na, not available.

[a]Values in parentheses are estimates.

the diet. Lysine deficiency in Label males is of great concern, as it leads to feather pecking and cannibalism; lysine requirements of 0.76% (as opposed to 0.68%) were recommended (Quentin *et al.*, 2005).

Diets low in energy tend to increase food intake (Quentin *et al.*, 2003; Brickett *et al.*, 2007b) as birds eat to energy value; they then attain the same BW with a poorer FCR. This may not be the case however when diets contain the same fat levels (Plumsted *et al.*, 2007). Diets low in energy and protein reduce BW and breast percentage in fast growing breeds (Gordon, 2002; Fanatico *et al.*, 2008), reduce fat levels (Fanatico *et al.*, 2007) and increase activity levels (T. Jones, unpublished results).

Rations are largely pelleted which increases food intake and BW and improves FCR in fast but not slow GR broilers. Additionally, increasing fine particles in the feed linearly reduces food intake and worsens FCR but the slope is five times greater in fast than slow breeds (Quentin *et al.*, 2004), indicating selection has limited the ability of the chicken to adapt to feed characteristics.

Sustainable cost of production

Consumers will pay for legislative improvements in animal welfare, especially for eggs (Bennett, 1996) where consumers are better informed about initiatives related to welfare and there is no substitute for the product (Vanhonacker and Verbeke, 2009). Apparent willingness to pay however, does not always translate for meat products (Castellini *et al.*, 2008), which are more expensive than eggs and may be double the price of standard produced chicken, as shown in Fig. 14.5. Price differential is a major barrier to purchasing high-welfare chicken meat with consumers primarily choosing on sell by date and appearance as a

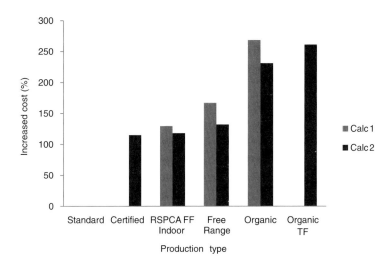

Fig. 14.5. The relative increased cost of producing broilers from alternative systems (FF, Freedom Food; TF, Total Freedom) compared with standard production: Calc 1, calculation 1 (P. Cook, unpublished data); Calc 2, calculation 2 (from Hubbard, 2008).

mark of quality and freshness (Hall and Sandilands, 2007). For consistent pro-welfare purchasing, price is often substituted for reduced quantity and there is a strong connection between bird welfare, human health, and product taste and quality (Vanhonacker and Verbeke, 2009); such consumers also believe they can make a difference to welfare through their purchasing choices. Most consumers believe better education of the issues and more informative labelling would make a positive impact on purchasing choices (Hall and Sandilands, 2007; Vanhonacker and Verbeke, 2009).

On one hand, consumers require evidence of genuine higher standards (Hovi *et al.*, 2003) and need to be convinced by product quality (Castellini *et al.*, 2008); on the other hand, producers need to be paid a fair sustainable price, which allows future investment in the enterprise. The retailer dictates both ends of this pricing structure and it is unclear whether a sustainable price is paid for the product. Indications from standard production are that producers are in fact losing money on each bird they rear. In April 2006 in the UK, the cost of production in standard systems was 54.69 p kg^{-1} live weight, while producers received 48.5 p kg^{-1}; there was then a call for an additional 12 p kg^{-1} to sustain the industry (NFU, 2006). Typical costs in May 2008 (NFU, 2008) were 69.88 p kg^{-1} live weight with producers receiving 68.08 p kg^{-1}. Similar data are not available for free range and organic chicken.

Health

The effects of alternative systems on disease and health of poultry, and the risk of human pathogens, are reviewed respectively by Lister and van Nijhuis (Chapter 4) and Van Hoorebeke *et al.* (Chapter 5, this volume).

Flock contamination with *Campylobacter* is the major concern for free range and organic systems as incidence levels are higher than from indoor systems (Heuer *et al.*, 2001; Avrain *et al.*, 2003). Broilers test positive for *Campylobacter* between 2 and 4 weeks of age (Herman *et al.*, 2003; El-Shibiny *et al.*, 2005), just before (Rivoal *et al.*, 2005; Huneau-Salaün *et al.*, 2007) or immediately after they go out on to the range (Colles *et al.*, 2008a). Since *Campylobacter* spp. are ubiquitous in the environment around chicken sheds (Herman *et al.*, 2003) and in the gastrointestinal tract and faeces of many wild animals and birds, exposure to the outdoor environment is often considered the cause of infection. There is little evidence to support this however (Newell and Fearnley, 2003), since the genotype of *Campylobacter* spp. found in rodents (Meerburg *et al.*, 2006) and wild birds on the same site (Colles *et al.*, 2008b) are different from those isolated from the chickens. In addition, there was no succession of *C. jejuni* genotypes between successive free range flocks and there was no association of ranging behaviour with *Campylobacter* shedding (Colles *et al.*, 2008a).

Standard biosecurity measures are considered to reduce the risk of *Campylobacter* colonization (Evans and Sayer, 2000; Newell and Fearnley, 2003; Huneau-Salaün *et al.*, 2007), but are limited because of the different physiology, epidemiology and ecology of the *Campylobacter* organism (Newell and Fearnley, 2003). Increasing temperatures (Heuer *et al.*, 2001; Huneau-

Salaün *et al.*, 2007), short turnaround times between flocks (Newell and Fearnley, 2003), moving birds between brood and rearing houses (Colles *et al.*, 2008a), thinning (Humphrey, 2006), faecal contamination of transport crates (Herman *et al.*, 2003) and lairage (associated with a prolonged period with no feed and water) (McCrea *et al.*, 2006a) are all associated with increased rates of colonization and shedding. Recent evidence suggests that neuro-transmitters (particularly noradrenaline), which are elevated in stressed and diseased animals, increase *C. jejuni* growth and motility (Humphrey, 2006).

Welfare

Access outdoors allows for foraging and exploration, and increases the range of environments, food sources and activity, creating the potential for improved welfare. Welfare criticism (excluding health) centres on the lack of ranging observed in flocks outdoors and the higher incidence of FPD.

Activity and ranging behaviour

Activity in fast GR broiler breeds drops off at 2–3 weeks, when they are able to thermoregulate body temperature physiologically (Rovee-Collier *et al.*, 1993) and energy is partitioned more to production than activity. Slow GR breeds are more active, performing more walking, perching and pecking behaviours than fast GR breeds (Castellini *et al.*, 2002c; Bokkers and Koene, 2003). Access outdoors can increase the activity of fast GR birds. Bird activity increased by a factor of 1.8 and rest was reduced by 0.8 when given outdoor access (to 81 days) (Castellini *et al.*, 2002b), and the percentage of time spent standing walking and pecking was significantly higher when outdoors than indoors (Jones *et al.*, 2007). On average birds walked for 98 strides per walking bout when outdoors compared with 7.2 strides per bout indoors (Jones *et al.*, 2007). Access outdoors did not affect the behaviour of fast GR birds when walking ability was poor (Weeks *et al.*, 2000).

Ranging in fast GR breeds, however, is generally low. Studies of large and small commercial flocks found on average 14% (Dawkins *et al.*, 2003) and 11% (range 0.2–51.4%) (Jones *et al.*, 2007) of the birds outside near the end of the growth cycle, respectively. Figure 14.6 shows that slow GR breeds range better than fast, and both range more when fed a diet of moderate rather than low energy (Nielsen *et al.*, 2003); slow breeds also spend more of their time outdoors (60% for Kabir versus 35% for Ross 208) (Castellini *et al.*, 2002c). Chickens have a diurnal rhythm to their ranging, with more birds outside in the morning and before dusk (Dawkins *et al.*, 2003; Nielsen *et al.*, 2003; Jones *et al.*, 2007). They tend to stay close to the house (Weeks *et al.*, 1994; Christensen *et al.*, 2003) and ranging behaviour, in terms of percentage outside and distance covered, increases with age (Mirabito and Lubac 2001; Mirabito *et al.*, 2001; Jones *et al.*, 2007).

Weather and the suitability of the outdoor environment greatly affect ranging behaviour. Chickens range more in summer (Jones *et al.*, 2007) and are negatively affected by low temperatures, wind and rain (Gordon and Forbes,

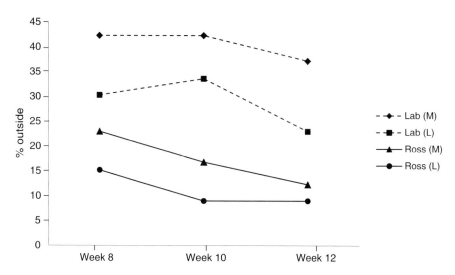

Fig. 14.6. The percentage of Labresse cross (Lab) and Ross 208 birds observed outside when fed diets with moderate (M) and low (L) energy (from Nielsen *et al.*, 2003).

2002). They prefer overcast days, and bush and tree cover over short grass (Dawkins *et al.*, 2003), although conifer wigwams were also found to be very attractive (Gordon and Forbes, 2002). Artificial shelter and outdoor drinkers and dust bathing areas are required by assurance schemes in an attempt to make the outdoor environment more attractive to the birds (Table 14.2). Tree provision further enhanced the outdoor environment and increased the ranging of Label birds over ryegrass provision at 11 weeks (77.2% versus 49.0%, respectively) (Mirabito *et al.*, 2001), while trees in their third year of growth increased the ranging of Ross broilers at 56 days over grass provision (22.4% (range 6.0–40.8%) versus 16.3% (range 3.1–30.5%), respectively) (Jones *et al.*, 2007).

Foot pad dermatitis

Wet litter, stocking density, feed and genetic line have all been associated with foot pad lesions and swelling which develop from the mechanistic raising and destruction of epithelial cells (e.g. Martland, 1985). Lesions are at least uncomfortable and most likely painful; they may also be a gateway for bacteria leading to walking and carcass impairment. Organic systems have been criticized for higher rates of FPD.

In one study, the incidence and severity of FPD was significantly worse in broilers from organic (98.1% incidence, 89.0% severe) than free range systems (32.8% incidence, 13.0% severe) (Pagazaurtundua and Warriss, 2006). The authors found no difference between RSPCA FF indoor, standard or corn-fed chickens (incidence 9.6%, 14.8% and 19.0%, respectively, with low rates of severe cases at 1–4%). In another study, slow GR broilers (Labresse cross) had significantly less FPD than fast GR females (Ross 208), particularly when fed a low-energy diet (2.3% versus 78.5%, respectively) (Nielsen *et al.*, 2003). High-protein and/or high-soybean meal can lead to high FPD when litter is good, while all-vegetable diets lead to the excretion of wet faeces; both aid litter deterioration (van der Sluis, 2010). The author also reported that the addition of complex zinc in the diet reduced the prevalence and severity of foot lesions.

Walking ability

Differences in the gaits of 13 genotypes of broilers with a wide range of growth profiles were entirely due to live weight and GR (Kestin *et al.*, 2001). Table 14.7 shows the incidence of gait scores found in free range and organic studies. As well as showing the poorer walking of fast GR breeds, the studies also highlighted the negative effect of average weekly temperatures <18°C (Jones *et al.*, 2007) and the positive effect of the outdoor environment (when night temperature was maintained above 15°C) on moderate gait problems (Fanatico *et al.*, 2008).

Breast blisters

The incidence of breast blisters in organic broilers can be high. An average incidence of 7% (range 1–17%) was found in Denmark (Fisker, 1999 cited by Nielsen, 2004), which was determined more by strain and sex differences than by perching behaviour (Nielsen, 2004). There was a higher incidence of breast blisters in Labresse birds than IS/A657 and in males more than females.

Table 14.7. The incidence (%) of broilers with different gait scores in three free range/organic studies.

| Reference | Slow breed | | | | Fast breed | | | | | |
	Weight (g) (age)	GS=0	GS=1	GS=2	Weight (g) (age)	GS=0	GS=1	GS=2	GS=3	GS=4
Nielsen *et al.* (2003)	2698 (84 days)	78	21	1	3882 (84 days)	0	13	60	25	2
Jones *et al.* (2007)					2000 (52 days)	68.6	25.6 (0–56.7)[a]		5.8 (0–20.0)	
Fanatico *et al.* (2008)	In: 2105 (91 days)	100	0	0	In: 3389 (63 days)	2.9	8.6	77.1	10	1.4
	Out: 2254 (91 days)	100	0	0	Out: 3370 (63 days)	8.3	30.6	52.8	8.3	0

GS, gait score; In, reared indoors; Out, reared with access outdoors.
[a]Value in parentheses is range.

MEAT TURKEYS

Production systems, legislation and assurance schemes

Alternative systems exist for both indoor and outdoor access turkeys. Legislative and assurance scheme requirements are given in Table 14.8 for indoor systems and in Table 14.9 for free range and organic systems. Quality British Turkey (QBT, 2009) is the industry Red Tractor Farm Assurance Scheme adopted for standard production in the UK. Stocking density is calculated from the equation $A = 0.0459W^{2/3}$, where A is the area required for a given final body weight, W, of turkey. RSPCA FF Indoor and Barn turkey standards reduce the permissible stocking density to 25 kg m^{-2}, and barn turkeys must not be sent for slaughter before 70 days of age. RSPCA FF requires natural daylight from January 2012, stipulates environmental enrichment (straw bales, rope, perches, objects) and requires turkeys with GS=3 and above to be culled.

Outdoor access systems stipulate a minimum slaughter age of between 70 and 140 days, with organic schemes reducing indoor maximum stocking density to as low as 16 kg m^{-2}. Schemes require outdoor access for half to two-thirds of the turkey's life at a rate of 4–12.5 m^2 per bird. Slow growing breeds are required, but not defined, by some standards but not all, and flock sizes are limited under organic regulations in an attempt to minimize nitrogen impact on the land.

Around 10% of UK turkey production comes from semi-intensive pole barn, free range or organic systems (Defra, 2007). This includes 1 14 million turkeys from seasonal or traditional farm fresh producers for the Christmas market; approximately 6.7% of total production (17 million turkeys) (BPC, 2006). In France, Label accounts for only 1% of turkey production; standard and certified account for 88% and 10%, respectively. There were 150,000 organic turkeys reared in Germany in 2005 (Statista, 2010) and 305,000 turkeys in 2009 (MEG, 2011), representing 1.41 and 2.48% of total German turkey production, respectively.

Safeguards for welfare

Turkeys will benefit from limits set on maximum stocking density, breeding for improved leg health and low pecking behaviour, and through the provision of natural light and environmental enrichment.

There are no legislative limits to the maximum stocking density permitted for turkeys reared in standard indoor systems. Limits to stocking density recommended by FAWC (1995) are followed by the Red Tractor Scheme in the UK (QBT, 2009); some voluntary country codes also set maximum levels, for example in Germany the maximum density for males and females is 52 and 58 kg m^{-2}, respectively (BML, 1999). Where there is no limit turkeys may be reared to much higher densities. There were no differences in FPD, breast blisters and feather pecking levels between commercial flocks at high densities

Table 14.8. European Union legislative and assurance scheme (UK and France) requirements for standard and alternative indoor turkey production systems.

Criteria	Legislation	Assurance Scheme		
	Barn (Commission Regulation, 2008)	Quality British Turkey UK (QBT, 2009)	RSPCA Freedom Food Indoor UK (RSPCA, 2010b)	Certified France (PMAF, 2010)
Min. age (days)	70	–	–	70
Max. area (birds m^{-2})		10 kg: 4.69 20 kg: 2.95		Variable
Max. SD (kg m^{-2})	25	10 kg: 46.9 20 kg: 59.1	25 (no thinning)	
Breed		No limitations	No limitations	Intermediate
Beak trimming		Permitted	Only in controlled housing on advice of vet	
Feed				100% vegetal
Light		Min. 10 lux Min. 8 h continuous dark or natural dark period in naturally lit houses	Min 12 h light Min. 8 h continuous dark, except naturally lit systems Av. illumination of at least 20 lux over at least half the floor area (incl. feeders and drinkers) No area <6 lux Gradual on/off light over 30 min period Natural light must be provided via openings min. 3% total floor area From 1 January 2012. natural daylight to be provided in all systems from 35 days; if pecking a problem entire building to be lit with UV light Shutters must be provided, opening min. 0.56 m^2	
Ventilation		Min./max. temperatures recorded daily	50–70% RH Not exceed: 5 ppm NH$_3$/5000 ppm CO$_2$/10 mg dust m^{-3}/50 ppm	

Table 14.8. – Continued

Criteria	Legislation		Assurance Scheme	
	Barn (Commission Regulation, 2008)	Quality British Turkey UK (QBT, 2009)	RSPCA Freedom Food Indoor UK (RSPCA, 2010b)	Certified France (PMAF, 2010)
Enrichment		Recommend: – perches – straw bales – pecking objects – suspended vegetables	For every 500 birds: – min. 2 m perch – 1 large, 2 small straw bales – 2 lengths rope Recommend: – additional 40 cm perch per bird – objects (brassicas, CDs, plastic bottles)	
Outcome measures		Record culls & mortality on daily basis; DOA & post-mortem inspection; if DOA exceeds 0.25% written report required	Record: – daily mortality, culls & reason for cull – FPD – breast blisters – back scratches – dirty feathers – transport deaths & injuries Inspection required if mortality 0.5% in 24 h	
Other			Thinning prohibited Birds with GS=3 or above must be humanely killed Management must prevent chronic joint disease or leg deformation Visual barriers to escape from others	

Min., minimum; max., maximum; SD, stocking density; DOA, dead on arrival; av., average; UV, ultraviolet; RH, relative humidity; FPD, foot pad dermatitis; GS, gait score; –, no standard requirement for this criterion.

Table 14.9. European Union legislative and assurance scheme requirements for standard and alternative outdoor turkey production in the UK, France and Germany.

Criteria	Legislation		Assurance Scheme			
	EU free range (Commission Regulation, 2008)	EU traditional free range (Commission Regulation, 2008)	RSPCA FF Free Range UK (RSPCA, 2010b)	Label Rouge France (LabelRouge, 2010)	Organic Soil Association, UK (Soil Association, 2010)	High Welfare & Organic Germany[a]
Min. age (days)	70	Whole: 140 Cut up: 126 (M) 98 (F)	70	126 (M) 98 (F) 140 (Christmas market)	Organic parents: – slow GR: any age – fast GR: 140 Non-organic parents: – slow GR: 70 – fast GR: 140	140 Bioland: 140 (M)/100 (F) Naturland: 140
Max. area (birds m^{-2})	6.25 (up to 7 weeks, 10 birds)		No thinning – restricts no. of birds	6.25 (up to 7 weeks, 10 birds)	No thinning – restricts no. of birds	Neuland: 10 Naturland, Bioland: – fixed: 10 – mobile: 16
Max. SD (kg m^{-2})	25		25	21	21	Neuland, Bioland: – fixed: 21 – mobile: 30 Demeter: – fixed: 16 – mobile: 18
Min. outdoor access	1/2 of life	>8 weeks continuous daytime access	Min. 1/2 of life Daily access >8 h	From 7 weeks	2/3 of life	Neuland: from 3 weeks Demeter: by 50 days Bioland: 1/3 of life
Min. outdoor space (m^2 per bird)	4	6	4	6	12.5 (800 birds ha^{-1})	Demeter, Neuland: 10 Naturland, Bioland: – fixed: 10 – mobile: 2.5

Table 14.9. – Continued

	Legislation		Assurance Scheme			
Criteria	EU free range (Commission Regulation, 2008)	EU traditional free range (Commission Regulation, 2008)	RSPCA FF Free Range UK (RSPCA, 2010b)	Label Rouge France (LabelRouge, 2010)	Organic Soil Association, UK (Soil Association, 2010)	High Welfare & Organic Germany[a]
Breed	No restrictions	Slow	No restrictions	Hardy, slow GR (black feathers required)	Slow GR defined as <35 g day⁻¹ over growth cycle, or max.: – male <105 g day⁻¹ – female <75 g day⁻¹ at any time	Slow (Bronze turkeys or females of white lines, common)
Beak trimming	–	–	Permitted in naturally lit systems	Prohibited	Prohibited	Prohibited (Bioland, Demeter, Naturland)
Feed	Min. 70% cereals			Min. 75% cereals or cereal-based products from 64 days	Fishmeal allowed No synthetic AA Synthetic vitamins permissible	
Light			As RSPCA FF Indoor		Artificial lighting may be used to extend day length to max. of 16 h, must end with dusk	Bioland: 5% window area
Ventilation			As RSPCA FF Indoor			
Enrichment			As RSPCA FF Indoor Min. 10 m² natural or artificial overhead shade per 1000 birds	Vegetation on 2/3 of outdoor area	Access to shelter at all times, protection from predators; natural and artificial cover to encourage ranging Woodland recommended	Combinations of winter garden (roofed shelter), veranda 1/3 useable space of indoor barn, straw bales, perches, dust bathing, depending on scheme

Note: In the Breed row, Organic Soil Association defines Slow GR as <35 g day⁻¹.

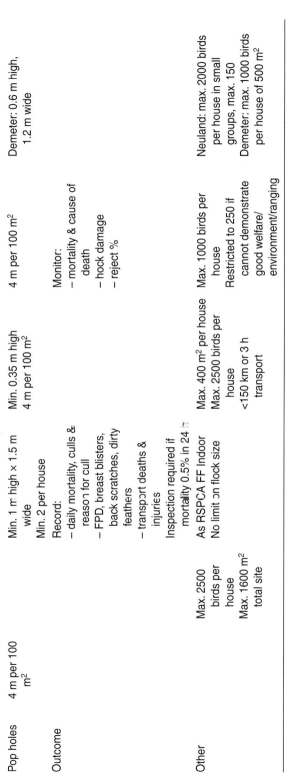

Pop holes	4 m per 100 m²	Min. 1 m high × 1.5 m wide Min. 2 per house	Min. 0.35 m high 4 m per 100 m²	4 m per 100 m²	Demeter: 0.6 m high, 1.2 m wide
Outcome		Record: – daily mortality, culls & reason for cull – FPD, breast blisters, back scratches, dirty feathers – transport deaths & injuries Inspection required if mortality 0.5% in 24 h		Monitor: – mortality & cause of death – hock damage – reject %	
Other	Max. 2500 birds per house Max. 1600 m² total site	As RSPCA FF Indoor No limit on flock size	Max. 400 m² per house Max. 2500 birds per house <150 km or 3 h transport	Max. 1000 birds per house Restricted to 250 if cannot demonstrate good welfare/environment/ranging	Neuland: max. 2000 birds per house in small groups, max. 150 Demeter: max. 1000 birds per house of 500 m²

FF, Freedom Food; min., minimum; max., maximum; SD, stocking density; M, male; F, female; FPD, foot pad dermatitis; GR, growth rate; AA amino acids; –, no standard requirement for this criterion.

aBioland (2010); Demeter (2009); Naturland (2010); Neuland (2008).

(8.5 and 7 birds m^{-2}, equivalent to 69.1 and 62.2 kg m^{-2}) (Mirabito *et al.*, 2002); significant effects of density have been shown however when investigating a wider range. Martrenchar (1999) found higher rates of bird disturbance, hip and foot pad lesions at stocking densities of 8 birds m^{-2} (equivalent to commercial levels and calculated at 61 kg m^{-2}) than at 6.5 birds m^{-2} (the free range level calculated at 52 kg m^{-2}) and 5 birds m^{-2} (recommended by FAWC, 1995; calculated as 40 kg m^{-2}). Additionally and similar to broilers (Dawkins *et al.*, 2004), gait deteriorated and growth rate was reduced as stocking density increased, particularly at the higher level.

Weight gain is rapid in modern turkey breeds, and issues arise with leg disorders. Fattening turkeys reduce the time they put weight on their legs, which may be associated with pain (Buchwalder and Huber-Eicher, 2004). Turkeys are more responsive to their environment and are more socially interactive than broiler chickens (Huff *et al.*, 2007); they therefore lie inactive for much less time than broilers, 40–60% (Martrenchar, 1999; Mirabito *et al.*, 2002) versus 80–85%, but are prone unfortunately to injurious pecking. Despite genetic selection for faster growth they are able to maintain higher activity levels, and since their feed intake per kilogram gain has not increased significantly (Ferket, 2002 cited by Flock *et al.*, 2006) high rates of growth in the parents (except for the fastest GR males) are largely controlled by diet shifting through a series of rations with differing CP levels (Aviagen, 2009b). Effective breeding strategies that include selection for leg health and walking ability, and reduced injurious pecking behaviour (Bentley, 2002, cited by Flock *et al.*, 2006), will benefit turkey welfare.

Turkeys prefer fluorescent over incandescent lighting; it emits an ultraviolet (UV) spectrum, following daylight more closely, and elicits dustbathing and preening behaviours (Sherwin, 1999). Turkeys are sensitive to UV-A, which enhances the ability to see seeds, berries, insects, etc., as well as visual markings in feathers for social recognition (Barber *et al.*, 2006); the unnatural appearance of markings under conventional lighting is thought to attract injurious pecking (Sherwin and Devereux, 1999). UV supplementation plus regular straw provision and visual barriers significantly reduced culling due to injurious pecking (Lewis *et al.*, 2000). Turkeys prefer to conduct all behaviours except resting and perching under higher light levels (20 to 200 lux) (Barber *et al.*, 2004); low light level is often used to control pecking behaviour, but also results in abnormal eye development (Thomson and Forbes, 1999 cited by Barber *et al.*, 2004).

Production, health and welfare – free range and organic systems

Mortality

Typically turkey hens and toms in Europe experience 3.5–5.0% and 8.0–12.0% mortality, respectively (Flock *et al.*, 2006; Damme, 2010), with higher rates recorded in the USA (Flock *et al.*, 2006). There was no difference in the mortality of non-beak trimmed free range BUT 6 and Kelly Bronze turkeys (Platz *et al.*, 2003). Intermediate and fast GR strains did however exhibit higher mortality than slow strains (3.1 and 4.6% versus 0%, respectively) with no effect of production system (mortality in barn and free range systems was ~2.4%) (Sarica *et al.*, 2009). Although not significantly different, mortality of BUT 6 turkeys in houses with veranda access ranged from 5.1 to 7.0% and in conventional houses ranged from 7.0 to 13.4% (Berk *et al.*, 2004), whereas for a slow GR strain only, mortality was significantly less in an enriched system (1.6%, in houses with elevated platforms, veranda access and outdoor run) than in a conventional system (5.8%) (Berk and Cottin, 2004).

Breed suitability

Examples of available genetic lines of turkeys are given in Table 14.10. Heavy strains tend to supply the portioned and processed markets, while slower strains are used as whole birds and heavy to medium strains service both markets. Selection for increased GR and reduced FCR over the 37 years from 1966 led to GR doubling, and improvements in feed conversion and breast yield of 50% and 5.4%, respectively, when fed the same diet (Havenstein *et al.*, 2007); interestingly increased GR did not lead to increased mortality. Slow GR breeds were however better able to cope with challenges (Huff *et al.*, 2007), showing little behavioural response to transport (they continued to eat/drink, etc. whereas fast GR strains remained prostrate and did not eat or drink for several hours) and better immune status following a disease challenge.

Product quality and sensory acceptability

There was no difference in carcass composition of different strains of fast GR turkeys (BUT 6, Hybrid Convertor and Nicholas 700), with less than 1% fat in all strains (Roberson *et al.*, 2003). Unlike broilers, there were no clear differences between the meat quality from fast (BUT 6 and Kelly Broad Breasted Bronze) and slow strains (Kelly Wrolstad and Kelly Super Mini) (Werner *et al.*, 2008); slow GR strains had more protein in the breast muscle, but similar fat and ash, and fast GR strains had larger muscle fibre diameter and higher shear force. Differences between barn and free range production systems for slow (Bronze), intermediate (Hybrid×Bronze) and fast GR (Hybrid) turkeys to 21 weeks were also negligible (Sarica *et al.*, 2009); there was however less carcass yield and abdominal fat outdoors, also shown by Le Bris (2005). Recently however Sarica *et al.* (2011) showed that outdoor access gave the breast muscle of slow, intermediate and fast GR turkeys a redder colour and higher protein content; differences in the colour, water-holding capacity and pH of the breast and thigh muscles between the genotypes were found, indicating that producers should choose the genotype appropriate for their production system (BW, age at slaughter, market goals, etc.).

Protein and nutrient balance

Nutrient requirements for turkeys are given precisely by the breeding companies (Aviagen, 2009b). Turkey poults have high amino acid requirements to meet

Table 14.10. The genetic potential of current turkey breeds (males and females) from breeding company targets (upper table) and cross-breeds (lower table) with regard to body weight (BW) and feed conversion ratio (FCR).

Company	Breed	Type	Males BW (kg)	Males FCR	Females BW (kg)	Females FCR
BUT (Avaigen)[a]	Big 6, 9, 10; T9, T8	Heavy to Medium heavy	15.30–12.59[c] 22.80–18.60[d]	2.17–2.18 2.69–2.71	7.32–6.03[f] 10.74–8.49[c]	2.09–2.12 2.45–2.52
Nicholas (Aviagen)[a]	N700 & N300	Heavy to Medium heavy	17.85–15.37[e] 22.45[d]	2.35–2.60 2.75	8.31–8.14[g] 9.68–9.39[c]	2.12–2.26 2.29–2.43
Hybrid (Hendrix Genetics)[b]	XL, Converter, Grade Maker	Heavy to Medium heavy	15.87–13.86[c] 23.14–17.86[d]	2.03–2.22 2.66–2.64	7.81–6.89[f] 10.88–9.46[c]	1.94–2.02 2.29–2.43

From Meier (2010) Breed	Colour	Males BW (kg) 140 days	Males % of Big 6	Females BW (kg) 112 days	Females % of Big 6
Big 6	White	20.39		10.74	
Big 9	White	19.45	95	10.25	95
BUT 9	White	17.54	86	9.12	85
BUT 8	White	16.65	82	8.49	79
N30 x B6FLX	Black/bronze	13.83	68	7.43	69
N30 x B5FLX	Black/bronze	11.99	59	6.44	60
A30 x 5FLX	Red/brown	11.99	59	6.44	60
N30 x T5FLX	Black/bronze	10.80	53	5.81	54
Kelly	Black/bronze/white	10.00	49	5.10	47

[a]BUT and Nicholas, see http://www.aviagen.com/ss/turkeys
[b]Hybrid, see http://www.hybridturkeys.com
[c]112 days (16 weeks).
[d]154 days (22 weeks).
[e]126 days (18 weeks).
[f]84 days (12 weeks).
[g]98 days (14 weeks).

the demands of their rapid growth, requiring 0.55% methionine and 1.05% sulfur amino acids from 0 to 4 weeks (Fanatico, 2010). Feeding sufficient indispensible amino acids in the start phase is a major problem in organic systems (Bellof and Schmidt, 2007). Both BW and mortality were negatively affected by nutritional deficiencies in a home mixed ration: male Bronze turkeys on two organic rearing sites over a 3-year period (one with home mixed and one with bought-in feed) averaged 9.9 and 15.4 kg (at 20 weeks) with mortality rates of 16.0 and 2.2%, respectively (Golze *et al.*, 2009).

Sustainable cost of production

Largely, cost of production figures are difficult to find. Defra (2007) estimated the UK cost of production for year-round BUT T8 female (20 weeks) and male (24 weeks) turkeys as £9.15 and £15.01 per bird, respectively, compared with seasonal turkeys from pole barns at £14.22 per bird (various white breeds) and free range bronze turkeys as £17.17 per bird (24 weeks). The figures equate to a 14% increase from standard male to free range production. For a similar time period, German estimates for organic turkey production were 29.5€ per bird (Deerberg, 2007), over 70% more expensive than standard production. Theuvsen *et al.* (2005) estimated the cost of adding perches (for 40% of the flock), reducing stocking density (to 36.5 kg m^{-2}), providing an outdoor climate (veranda) and free range conditions to be 0.012, 0.08, 0.03 and 0.35€ kg^{-1} slaughter weight, respectively. German consumer conjoint analysis by the same authors indicated consumers were prepared to pay 0.2€ kg^{-1} for perches, 1.17€ kg^{-1} for reduced stocking density and 2.63€ kg^{-1} for free range, suggesting no gap between ethics and economics. However, survey answers did not translate into willingness to pay, as commercial ventures which produced turkeys at reduced stocking densities and increased retail price by 20%, failed. Securing the market for higher-welfare higher turkey meat at an elevated sales price can pay dividends however: Heritage turkey production in the USA generated US$4.80 per bird more income, despite a higher cost of production to 26–28 weeks (14 lb/6.36 kg slaughter weight) compared with white broad-breasted strains at 16–18 weeks (16 lb/7.27 kg) (Born *et al.*, 2007).

Health

The effects of alternative systems on disease and health of poultry are reviewed by Lister and van Nijhuis (Chapter 4, this volume).

Activity and ranging behaviour

Environmental enrichment in the form of raised platforms and outdoor access via verandas (roofed outside runs with mesh sides) reduced mortality and improved health and carcass quality in turkeys without reducing production performance (Berk, 2000, 2001). Behavioural expression in the veranda was enhanced, with turkeys performing more wing flapping, wing stretching, dust bathing, aggression, ground pecking and feather pecking (Berk *et al.*, 2004)

(Fig. 14.7). Veldkamp and Kiezebrink (2006) also found higher rates of feather pecking in the veranda (5.0 pecks every 30 min versus only 0.3 pecks every 30 min indoors), and significantly higher rates of feed intake in veranda systems for comparable BW (385 g day^{-1} versus 360 g day^{-1} for indoors).

There is evidence that fast and intermediate breeds do not range as well as slow (Sarica *et al.*, 2007, cited by Sarica *et al.*, 2009). In winter, 56.2% of Kelly Bronze turkeys were observed outside on the range compared with 35.7% of BUT 6, which tended to stay close to the house (Bergmann, 2006); Bronze turkeys were more active on the range with fewer observed to rest on the range (11.7% versus 19.0% of BUT 6). American Bronze and Californian white turkeys raised on pasture, with wheat provided *ad libitum* in the house when the birds returned off pasture, were observed to range at frequencies of between 60 and 74% when scan sampled between 10 am and 1 pm (Karabayir *et al.*, 2008); they ranged most frequently in cooler weather.

Beak trimming and feather pecking

Beak trimming of turkeys is conducted by various methods to avoid damage caused by feather pecking and cannibalism. The procedure can be traumatic and the age at trimming affects the pain duration and healing level of the beak (Hughes and Gentle, 1995). When turkeys were beak trimmed by various methods at a young age (1 or 21 days of age) there was no evidence of neuroma formation at 42 days (Gentle *et al.*, 1995); the Bio-beaker (which passes an electrical current through the premaxilla) led to most tissue damage, while secateurs led to most precision but least denervated area. The heated blade debeaker led to additional tissue damage (Gentle *et al.*, 1995) which is likely to cause chronic pain (Mench and Siegel, 1997), while the arc trimmer resulted in a slight modification of behaviour, but showed no lasting effects (Noble *et al.*, 1996). Since 2000, the infrared method is generally applied in hatcheries to

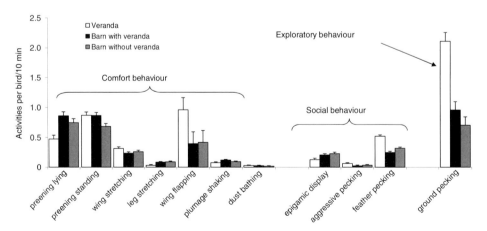

Fig. 14.7. Behaviour of tom turkeys in barns with or without veranda access, as well as in the veranda itself. Values are means with their standard errors represented by vertical bars (Berk *et al.*, 2004).

trim the upper beak, but the lower beak tip is damaged including the bill tip organ (Fiedler and König, 2006). Beak trimming is generally not permitted in organic flocks and is also not accepted as a routine procedure in many commercial flocks in Europe.

Factors affecting the development of feather pecking and cannibalism are multifactorial, such as breed, feed composition, rearing environment and other housing and management factors. Light, especially light intensity and light source, is very important. Turkeys can be reared in standard housing without beak trimming, where control of pecking behaviour (and general suppression of behaviour) can be limited via reducing light levels when necessary. However, controlling light intensity in pole barns, verandas and free range systems is almost impossible. Selection against feather pecking may be possible (Martrenchar, 1999), and redirecting the pecking behaviour in non-trimmed turkeys with appropriate environmental enrichment (Lewis *et al.*, 2000) or a free range environment (Platz *et al.*, 2003) have proved effective. Production system did not affect the incidence of skin injuries in 16-week-old males (12.8%) or females (13.8%) reared on 24 commercial farms (Mitterer-Istyagin *et al.*, 2011); injuries occurred mainly on the head region, snood and back.

Foot pad dermatitis

Turkeys with severe FPD have slower weight gain and flocks with high FPD incidence often show a high prevalence of other types of contact dermatitis (Gonder and Barnes, 1987). Severity of FPD tends to increase with age and females were found to be affected more than males (60% versus 33%) (Krautwald-Junghanns *et al.*, 2011). Wet litter is a recognized major cause of FPD in turkeys (Martland, 1984). Lesion prevalence was higher on straw litter than wood shavings, and with bell drinkers rather than cups, while the addition of dry litter reduced the prevalence of FPD in turkeys (Ekstrand and Algers, 1997). Lignocellulose as a litter substrate was found to be an effective absorber of water leading to low levels of FPD (Berk, 2009), and litter moisture should be maintained at 30% or less for good foot condition (Youssef *et al.*, 2009; Wu and Hocking, 2011).

In white turkey poults FPD has been associated with methionine deficiency (Chavez and Kratzer, 1972, 1974) and breed effects have been shown; free range Heritage Bourbon Red turkeys had fewer FPD lesions at 4 and 17 weeks than a commercial broad-breasted white breed (McCrea *et al.*, 2006b).

Walking ability

Over 5% of all commercial turkeys exhibit leg disorders (Ferket, 2009), and almost 50% of total flock mortality in toms can occur in the last 3–5 weeks of fattening due to leg problems (Powell, 2007). Leg disorders are mostly accompanied with reduced growth, a predisposition to cannibalism and carcass downgrades at processing (Ferket, 2009). Causes can be infectious (viral and bacterial agents) and non-infectious (genetic, nutrition, management), with the incidence of leg problems significantly higher in poorly ventilated houses, with poor litter quality, high dust concentration and poorly adjusted drinkers and feeders (Hafez, 2001). Woodward (2004) categorized leg disorders in turkeys as those associated with nutritional deficiencies, poor litter or high GR, while Powell (2007) and Ferket (2009) categorized leg problems into three critical periods: week 1, influenced by maternal and incubation effects; weeks 2–10, induced by malabsorption; and weeks 10–18, related to high GR. Controlling early growth by feed (diets low in energy/protein) or light restriction programmes can reduce the incidence of leg disorders, often at the expense of meat yield (Powell, 2007; Ferket, 2009); inadequate dietary levels of minerals, vitamins and protein should be avoided as well as enteric diseases that contribute to poor nutrient absorption (Ferket, 2009).

A study comparing Bronze and BUT 6 turkeys with free range access showed that reduced walking ability was significantly higher in BUT 6 turkeys (56.3% versus 35.3% in Bronze), as was incorrect leg position (87.5% versus 67.7% in Bronze) (Bergmann, 2006). However a previous study found no significant difference between the two breeds with regard to reduced walking ability (5.2% Kelly Bronze and 3.2% Big 6) under outdoor rearing conditions at the end of fattening (Le Bris, 2005).

Breast blisters

Breast blisters in turkeys are associated with high BW and poor feathering at 8 weeks, as well as cool temperatures, coarse rather than fine sawdust and use of newspaper as litter (Newberry, 1993). Breast blisters and buttons in free range trials occurred in summer only, when growth rate was highest, and occurred at higher levels in BUT 6 turkeys (37.8% blisters, 16.1% buttons) than in Kelly Bronze (14.8% blisters, 33.3% buttons) (Bergmann, 2006).

Perching is performed by high BW turkeys and is often considered to be a cause of breast blisters and buttons. Nicholas N700 turkeys perched more frequently and had higher rates of breast blisters (34.9%) than BUT 6 turkeys (12.3%), both medium to heavy breeds (Berk, 2000). Perch design is therefore of great importance; raised platforms with a ramp or bales of straw are therefore recommended for perching (Berk, 2003; Spindler and Hartung, 2007; Letzguß and Bessei, 2009). Finally, the number and severity of breast blisters and buttons in turkeys with veranda access were lower than in indoor systems (Veldkamp and Kiezebrink, 2006).

CONCLUSIONS

Alternative systems have the potential to deliver good production, health and welfare to broilers and turkeys, with successful delivery dependent on key factors such as breed suitability, quality of the environment and provision of a balanced ration (particularly in relation to organic diets). In order to be sustainable, meat from chickens and turkeys reared in alternative systems must be acceptable to consumers in terms of both product quality and price. From a human health perspective, further work is required to understand and reduce the rates of *Campylobacter* infection in all systems.

Breed suitability is important for both broilers and turkeys, as the birds must be able to cope with a wide range of environmental conditions (especially temperature fluctuations and wet weather) and maintain health over a longer growth period. Slower growing breeds are often used and generally have lower rates of mortality, foot pad lesions and hock burn than fast growing breeds. Slow growing breeds also have better walking ability and are more active, thus able to conduct more natural behaviours and range further and for longer periods than fast breeds. Lack of activity and poor walking ability in relation to growth rate is particularly problematic for broiler chickens, and lack of outdoor ranging is a criticism of free range broiler systems.

The quality of the outdoor environment is important in encouraging ranging behaviour. Natural shade and shelter in the form of hedges and trees are particularly effective at encouraging broilers out of the house throughout the day. Environmental enrichment in the form of natural light, straw bales and pecking objects increases activity in extensive indoor broiler systems and reduces injurious feather pecking in turkeys that have not been beak trimmed. Breast blisters are problematic in the slowest broiler breeds which have very narrow breasts, and further work on perch design to alleviate these problems is needed. Despite their size, turkeys are more active than broiler chickens and perform high levels of perching behaviour, and raised platforms are effective at eliminating breast blisters in heavy broad-breasted breeds under all production systems.

Dry litter is essential for healthy feet in all production systems. Maintenance of litter quality is difficult in wet weather conditions particularly for free range and organic systems, and is further exacerbated by diet in organic systems. The balance of organic rations needs to be improved for both broilers and turkeys, especially in relation to indispensible amino acids and trace elements (such as zinc), which are required for efficient utilization of the diet and the reduction of foot pad lesions. High incidence of foot lesions is a particular criticism of organic systems, although it can occur in all systems.

Outdoor access reduces fat and improves protein and n–3 polyunsaturated fatty acids in the meat, improving the nutritional quality of the product. For broilers however, the meat from slower breeds appears to be drier in texture and may be better suited to the whole bird market. Cost of production in alternative systems is higher than in conventional, and the consumer must be willing to pay the difference if the market is to be sustainable. Better labelling, consumer education and a fair distribution of cost throughout the supply chain are also needed.

REFERENCES

ACP (2010a) Poultry Standards – Broilers and Poussin. Assured Chicken Production, Red Tractor Farm Assurance, Assured Food Standards. http://www.assuredchicken.org.uk/resources/000/471/416/Poultry_Standards_BroilersPoussin_web.pdf (accessed 1 August 2010).

ACP (2010b) Poultry Standards – Free Range. Assured Chicken Production, Red Tractor Farm Assurance, Assured Food Standards. http://www.assuredchicken.org.uk/resources/000/471/417/Poultry_Standards_Free_Range_web.pdf (accessed 1 September 2010).

Appleby, M.C., Hughes, B.O. and Savory, C.J. (1994) Current state of poultry welfare: progress, problems and strategies. *British Poultry Science* 35, 467–475.

Aviagen (2009a) Ross 308 Broiler Performance Objectives. http://en.aviagen.com/assets/Tech_Center/Ross_Broiler/Ross_308_Broiler_Performance_Objectives.pdf (accessed 1 September 2010).

Aviagen (2009b) Feeding programmes for BUT breeding turkeys – Key points. http://en.aviagen.com/assets/Tech_Center/Turkeys_breeders/ATL/nutrition/ATL_Breeder_Nutrition_0048.pdf (accessed 1 September 2010).

Avrain, L., Humbert, F., L'Hospitalier, R., Sanders, P., Vernozy-Rozand, C. and Kempfa, I. (2003) Antimicrobial resistance in *Campylobacter* from broilers: association with production type and antimicrobial use. *Veterinary Microbiology* 96, 267–276.

AVEC (2010) Association of Poultry Processors and Poultry Trade in the EU Countries – Annual Report 2010. http://www.avec-poultry.eu/Default.aspx?ID=6379 (accessed 1 September 2010).

Barber, C.L., Prescott, N.B., Wathes, C.M., Le Sueur, C. and Perry, G.C. (2004) Preferences of growing ducklings and turkeys poults for illuminance. *Animal Welfare* 13, 211–244.

Barber, C.L., Prescott, N.B., Jarvis, J.R., Le Sueur, C., Perry, G.C. and Wathes, C.M. (2006) Comparative study of the photopic spectral sensitivity of domestic ducks (*Anas platyrhynchos domesticus*), turkeys (*Meleagris gallopavo gallopavo*) and humans. *British Poultry Science* 47, 365–374.

Bellof, G. and Schmidt, E. (2007) Ökologische Geflügelmast Lösungsmöglichkeiten für eine 100% Bio-Fütterung. http://orgprints.org/13840 (accessed 1 October 2010).

Bennett, R.M. (1996) Willingness-to-pay measures of public support for farm animal welfare legislation. *The Veterinary Record* 139, 320–321.

Bergmann, S.M. (2006) Vergleichende Untersuchung von Mastputenhybriden (B.U.T. Big 6) und einer Robustrasse (Kelly Bronze) bezüglich Verhalten, Gesundheit und Leistung in Freilandhaltung. Inaugural dissertation, Tierärztlichen Fakultät der Ludwig-Maximilians-Universität München, Germany.

Berk, J. (2000) Modified husbandry, productivity, health and animal behaviour. In: Hafez, H.M. (ed.) *Proceedings of International Meeting of the Working Group 10 (Turkey) on Turkey Production in Europe in the New Millennium*, Berlin, Germany, 24–25 November 2000. Eugen Ulmer, Stuttgart, Germany, pp. 103–109.

Berk, J. (2001) The enrichment of environment – a possible way to improve the health and welfare of tom turkeys? In: Oester, H. and Wyss, C. (eds) *Proceedings of the 6th European Symposium on Poultry Welfare*, Zollikofen, Switzerland, 1–4 September 2001. Swiss Branch of the World's Poultry Science Association, pp. 278–280.

Berk, J. (2003) Can alternative housing systems improve the performance and health of tom turkeys? In: Hafez, H.M. (ed.) *Turkey Production: Balance Act Between Consumer Protection, Animal Welfare and Economic Aspects*. Ulmer Verlag, Stuttgart, Germany, pp. 103–114.

Berk, J. (2009) Effects of different types of litter on performance and pododermatitis in male turkeys. In: Hafez, H.M. (ed.) *Turkey Production: Toward Better Welfare and Health. Proceedings of the 5th International Symposium on Turkey Production Meeting of the Working Group 10 (Turkey)*, 28–30 May 2009. Mensch und Buch Verlag, Berlin, pp. 29–30

Berk, J. and Cottin, E. (2004) Final Report: The roles of selection and husbandry in the development of locomotory dysfunction in turkeys. QLRT-1999-01549. http://ec.europa.eu/research/agriculture/projects/qlrt_1999_01549_en.htm (accessed 24 October 2011).

Berk, J., Wartemann, S., Hinz, T. and Linke, S. (2004) Einsatz eines Außenklimabereiches in der Putenmast als Möglichkeit der Strukturierung der Haltungsumwelt zur Verbesserung der Tiergesundheit, des Wohlbefindens und der Ökonomie unter Beachtung umweltrelevanter Aspekte. *Abschlußbericht zum FuE-Vorhaben 99UM019.* Institut für Tierschutz und Tierhaltung, Celle, Germany.

Berri, C., Le Bihan-Duvala, E., Baézaa, E., Chartrina, P., Picgirardb, L., Jehlc, N., Quentina, M., Picarda, M. and Duclosa, M.J. (2005) Further processing characteristics of breast and leg meat from fast-, medium- and slow-growing commercial chickens. *Animal Research* 54, 123–134.

Bioland (2010) Bioland-Richtlinien: 4 Tierhaltung. http://www.bioland.de/bioland/richtlinien.html (accessed 1 March 2011).

Blatchford, R.A., Klasing, K.C., Shivaprasad, H.L., Wakenell, P.S., Archer, G.S. and Mench, J.A. (2009) The effect of light intensity on the behavior, eye and leg health, and immune function of broiler chickens. *Poultry Science* 88, 20–28.

BML (1999) Bundeseinheitliche Eckwerte für eine freiwillige Vereinbarung zur Haltung von Jungmasthühnern (Broiler, Masthähnchen) und Mastputen. BML – 321-3545/2; 4157/3659 (23 September 1999). Bundesministerium für Landwirtschaft und Ernährung, Bonn, Germany.

Bokkers, E.A.M. and Koene, P. (2003) Behaviour of fast and slow growing broilers to 12 weeks of age and the physical consequences. *Applied Animal Behaviour Science* 81, 59–72.

Born, H., Bender, M. and Beranger, J. (2007) Economics of Heritage turkey production on range. In: *How to Raise Heritage Turkeys on Pasture.* American Livestock Breeds Conservancy, Pittsboro, North Carolina, pp. 103–107.

BPC (2006) Annual turkey slaughter information calculated from: Defra UK Poultry Slaughterings monthly dataset at http://www.defra.gov.uk/statistics/foodfarm/food/poultry.

Bradshaw, R.H., Kirkden, R.D. and Broom, D.M. (2002) A review of the aetiology and pathology of leg weakness in broilers in relation to their welfare. *Avian and Poultry Biology Reviews* 13, 45–103.

Brickett, K.E., Dahiya, J.P., Classen, H.L., Annett, C.B. and Gomis, G. (2007a) The impact of nutrient density, feed form, and photoperiod on the walking ability and skeletal quality of broiler chickens. *Poultry Science* 86, 2117–2125.

Brickett, K.E., Dahiya, J.P., Classen, H.L. and Gomis, G. (2007b) Influence of dietary nutrient density, feed form, and lighting on growth and meat yield of broiler chickens. *Poultry Science* 86, 2172–2181.

Brown, S.N., Nute, G.R., Baker, A., Hughes, S.I. and Warriss, P.D. (2008) Aspects of meat and eating quality of broiler chickens reared under standard, maize-fed, free-range or organic systems. *British Poultry Science* 49, 118–124.

Buchwalder, T. and Huber-Eicher, B. (2004) Effect of increased floor space on aggressive behaviour in male turkeys (*Meleagris gallopavo*). *Applied Animal Behaviour Science* 89, 207–214.

Castellini, C., Mugnai, C. and Dal Bosco, A. (2002a) Meat quality of three chicken genotypes reared according to the organic system. *Italian Journal of Food Science* 14, 401–412.

Castellini, C., Mugnai, C. and Dal Bosco, A. (2002b) Effect of organic production system on broiler carcass and meat quality. *Meat Science* 60, 219–225.

Castellini, C., Dal Bosco, A., Mugnai, C. and Bernardini, M. (2002c) Performance and behaviour of chickens with different growing rate reared according to the organic system. *Italian Journal of Animal Science* 1, 291–300.

Castellini, C., Berri, C., Le Bihan-Duval, E. and Martino, G. (2008) Qualitative attributes and consumer perception of organic and free-range poultry meat. *World's Poultry Science Journal* 64, 500–512.

CCP (2010) Produit Certifié – Filière Viande – Volaille. http://www.produitcertifie.fr/index. php?option=com_content&view=article&id=74 (accessed 1 January 2011).

Chavez, E. and Kratzer, F.H. (1972) Prevention of foot pad dermatitis in poults with methionine. *Poultry Science* 51, 1545–1548.

Chavez, E. and Kratzer, F.H. (1974) Effect of diet on of foot pad dermatitis in poults. *Poultry Science* 53, 755–760.

Christensen, J.W., Nielsen, B.L., Young, J.F. and Noddergaard, F. (2003) Effects of calcium deficiency in broilers on the use of outdoor areas, foraging activity and production parameters. *Applied Animal Behaviour Science* 83, 229–240.

Classen, H.L., Riddell, C. and Robinson, F.E. (1991) Effects of increasing photoperiod length on performance and health of broiler chickens. *British Poultry Science* 32, 21–29.

Colles, F.M., Jones, T.A., McCarthy, N.D., Sheppard, S.K., Cody, A.J., Dingle, K.E., Dawkins, M.S. and Maiden, C.J. (2008a) *Campylobacter* infection in broiler chickens in a free-range environment. *Environmental Biology* 10, 2042–2050.

Colles, F.M., Dingle, K.E., Cody, A.J. and Maiden, M.C.J. (2008b) Comparison of *Campylobacter* populations in wild geese with those in starlings and free-range poultry on the same farm. *Applied and Environmental Microbiology* 74, 3583–3590.

Commission Regulation (2008) Commission Regulation (EC) No 543/2008 of 16 June 2008 laying down detailed rules for the application of Council Regulation (EC) 1234/2007 as regards the marketing standards for poultry meat. *Official Journal of the European Union* L 157, 17/06/2008, pp. 46–87; available at: http://www.fsai.ie/uploadedFiles/ Legislation/Legislation_Update/Reg543_2008.pdf (accessed 1 August 2010).

Cooper, M.D., Allanson-Bailey, S., Gauthier, R. and Wrathall, J. (2008) Higher welfare standards and broiler welfare. *World Poultry* 18, 20–21.

Corr, S.A., Gentle, M.J., McCorquodale, C.C. and Bennett, D. (2003) The effect of morphology on walking ability in the modern broiler: a gait analysis. *Animal Welfare* 12, 159–171.

Council Directive (2007) Council Directive 2007/43/EC of 28 June 2008 laying down minimum rules for the protection of chickens kept for meat production. *Official Journal of the European Union* L 182, 12/07/2008, pp. 19–28; available at: http://eur-lex. europa.eu/LexUriServ/LexUriServ.do?uri=OJ:L:2007:182:0019:0028:EN:PDF (accessed 1 August 2010).

Council Regulation (1999) Council Regulation EC No 1804/1999 of 19 July 1999 supplementing Regulation (EEC) No 2092/91 on organic production of agricultural products and indications referring thereto on agricultural products and foodstuffs to include livestock production. Official Journal of the European Communities L 222, 24/08/1999, pp. 1–28; available at: http://eur-lex.europa.eu/LexUriServ/LexUriServ.do?uri=OJ:L:199 9:222:0001:0028:EN:PDF (accessed 1 November 2010).

Damme, K. (2010) Faustzahlen zur Betriebswirtschaft. In: *Geflügeljahrbuch 2011*. Ulmer Verlag, Stuttgart, Germany, p. 70.

Davis, N.J., Prescott, N.B., Savory, C.J. and Wathes, C.M. (1999) Preferences of growing fowls for different light intensities in relation to age, strain and behaviour. *Animal Welfare* 8, 193–203.

Dawkins, M.S., Cook, P., Whittingham, M., Mansell, K., and Harper, A. (2003) What makes free-range broilers range? *In-situ* measurement of habitat preference. *Animal Behaviour* 66, 342–344.

Dawkins, M.S., Donnelly, C.A. and Jones, T.A. (2004) Chicken welfare is influenced more by housing conditions than by stocking density. *Nature* 427, 342–344.

Dawkins, M.S., Lee, H., Waitt, C.D. and Roberts, S.J. (2009) Optical flow patterns in broiler chicken flocks as automated measures of behaviour and gait. *Applied Animal Behaviour Science* 119, 203–209.

Deerberg, F. (2007) Aufbau eines bundesweiten Berater – Praxisnetzwerkes zum Wissensaustausch und Methodenabgleich für die Bereiche Betriebsvergleich (BV) und Betriebszweigauswertung (BZA) Arbeitskreis 4: Geflügel. http://orgprints.org/ 13358/5/13358-03OE495-soel-zerger-2007-BPN_AK4_Gefluegel.pdf (accessed 1 October 2010).

Defra (2007) The UK turkey and geese production industry: a short study. http://archive.defra. gov.uk/foodfarm/food/industry/sectors/eggspoultry/documents/turkey-geese-report.pdf (accessed 1 October 2010).

Defra (2010) Implementing Council Directive 2007/43 in England. http://archive.defra.gov. uk/foodfarm/farmanimal/welfare/onfarm/meatchks.htm (accessed 1 December 2010).

Demeter (2010) Demeter-Richtlinie Erzeugung: VII.2: I. Weisungen für die Geflügelhaltung. http://www.demeter.de/ebenenangleichung/zielgruppe/schnittmengen/leben-arbeiten/ das-ist-bio-dynamische-landwirtschaft/richtlinien-erzeugung/?MP=13-1491 EZ_12-10_ VII.2_Weisungen_GeflÃ¼gelhaltung[2].pdf (accessed 1 December 2010).

Ekstrand, C. and Algers, B. (1997) Rearing conditions and foot-pad dermatitis in Swedish turkey poults. *Acta Veterinaria Scandinavia* 38, 167–174.

El-Shibiny, A., Connerton, P.L. and Connerton, I.F (2005) Enumeration and diversity of campylobacters and bacteriophages isolated during the rearing cycles of free-range and organic chickens. *Applied and Environmental Microbiology* 71, 1259–1266.

Estevez, I. (2007) Density allowances for broilers: where to set the limits? *Poultry Science* 86, 1265–1272.

Evans, S.J. and Sayers, A.R. (2000) A longitudinal study of *Campylobacter* infection of broiler flocks in Great Britain. *Preventive Veterinary Medicine* 46, 209–223.

Fanatico, A. (2010) Organic Poultry Production: Providing Adequate Methionine. https://attra. ncat.org/attra-pub/summaries/summary.php?pub=336 (accessed 1 August 2010).

Fanatico, A. and Born, H. (2002) Label Rouge: pasture based poultry production in France. https://attra.ncat.org/attra-pub/summaries/summary.php?pub=224 (accessed 1 May 2010).

Fanatico, A.C., Pillai, P.B., Cavitt, L.C., Owens, C.M. and Emmert, J.L. (2005) Evaluation of slower-growing broiler genotypes grown with and without outdoor access: growth performance and carcass yield. *Poultry Science* 84, 1321–1327.

Fanatico, A.C., Pillai, P.B., Cavitt, L.C., Emmert, J.L., Meullenet, J.F. and Owens, C.M. (2006) Evaluation of slower-growing broiler genotypes grown with and without outdoor access: sensory attributes. *Poultry Science* 85, 337–343.

Fanatico, A.C., Pillai, P.B., Emmert, J.L. and Owens, C.M. (2007) Meat quality of slow- and fast-growing chicken genotypes fed low-nutrient or standard diets and raised indoors or with outdoor access. *Poultry Science* 86, 2245–2255.

Fanatico, A.C., Pillai, P.B., Hester, P.Y., Falcone, C., Mench, J.A., Owens, C.M. and Emmert, J.L. (2008) Performance, livability, and carcass yield of slow- and fast-growing chicken genotypes fed low-nutrient or standard diets and raised indoors or with outdoor access. *Poultry Science* 87, 1012–1021.

FAWC (1995) *Report on the Welfare of Turkeys*. PB 2033. Farm Animal Welfare Council, London; available at: http://www.fawc.org.uk/reports/turkeys/turkrtoc.htm (accessed 1 September 2010).

Ferket, P. (2009) Leg problems in turkeys. In: Hafez, H.M. (ed.) *Turkey Production: Toward Better Welfare and Health*. Mensch und Buch Verlag, Berlin, pp. 113–122.

Fiedler, H.H. and König, K. (2006) Assessment of beak trimming in day-old turkey chicks by infrared irradiation in view of animal welfare. *Archiv für Geflügelkunde* 70, 241–249.

Flock, D.K., Laughlin, K.F. and Bentley, J. (2006) Minimising losses in poultry breeding and production: how breeding companies contribute to poultry welfare. *Lohmann Information* 41, 20–28.

Gentle, M.J., Thorp, B.H. and Hughes, B.O. (1995) Anatomical consequences of partial beak amputation (beak trimming) in turkeys. *Research in Veterinary Science* 58, 158–162.

Golze, M., Wehlitz, R. and Jäckel, U. (2009) Öko-Puten Ökologische Putenerzeugung: Aufzucht und Produktqualität. *Rundschau für Fleischhygiene und Lebensmittelüberwachung* 8, 314–317.

Gonder, E. and Barnes, H.J. (1987) Focal ulcerative dermatitis ('breast buttons') in marketed turkeys. *Avian Diseases* 31, 52–58.

Gordon, S.H. (2002) Effect of breed suitability, system design and management on welfare and performance of traditional and organic table birds. Defra Project OF0153. http://www.orgprints.org/8104/01/OF0153_2552_FRP.pdf (accessed 1 July 2010).

Gordon, S.H and Forbes, M.J. (2002) Management factors affecting the use of pasture by table chickens in extensive production systems. In: Powell, J., *et al.* (eds) *Proceedings of the UK Organic Research 2002 Conference*, Organic Centre Wales, Institute of Rural Studies, University of Wales Aberystwyth, 26–28 March 2002, pp. 269–272; available at: http://orgprints.org/8257 (accessed 1 July 2010).

Hafez, H.M. (2001) Turkey health disorders: causes and control approaches. In: Hafez, H.M. (ed.) *Turkey Production in Europe in the New Millennium*. Ulmer Verlag, Stuttgart, Germany, pp. 137–156.

Hall, C. and Sandilands, V. (2007) Public attitudes to the welfare of broiler chickens. *Animal Welfare* 16, 499–512.

Havenstein, G.B., Ferket, P.R. and Qureshi, M.A. (2003) Growth, livability, and feed conversion of 1957 versus 2001 broilers when fed representative 1957 and 2001 broiler diets. *Poultry Science* 82, 1500–1508.

Havenstein, G.B., Ferket, P.R., Grimes, J.L., Qureshi, M.A. and Nestor, K.E. (2007) Comparison of the performance of 1966- versus 2003-type turkeys when fed representative 1966 and 2003 turkey diets: growth rate, liveability, and feed conversion. *Poultry Science* 86, 232–240.

Herman, L., Heyndrickx, M., Grijspeerdt, K., Vandekerchove, D., Rollier, I. and De Zutter, L. (2003) Routes for *Campylobacter* contamination of poultry meat: epidemiological study from hatchery to slaughterhouse. *Epidemiology and Infection* 131, 1169–1180.

Heuer, O.E., Pedersen, K., Andersen, J.S. and Madsen, M. (2001) Prevalence and antimicrobial susceptibility of thermophilic *Campylobacter* in organic and conventional broiler flocks. *Letters in Applied Microbiology* 33, 269–274.

Horsted, K., Henning, J. and Hermansen, J.E. (2005) Growth and sensory characteristics of organically reared broilers differing in strain, sex and age at slaughter. *Acta Agriculturae Scandinavica, Section A – Animal Science* 55, 149–157.

Hovi, M., Sundrum, A. and Thamsborg, S.M. (2003) Animal health and welfare in organic livestock production in Europe: current state and future challenges. *Livestock Production Science* 80, 41–53.

Hubbard (2008) French broiler market and French and UK Quality Products. Industry presentation by Claude Toudic, Hubbard Headquarters, Quintin, Brittany, France, 22nd September 2008.

Huff, G., Huff, W., Rath, N., Donoghue, A., Anthony, N. and Nestor, K. (2007) Differential effects of sex and genetics on behavior and stress response of turkeys. *Poultry Science* 86, 1294–1303.

Hughes, B.O. and Gentle, M.J. (1995) Beak trimming of poultry: its implications for welfare. *World's Poultry Science Journal* 51, 51–61.

Humphrey, T (2006) Are happy chickens safer chickens? Poultry welfare and disease susceptibility. *British Poultry Science* 47, 379–391.

Huneau-Salaün, A., Denis, M., Balaine, L. and Salvat, G. (2007) Risk factors for *Campylobacter* spp. colonization in French free-range broiler-chicken flocks at the end of the indoor rearing period. *Preventive Veterinary Medicine* 80, 34–48.

Jahan, K., Paterson, A. and Piggott, J.R. (2005) Sensory quality in retailed organic, free range and corn-fed chicken breast. *Food Research International* 38, 495–503.

Jones, T.A., Donnelly, C.A. and Dawkins, M.S. (2005) Environmental and management factors affecting the welfare of chickens on commercial farms in the UK and Denmark stocked at five densities. *Poultry Science* 84, 1155–1165.

Jones, T.A., Feber, R., Hemery, G., Cook, P., James, K., Lamberth, C. and Dawkins, M.S. (2007) Welfare and environmental benefits of integrating commercially viable free-range broiler chickens into newly planted woodland: a UK case study. *Agricultural Systems* 94, 177–188.

Karabayir, A., Tolu, C. and Ersoy, E. (2008) Some behavioural traits of American Bronze and White (California) turkeys grazing on pasture. *Journal of Animal and Veterinary Advances* 7, 1113–1116.

Kennedy, O.B., Stewart-Knox, B.J., Mitchell, P.C. and Thurnham, D.I. (2005) Flesh colour dominates consumer preference for chicken. *Appetite* 44, 181–186.

Kestin, S.C. and Knowles, T.G. (2004) Estimating the number of broilers to sample to determine the prevalence of lameness. In: Weeks, C.A. and Butterworth, A. (eds) *Measuring and Auditing Broiler Welfare*. CABI Publishing, Wallingford, UK, pp. 295–298.

Kestin, S.C., Knowles, T.G., Tinch, A.E. and Gregory, N.G. (1992) Prevalence of leg weakness in broiler chickens and its relationship with genotype. *The Veterinary Record* 131, 190–194.

Kestin, S.C., Gordon, S., Su, G. and Sorensen, P. (2001) Relationships in chickens between lameness, liveweight, growth rate and age. *The Veterinary Record* 148, 195–197.

Knowles, T.G., Kestin, S.C., Haslam, S.M., Brown, L.E., Butterworth, A., Pope, S.J., Pfeiffer, D. and Nicol, C.J. (2008) Leg disorders in broiler chickens: prevalence, risk factors and prevention. *PLoS ONE* 3(2), e1545; doi:10.1371/journal.pone.0001545.

Koreleski, J. and Świątkiewicz, S. (2008) Effect of protein and methionine levels in a semiorganic diet for dual-purpose type chickens on slaughter performance and nitrogen balance. *Journal of Animal and Feed Sciences* 17, 381–391.

Krautwald-Junghanns, M.-E., Ellerich, R., Mitterer-Istyagin, H., Ludewig, M., Fehlhaber, K., Schuster, E., Berk, J., Petermann, S. and Bartels, T. (2011) Examinations on the prevalence of footpad lesions and breast skin lesions in British United Turkeys Big 6 fattening turkeys in Germany. Part I: Prevalence of footpad lesions. *Poultry Science* 90, 555–560.

Kristensen, H.H., Aerts, J.M., Leroy, T., Wathes, C.M., Berckmans, D. (2006a) Modelling the dynamic activity of broiler chickens in response to step-wise changes in light intensity. *Applied Animal Behaviour Science* 101, 125–143.

Kristensen, H.H., Perry, G.C., Prescott, N.B., Ladewig, J., Ersbøll, A.K. and Wathes, C.M. (2006b) Leg health and performance of broiler chickens reared in different light environments. *British Poultry Science* 47, 257–263.

LabelRouge (2010) Label-Rouge-Broad range of Label Rouge poultry – Turkey cuts. http://www.poultrylabelrouge.com//014_differentes_volailles_dinde.php (accessed 1 July 2010).

Lawlor, J.B., Sheehan, E.M., Delahunty, C.M., Kerry, J.P. and Morrissey, P.A. (2003) Sensory characteristics and consumer preference for cooked chicken breasts from organic, corn-fed, free-range and conventionally reared animals. *International Journal of Poultry Science* 2, 409–416.

Le Bris, J. (2005) Gesundheit, Leistung und Verhalten konventioneller Mastputenhybriden unter den Bedingungen ökologischer Haltungsanforderungen. Inaugural dissertation, Tierärztliche Fakultät der Ludwig-Maximilians-Universität München, Germany.

Leterrier, C., Rose, N., Constantin, P. and Nys, Y. (1998) Reducing growth rate of broiler chickens with a low energy diet does not improve cortical bone quality. *British Poultry Science* 39, 24–30.

Letzguß, H. and Bessei, W. (2007) Tiergerechte Mastputenhaltung mit Beschäftigungs- und Strukturelementen. *Versuchsbericht KTBL-Projekt „Landwirtschaftliches Bauen 2005– 2007"*. Kuratorium für Technik und Bauwesen in der Landwirtschaft, Stuttgart, Germany.

Letzguß, H. and Bessei, W. (2009) Effects of environmental enrichment on the locomotor activity of turkeys. *World's Poultry Science Journal, Book of Abstracts, 8th European Symposium on Poultry Welfare*, Cervia, Italy, 18–22 May 2009, p. 46.

Lewis, P.D., Perry, G.C., Sherwin, C.M. and Moinard, C. (2000) Effect of ultraviolet radiation on the performance of intact male turkeys. *Poultry Science* 79, 850–855.

Lund, V. (2006) Natural living – a precondition for animal welfare in organic farming. *Livestock Science* 100, 71– 83.

Malleau, A.E., Duncan, I.J.H., Widowski, T.M. and Atkinson, J.L. (2007) The importance of rest in young domestic fowl. *Applied Animal Behaviour Science* 106, 52–69.

Martland, M.F. (1984) Wet litter as a cause of plantar pododermatitis, leading to foot ulceration and lameness in fattening turkeys. *Avian Pathology* 13, 241–252.

Martland, M.F. (1985) Ulcerative dermatitis in broiler chickens: the effects of wet litter. *Avian Pathology* 14, 353–364.

Martrenchar, A. (1999) Animal welfare and intensive production of turkey broilers. *World's Poultry Science Journal* 55, 143–152.

Mascetti, G.G., Bobbo, D., Rugger, M. and Vallortigara, G. (2004) Monocular sleep in male domestic chicks. *Behavioural Brain Research* 153, 447–452.

McCrea, B.A., Tonooka, K.H., VanWorth, C., Boggs, C.L., Atwill, E.R. and Schrader, J.S. (2006a) Prevalence of *Campylobacter* and *Salmonella* species on farm, after transport, and at processing in specialty market poultry. *Poultry Science* 85, 136–143.

McCrea, B., Leslie, M., Stevesen, L., Macklin, K., Bauermeister, L. and Hess, J. (2006b) Foot pad lesions, pasture condition, and bacterial pathogens in free range Heritage vs. commercial turkey varieties. *Poultry Science* 85, 24.

McKay, J.C. (2009) The genetics of modern commercial poultry. In: Hocking, P.M. (ed.) *Biology of Breeding Poultry.* CABI Publishing, Wallingford, UK, pp. 3–9.

Meerburg, B.G., Jacobs-Reitsma, W.F., Wagenaar, J.A. and Kijlstra, A. (2006) Presence of *Salmonella* and *Campylobacter* spp. in wild small mammals on organic farms. *Applied and Environmental Microbiology* 72, 960–962.

MEG (2011) MEG Marktinfo Eier & Geflügel. *Geflügel kompakt* (3), 02/2011, 1-4.

Meier, H. (2010) Die Putenzuchtunternehmen im Wandel. In: *Geflügeljahrbuch 2010*. Eugen Ulmer KG, Stuttgart, Germany, pp 103–110.

Mench, J.A. and Siegel, P.B. (1997) Animal welfare issues: poultry. In: Reynnells, R.D. and Eastwood, B.R. (eds) *Animal Welfare Issues Compendium: A Collection of 14 Discussion Papers*. US Department of Agriculture Cooperative State Research, Education and Extension Service Plant and Animal Production, Protection and Processing, pp. 100- 107.

Mirabito, L. and Lubac, L. (2001) Descriptive study of outdoor run occupation by 'Red Label' type chickens. *British Poultry Science* 42, S16–S17.

Mirabito, L., Joly, T., and Lubac, L. (2001) Impact of the presence of peach orchards in the outdoor hens run on the occupation of the space by Red Label type chickens. *British Poultry Science* 42, S18–S19.

Mirabito, L., Berthelot, A., Baron, F., Bouvarel, I., Aubert, C., Bocquier, C., Dalibard, F., Sante, V. and Le Pottier, G. (2002) Influence of reducing the stocking density on the performance, behaviour and physical integrity of meat turkey (abstract). In: *Proceedings of the 11th European Poultry Conference*, Bremen, Germany, 6–10 August 2002 (CD ROM) World's Poultry Science Association, German Branch.

Mitterer-Istyagin, H., Ludewig, M., Bartels, T., Krautwald-Junghanns, M.E., Ellerich, R., Schuster, E., Berk, J., Petermann, S. and Fehlhaber, K. (2011) Examinations on the prevalence of footpad lesions and breast skin lesions in B.U.T. Big 6 fattening turkeys in Germany. Part II: Prevalence of breast skin lesions (breast buttons and breast blisters). *Poultry Science* 90, 775–780.

Moberly, R.L., White, P.C.L. and Harris, S. (2004) Mortality due to fox predation in free-range poultry flocks in Britain. *The Veterinary Record* 155, 48–52.

Mushtaq, T., Sarwar, M., Ahmad, G., Mirza, M.A., Ahmad, T., Noreen, U., Mushtaq, M.M.H. and Kamran, H. (2009) Influence of sunflower meal based diets supplemented with exogenous enzyme and digestible lysine on performance, digestibility and carcass response of broiler chickens. *Animal Feed Science and Technology* 149, 275–286.

Naturland (2010) Naturland Richtlinien Erzeugung: Teil B: II Viehwirtschaft. http://www.naturland.de/richtlinien.html (accessed 1 May 2010).

Neuland (2008) Richtlinien für die artgerechte Mastgeflügelhaltung. http://www.neuland-fleisch.de/landwirte/richtlinien-mastgefluegelhaltung.html (accessed 1 July 2010).

Newberry, R.C. (1993) The role of temperature and litter type in the development of breast buttons in turkeys. *Poultry Science* 72, 467–474.

Newell, D.G. and Fearnley, C. (2003) Sources of *Campylobacter* colonization in broiler chickens. *Applied and Environmental Microbiology* 69, 4343–4351.

NFU (2006) NFU calls for poultry price rise. www.nfu*online*.com, 25 April; available at http://www.nfuonline.com/Media_centre/2006/NFU_calls_for_poultry_price_rise/ (accessed 1 February 2011).

NFU (2008) Consumers want British chicken but producers need a fair price – NFU. www.nfu*online*.com, 13 May; available at: http://www.nfuonline.com/Media_centre/2008/Consumers_want_British_chicken_but_producers_need_a_fair_price_-_NFU/ (accessed 1 February 2011).

Nielsen, B.L. (2004) Breast blisters in groups of slow-growing broilers in relation to strain and the availability and use of perches. *British Poultry Science* 45, 306–315.

Nielsen, B.L., Thomsen, M.G., Sørensen, P. and Young, J.F. (2003) Feed and strain effects on the use of outdoor areas by broilers. *British Poultry Science* 44, 161–169.

Noble, D.O., Nestor, K.E. and Krueger, K.K. (1996) The effect of beak trimming on two strains of commercial tom turkeys. *Poultry Science* 75, 1468–1471.

Pagazaurtundua, A. and Warriss, P.D. (2006) Levels of foot pad dermatitis in broiler chickens reared in 5 different systems. *British Poultry Science* 47, 529–532.

Platz, S., Berger, J., Ahrens, F., Wehr, U., Rambeck, W., Amselgruber, W. and Erhard, M.H. (2003) Health, productivity and behaviour of conventional turkey breeds under ecological outdoor rearing conditions. In: Hafez, H.M. (ed.) *Turkey Production: Balance Act Between Consumer Protection, Animal Welfare and Economic Aspects.* Ulmer Verlag, Suttgart, Germany, pp. 115–121.

Plumstead, P.W., Romero-Sanchez, H., Paton, N.D., Spears, J.W. and Brake, J. (2007) Effects of dietary metabolizable energy and protein on early growth responses of broilers to dietary lysine. *Poultry Science* 86, 2639–2648.

PMAF (2010) Label Rouge & Agriculture Biologique. http://pmaf.org/pdf/labels/fiche_poulets.pdf (accessed 1 October 2010).

Polak, M., Przybylska-Gornowicz, B. and Faruga, A. (2010) The effect of two different rearing conditions on muscle characteristics in broilers of two commercial lines – a light microscope study. *Japan Poultry Science* 47, 125–132.

Ponte, P.I.P., Rosado, C.M.C., Crespo, J.P., Crespo, D.G., Mourão, J.L., Chaveiro-Soares, M.A., Brás, J.L.A., Mendes, I., Gama, L.T., Prates, J.A.M., Ferreira, L.M.A. and Fontes, C.M.G.A. (2008a) Pasture intake improves the performance and meat sensory attributes of free-range broilers. *Poultry Science* 87, 71–79.

Ponte, P.I.P., Prates, J.A.M., Crespo, J.P., Crespo, D.G., Mourão, J.L., Alves, S.P., Bessa, R.J.B., Chaveiro-Soares, M.A., Gama, L.T., Ferreira, L.M.A. and Fontes, C.M.G.A.

(2008b) Restricting the intake of a cereal-based feed in free-range-pastured poultry: effects on performance and meat quality. *Poultry Science* 87, 2032–2042.

Powell, K.C. (2007) Current skeletal issues in turkeys. In: Hafez, H.M. (ed.) *Turkey Production: Current Challenges*. Mensch und Buch Verlag, Berlin, pp. 90–109.

Proplanta (2010) EU: Putenerzeugung schrumpft. http://www.proplanta.de/Agrar-Nachrichten/Tier/EU-Putenerzeugung-schrumpft_article1279073552.html (accessed 1 July 2010).

QBT (2009) Quality British Turkey, Assured Food Standards in association with Red Tractor. Agricultural standards: breeder replacements, laying birds, hatchery, growing farms, free range. http://www.redtractor.org.uk/site/REDT/UploadedResources/QBT_Agricultural_Standards_Oct2009.pdf (accessed 1 October 2010).

Quentin, M., Bouvarel, I., Berri, C., Le Bihan-Duval, E., Baéza, E., Jégo, Y. and Picard, M. (2003) Growth, carcass composition and meat quality response to dietary concentrations in fast-, medium and slow-growing commercial broilers. *Animal Research* 52, 65–77.

Quentin, M., Bouvarel, I. and Picard, M. (2004) Short- and long-term effects of feed form on fast- and slow-growing broilers. *Journal of Applied Poultry Research* 13, 540–548.

Quentin, M., Bouvarel, I. and Picard, M. (2005) Effects of crude protein and lysine contents of the diet on growth and body composition of slow-growing commercial broilers from 42 to 77 days of age. *Animal Research* 54, 113–122.

Rivoal, K., Ragimbeau, C., Salvat, G., Colin, P. and Ermel, G. (2005) Genomic diversity of *Campylobacter coli* and *Campylobacter jejuni* isolates recovered from free-range broiler farms and comparison with isolates of various origins. *Applied and Environmental Microbiology* 71, 6216–6227.

Roberson, K.D., Rahn, A.P., Balander, R.J., Orth, M.W., Smith, D.M., Booren, B.L., Booren, A.M., Osburn, W.N. and Fulton, R.M. (2003) Evaluation of the growth potential, carcass components and meat quality characteristics of three commercial strains of tom turkeys. *Journal of Applied Poultry Research* 2, 229–236.

Rovee-Collier, C., Collier, G., Egert, K. and Jackson, D. (1993) Developmental consequences of diet and activity. *Physiology & Behavior* 53, 353–359.

Rozenboim, I., Robinzon, B. and Rosenstrauch, A. (1999) Effect of light source and regimen on growing broilers. *British Poultry Science* 40, 452–457.

RSPCA (2008) *RSPCA Welfare Standards for Chickens*. Royal Society for the Prevention of Cruelty to Animals, Horsham, UK.

RSPCA (2010a) The welfare state: measuring animal welfare in the UK. http://www.rspca.org.uk/in-action/whatwedo/animalwelfareindicators (accessed 1 February 2011).

RSPCA (2010b) *RSPCA Welfare Standards for Turkeys*. Royal Society for the Prevention of Cruelty to Animals, Horsham, UK.

Sanotra, G.S., Lund, J.D. and Vestergaard, K.S. (2002) Influence of light–dark schedules and stocking density on behaviour, risk of leg problems and occurrence of chronic fear in broilers. *British Poultry Science* 43, 344–354.

Santos, A.L., Sakomura, N.K., Freitas, E.R., Fortes, C.M.S. and Carrilho, E.N.V.M. (2005) Comparison of free range broiler chicken strains raised in confined or semi-confined systems. *Brazilian Journal of Poultry Science* 7, 85–92.

Sarica, M., Ocak, N., Karacay, N., Yamak, U., Kop, C. and Altop, A. (2009) Growth, slaughter and gastrointestinal tract traits of three turkey genotypes under barn and free-range housing systems. *British Poultry Science* 50, 487–494.

Sarica, M., Ocak, N., Turhan, S., Kop, C. and Yamak, U.S. (2011) Evaluation of meat quality from three turkey genotypes reared with or without outdoor access. *Poultry Science* 90, 1313–1323.

SCAHAW (2000) *The Welfare of Chickens Kept for Meat Production (Broilers)*. Report of the Scientific Committee on Animal Health and Welfare. SANCO.B.3/AH/R15/2000. European Commission, Health and Consumer Protection Directorate-General, Brussels.

Schwean-Lardner, K. and Classen, H. (2010) Lighting for Broilers. Aviagen technical report. 0210-AVN-024. http://en.aviagen.com/assets/Uploads/RossTechLightingforBroilers.pdf (accessed 1 August 2010).

Sherwin, C.M. (1999) Domestic turkeys are not averse to compact fluorescent lighting. *Applied Animal Behaviour Science* 64, 47–55.

Sherwin, C.M. and Devereux, C.L. (1999) Preliminary investigations of ultraviolet-induced markings on domestic turkey chicks and a possible role in injurious pecking. *British Poultry Science* 40, 429–433.

Soil Association (2010) Standards for producers. http://www.soilassociation.org/organicstandards (accessed 1 September 2010).

Spindler, B. and Hartung, J. (2007) Tiergerechte Mastputenhaltung mit Beschäftigungs- und Strukturelementen. *Abschlussbericht Modellvorhaben „Landwirtschaftliches Bauen 2005–2007".* Tierärztliche Hochschule Hannover, Germany.

Statista (2010) It must be pay. http://de.statista.com/statistik/printstatsek/studie/75968/ (accessed 1 May 2011).

Sundrum, A. (2001) Organic livestock farming – a critical review. *Livestock Production Science* 67, 207–215.

Theuvsen, L., Essmann, S. and Brand-Sassen, H. (2005) Livestock husbandry between ethics and economics: finding a feasible way out by target costing? Presented at the *XIth International Congress of the EAAE (European Association of Agricultural Economists), 'The Future of Rural Europe in the Global Agri-Food System',* Copenhagen, Denmark, 24–27 August 2005.

van der Sluis, W. (2010) Zinc provides key to more valuable feet. *Poultry World* July issue, 30.

Vanhonacker, F. and Verbeke, V. (2009) Buying higher welfare poultry products? Profiling Flemish consumers who do and do not. *Poultry Science* 88, 2702–2711.

Veldkamp, T. and Kiezebrink, M. (2006) Effects of a turkey house with veranda on behaviour, performance, carcass quality, and health of male turkeys. *World's Poultry Science Journal, Book of Abstracts, 12th European Poultry Conference*, Verona, Italy, 10–14 September 2006, pp. 443–444.

Walker, A. and Gordon, S. (2003) Intake of nutrients from pasture by poultry. *Proceedings of the Nutrition Society* 62, 253–256.

Weeks, C.A., Nicol, C.J., Sherwin, C.M. and Hunt, S.C. (1994) Comparison of the behavior of broiler chickens in indoor and free-range environments. *Animal Welfare* 3, 179–192.

Weeks, C.A., Danbury, T.D., Davies, H.C., Hunt, P. and Kestin, S.C. (2000) The behaviour of broiler chickens and its modification by lameness. *Applied Animal Behaviour Science* 67, 111–125.

Welfare Quality (2010) Preventing lameness in broiler chickens (Popular Fact Sheet WQ Broilers 0303). http://www.welfarequality.net/everyone/41858/5/0/22 (accessed 1 April 2011).

Werner, C., Riegel, J. and Wicke, M. (2008) Slaughter performance of four different turkey strains, with special focus on the muscle fiber structure and the meat quality of the breast muscle. *Poultry Science* 87, 1849–1859.

Woodward, P. (2004) Nutritional strategies to reduce leg problems in turkeys. In: *Proceedings of the 31st Annual Carolina Poultry Nutrition Conference Carolina Feed Industry Association*, Research Triangle Park, North Carolina, pp. 67–74; available at: http://www.ces.ncsu.edu/depts/poulsci/conference_proceedings/nutrition_conference/2004/woodward_2004.pdf (accessed 1 October 2011).

Wu, K. and Hocking, P.M. (2011) Turkeys are equally susceptible to foot pad dermatitis from 1 to 10 weeks of age and foot pad scores were minimized when litter moisture was less than 30%. *Poultry Science* 90, 1170–1178.

Youssef, I., Beineke, A. and Kamphues, J. (2009) Effects of soybean meal and its constituents (potassium, oligosaccharides) and moisture content in litter on prevalence and severity of foot pad dermatitis in young turkeys. In: Hafez, H.M. (ed.) *Turkey Production: Toward Better Welfare and Health*. Mensch und Buch Verlag, Berlin, pp. 40–47.

CHAPTER *15*

Nutritional Challenges of Alternative Production Systems

M.G. MacLeod and J.S. Bentley

ABSTRACT

Many of the nutritional challenges posed by alternative systems can be addressed by application of existing scientific knowledge. However, regulations applied to alternative systems may limit the nutritionist's freedom of action, particularly with regard to the ingredients which can be used to formulate diets. Practical comments on meeting the nutrition-related stipulations of the various regulations are included in the present chapter. It is possible to formulate diets without animal protein, potentially genetically modified organisms (e.g. soybean and maize products) and synthetic amino acids, but it is difficult to attain nutritional optima. On the positive side, in free range systems, the bird s nutritional inputs may be enhanced by access to forage plants and animals. Also, there is clearly greater scope for the bird to be provided with food in ways that give greater opportunities for a repertoire of feeding behaviour, such as feedstuff choice. Some alternative systems may increase the bird's energy requirements because of increased expenditure of energy on physical activity and on thermoregulation in a cooler environment. Since there is so much scope for variation in environmental factors in alternative systems, nutritional decisions may have to be made on an iterative basis, meaning that cooperation between the producer and the nutritionist may be the key to success. This is particularly true where there are strict regulatory limitations on rate of growth or final body weight. There is a tendency for alternative systems to have a greater ecological impact than conventional systems, largely because of the lower efficiency of nutrient utilization. This chapter comments on nutritional methods of helping to reduce environmental impact.

INTRODUCTION

The nutritional challenges posed by alternative systems are likely to involve the application of existing knowledge rather than the development of totally novel scientific principles. Given that the aim of science is to produce generalizing theories that hold true across as wide as possible a range of conditions, it would be an admission of failure if every adjustment of husbandry and nutritional practices needed new nutritional research. Many of the general principles of poultry nutrition are applicable across breeds and production systems and are described in books such as Larbier and Leclercq (1994) and Leeson and Summers (2008). Feedstuffs and their evaluation are comprehensively covered in McNab and Boorman (2002). Nutritionists and producers should consult the existing literature before assuming that new research is needed to solve the perceived problems of alternative systems.

NUTRITION REGULATIONS FOR ALTERNATIVE SYSTEMS

The formulation of feed for alternative systems is governed by a large number of schemes. These, unlike nutritional recommendations, are not always evidence based and therefore carry the risk of being a 'moveable feast'. For instance, the rules on organic production within the European Union (EU) at the time of writing allow the use of fish meal as long as it is from sustainable fisheries. This may be very convenient in allowing the use of a methionine-rich protein source but depends on decisions on fishery sustainability. (A relatively recent previous version of the rules, (EC) No 834/2007, stated that 'The products of hunting and fishing of wild animals shall not be considered as organic production' (Council Regulation, 2007).) Why industrial fishing might be preferable to industrial production of methionine is perhaps a question for the organic consumer. Organic schemes are the most demanding nutritionally but there is a range of others, such as Freedom Food, Label Range and even individual certification schemes run by some of the larger retailers. Organic feed is governed by EU directives such as the latest update (EC) No 889/2008 (Commission Regulation, 2008).

A summary of the rules regarding poultry feed is:

(i) primarily obtaining feed for livestock from the holding where the animals are kept or from other organic holdings in the same region;
(ii) livestock shall be fed with organic feed that meets the animal's nutritional requirements at the various stages of its development. A part of the ration may contain feed from holdings which are in conversion to organic farming;
(iii) with the exception of bees, livestock shall have permanent access to pasture or roughage;
(iv) non organic feed materials from plant origin, feed materials from animal and mineral origin, feed additives, certain products used in animal nutrition and processing aids shall be used only if they have been authorised for use in organic production under Article 16;
(v) growth promoters and synthetic amino acids shall not be used.

A point of particular relevance from Article 16 mentioned above is that:

(ii) feed of mineral origin, trace elements, vitamins or provitamins shall be of natural origin. In case these substances are unavailable, chemically well-defined analogic substances may be authorised for use in organic production.

For poultry, modifications of the regulations to allow importation of organic materials from other countries have been made, which specify that the supplying country must have equivalent organic regulations and certification procedures. The critical example in the case of poultry is soybean meal. The UK Government provides a 'living document' giving guidance on EU organic regulations (Defra, 2010).

Practical nutrition guidance to meet the requirements of the various regulations is discussed under 'Practical Feeding Programmes', later in this chapter. It is possible to formulate diets without animal protein, potential genetically modified organisms (e.g. soybean and maize products) and synthetic amino acids, but it is difficult to attain all the nutritional optima. This may lead to performance below the birds' genetic potential and compromise the health and welfare of the birds. Hadorn *et al.* (2000) tested such diets and found significant effects on productivity, as would be predicted from nutritional theory. There was also a significant increase in mortality when synthetic amino acids were omitted. We can hypothesize that one possible factor may have been the importance of methionine in the immune system (Rama Rao *et al.*, 2003). Also, the omission of sulfur-containing amino acids resulted in poorer plumage condition later in the laying period. Nitrogen excretion may be higher on 'vegetable protein' diets if a larger amount of less well-balanced protein has to be used to approach standard nutritional requirements (Hadorn *et al.*, 2000). It would be unfortunate if a side effect of organic or 'vegetable protein only' production was to increase nitrogen pollution and other environmental burdens.

NUTRITIONAL PRINCIPLES FOR THE NON-SPECIALIST

Growth and egg production require the provision of the chemical components of the body or egg and also the energy required to convert these components into the substances deposited. Tables of nutrient requirements are produced by the feedstuffs and breeding industries and by organizations such as the National Research Council of the USA (NRC, 1994). Much of the dietary energy consumed by the bird is used for 'maintenance', i.e. to sustain all of the physiological and biochemical processes which keep the bird in a steady state. Maintenance energy requirement can be measured, or estimated in various ways, as the energy intake which leads to zero energy balance. Any 'alternative system' is likely to affect the bird's maintenance requirement, particularly if locomotor activity or the thermal environment is altered by the system of housing, husbandry or nutrition.

Costs of activity

Differences in activity contribute to between-breed differences in energy expenditure. Even in relatively confined conditions, about 12% of the energy expenditure of a light layer strain was attributable to locomotor activity, compared with about 5% for broilers (MacLeod *et al.*, 1982). Activity can also be predicted to produce differences in energy requirements and food intakes between different housing systems. Pre-oviposition behaviour increases heat production by about 60% over the resting value (MacLeod and Jewitt, 1985), similar to treadmill measurements of the cost of walking (van Kampen, 1976). The activity of feed intake has been shown to increase heat production by about 25% (MacLeod and Jewitt, 1985). Measurements such as these could be combined with activity diary observations to estimate the locomotory energy costs of different systems. However, a more direct integrated measurement of overall energy expenditure by the unrestrained bird can be made by methods such as the doubly labelled water technique (Ward and MacLeod, 1991, 1992, 1994).

Nutrition and production

Amino acids

Both number and size of eggs respond to the concentration of the first-limiting amino acid in the diet. Even if an amino acid is severely limiting, egg weight seldom falls below 90% of maximum, so any further response must be in rate of lay (Morris and Gous, 1988). Al-Saffar and Rose (2002) compiled the results from a large number of independent experiments to show that responses in both number and size could be approximated by an asymptotic exponential curve. Statistical treatments of such curves are given by Curnow (1973), Fisher *et al.* (1973) and France and Thornley (1984). Similar models have been derived for broilers. Such models allow calculation of the financially optimal intake of an individual amino acid; this occurs where the gradient of the response curve is equal to the ratio of the cost of an additional unit of amino acid to the financial return from an additional unit of growth or egg production. Pesti *et al.* (2009) compared different statistical methods for estimating amino acid requirements from experimental measurements. Requirements, expressed per unit of feed, can be altered by various factors related to the bird and even by the physical form of the diet, which is relevant to the current discussion (Lemme *et al.*, 2006).

Egg size can potentially be controlled by precise formulation on the first-limiting amino acid (usually methionine + cystine, lysine or tryptophan). Precision can be reinforced by rapid feedback of data on food intake, production and egg size. The survey of El-Saffar and Rose (2002) gave some support to the idea (Morris *et al.*, 1999) that amino acid concentrations are better described as proportions of dietary protein than as proportions of the entire diet.

Yolk colour

Yolk colour is an aspect of product quality that can be expected to improve with access to suitable pasture. Especially when diets are based on wheat or barley rather than maize, synthetic or concentrated xanthophyll supplements may be added to the feed to give the preferred intensity of yolk colour (Nys, 2000). The plant pigments are natural derivatives of β-carotene. They are present at high concentrations in marigold meal and some species of algae but are also present at practically useful concentrations in lucerne and grass meals.

Quantitative control of feed and nutrient intake

When birds are given a single compound diet, quantity of food eaten and time of eating are the only responses available to the bird. Under such conditions, food intake can only be a compromise among specific requirements for individual diet components. The requirements for energy and protein (or, more specifically, amino acids) are probably the strongest drivers of food intake. However, dietary energy concentration has the most influential effect under most conditions. Broadly speaking, this means that the intake of other nutrients in a compound diet will be inversely proportional to the energy concentration of the diet. Husbandry factors which increase energy intake (most commonly decreased ambient temperature or increased locomotor activity) will therefore increase the intake of other components of the diet unless their concentrations are reduced. Over a broad range of ambient temperature (e.g. 10°C to 28°C in layers), food intake increases as temperature declines. This is related to the greater heat production required to maintain body temperature. Factors such as the quality of feather insulation (Tullett *et al.*, 1980) and stocking density (Savory and MacLeod, 1980) influence this relationship. Over the course of a day, however, a temporal rhythm (e.g. in calcium requirement; Hughes, 1972) may override or conflict with the requirement for other major nutrients.

Qualitative control of feed and nutrient intake

Selecting among food sources so as to obtain the appropriate mixture of nutrients is of evolutionary advantage to birds living under natural or quasi-natural conditions. That wild birds have this ability is clear from field and laboratory studies. For example, adult red grouse feed mainly on heather shoots but their chicks supplement this diet with invertebrates, mainly insects (Park *et al.*, 2001). The insect 'supplement' is clearly supplying a growth-limiting nutrient, which we can describe as protein but which can be narrowed down to individual amino acids. This ability to select among foods is of such fundamental evolutionary advantage that it seems unlikely to have been eliminated from domestic poultry by generations of breeding on compound diets. The persistence of this ability has been tested many times in poultry, with variable results (Hughes, 1984; Rose and Kyriazakis, 1991; Forbes and Covasa, 1995; Henuk and Dingle, 2002), although choice feeding was common practice before requirements had been sufficiently well defined to allow the formulation of nutritionally complete diets. However, the re-development of free range poultry husbandry raises the possibility of birds obtaining a supplementary source of feed items (including invertebrate animals) from the range or pasture. This is an area that needs quantification under field conditions.

Feeding of whole grains

Feeding of whole grains is likely to occur as part of the nutritional strategy in alternative systems. This has several potential advantages: it provides a form of environmental enrichment for the bird (Picard *et al.*, 2002), it encourages muscular development of the gizzard and it reduces feed processing costs. Grain (e.g. wheat, barley, oats) can be provided separately in a choice feeding system, mixed with mash or fed at alternating times to a compound diet (Rose *et al.*, 1995). Starch digestibility is improved by the addition of whole wheat (Svihus and Hetland, 2001; Hetland *et al.*, 2002) and also by inclusion of oat hulls (Hetland and Svihus, 2001). The gizzard has a well-developed ability to grind down larger particles such as whole grains and increased gizzard size and activity may increase the opportunity for enzymatic digestion. However, not all whole grain systems have given positive results (Bennett and Classen, 2003). It should be noted that simply adding whole cereal grains 'on top of' an existing compound diet will dilute many nutrients. This may be advantageous if maintenance energy requirements have increased (e.g. under more extensive systems), since energy intake will be allowed to increase without excessive additional intake of the more expensive components of the diet. Umar Faruk *et al.* (2010) noted that loose-mix feeding of grain with a compound 'balancer' diet had no effect on intake of metabolizable energy (ME). However the loose-mix treatment reduced feed and protein intake due to lower balancer diet intake. It also resulted in lower egg production and lower egg and body weights than sequential feeding. Sequential feeding of whole grains and a concentrate resulted in similar egg laying performance to conventional feeding and thus could be used to advantage in situations where it is applicable.

Specific nutrient appetites

It may be possible to cater for specific nutrient appetites in some alternative systems. A calcium appetite is particularly clear in the laying hen (Mongin and Sauveur, 1979) and the effects of the onset of lay (Meyer *et al.*, 1970) and even the deposition of the individual eggshell (Hughes, 1972) are detectable. Separate feeding of a calcium source is one form of free choice feeding that is reliably successful. It has the advantage over feeding calcium only as part of a complete compound diet that the intake of calcium is dissociated from energy and protein intake and can occur at the time of maximum physiological demand.

A slight deficiency of an amino acid in a compound diet has been shown to lead to a compensatory increase in food intake. Gross deficiency or excess, which can be summarized as an amino acid imbalance, usually leads to a reduction in intake (d'Mello, 1994). When offered a choice between a diet adequate in methionine and one 65% below adequate (Hughes, 1979), laying hens selected about 60% of total intake in the form of the adequate diet. This gave an egg production only slightly lower than in controls fed only on the methionine-adequate diet.

Hughes (1979) described specific appetites for calcium, zinc, phosphorus and thiamine. Zuberbuehler *et al.* (2002) found that selenium-deficient hens preferentially selected a high-selenium diet, presumably in response to post-ingestional feedback, since inadequate selenium, often in combination with low vitamin E status, causes deficiency symptoms in many species. Post-ingestional feedback is likely to be the key to most specific appetites. However, post-ingestional effects must become associated with an identifiable characteristic of the food – smell, taste, appearance, location – before the bird can learn to select for or against it.

Supplementary range feeding

A potential advantage of access to outdoor areas is the availability of supplementary feed items, whether animal, vegetable or mineral. However, this advantage can be difficult to quantify since it depends on essentially ecological factors, such as the quality and biodiversity of the 'range' area and stocking density, and also on behavioural factors such as the readiness and ability of the birds to move over the area and select from its resources. Knowing the intake and composition of forage has the potential to allow fine-tuning of the main (farmer-provided) diet, although there is so much scope for variation between and within farms that it may not always be economically justifiable. Assessing the contribution of foraging to nutrient intake may have to rely on ecological methods, such as sampling of crop contents (Antell and Ciszuk, 2006). Horsted *et al.* (2007) used this technique to assess the intake of different forages when hens were given either a typical organic layer concentrate (184 g crude protein kg^{-1} dry matter) or a nutrient-restricted diet consisting of whole wheat (120 g crude protein kg^{-1} dry matter) and oyster shell grit. The latter diet was intended to encourage foraging and did indeed produce significant effects, being associated with greater crop contents of plant materials, oyster shell, insoluble grit and soil. There was no significant difference in intakes of animal matter, such as earthworms and larvae, which might have been expected if the birds were 'adjusting' their nutrient intake. However, the authors suggested that the range area had already been depleted of such items before the measurements started, illustrating a source of variation which can potentially be controlled if sufficient land is available.

UK government departments recently published the advice that:

> Article 23(5) of EC 889/2008 requires that when the production of each batch of poultry has been completed, runs must be left empty to allow vegetation to grow back but leaves Defra to determine the period during which runs must be left empty. In the UK the period for which runs shall be left empty between batches of layers must be not less than two months and in the case of poultry for meat production the total of the periods in any one year that runs are empty must be not less than two months per year. (Defra, 2010: 11.)

A suitably managed poultry pasture can be seen as a source of materials other than the obvious macronutrients. Ponte *et al.* (2008c) studied some of the effects of a legume-rich pasture (clover, etc.) on broiler meat quality and performance. The birds were kept in portable floorless pens and given a cereal-based diet at 50, 75 or 100% of *ad libitum* intake. Although growth rates were reduced by feed restriction, leguminous plant intake increased from 1.5% to 5% of total dry matter intake. Pasture intake decreased meat pH and improved skin pigmentation. It also had a small positive effect on meat vitamin E content and a small negative effect on meat cholesterol content. There were clear effects on fatty acid profile, with breast meat concentrations of $n-3$ polyunsaturated fatty acids (α-linolenic acid, eicosapentaenoic acid, doco-sapentaenoic acid and docosahexaenoic acid) being significantly greater in birds consuming leguminous plant material. However, some of the measured effects may have been associated with the different degrees of feed restriction, since Ponte *et al.* (2008b) did not find the same effects when unrestricted birds had access to similar pasture. Although there were no significant effects on single measures of tenderness, juiciness or flavour, a 30-person sensory assessment panel gave greater scores for overall appreciation to broilers that had consumed clover (Ponte *et al.*, 2008a).

Nutrient effects on behaviour

It has been asserted (FAWC, 1997) that the lack of animal protein in the diet makes pecking damage and 'cannibalism' more likely; this assertion has not been supported by controlled experiment (McKeegan *et al.*, 2001). However, an imbalanced diet (independently of whether animal protein is included) may induce such behavioural effects. As an example, Elwinger *et al.* (2008) compared the responses of three layer lines, kept in aviary pens with access to outdoor runs, to four different experimental diets. Diets based on feedstuffs suitable for organic production, and differing in methionine and cystine content, were tested against a control diet. The diet with lowest protein led to feather pecking in all lines, with one line being particularly susceptible. Severe feather pecking occurred in one of the lines and was worst on the low-protein diet. There was an incidence of cannibalism but only in one pen group fed on the diet with the lowest methionine content. Diets low in methionine influenced egg weight as well as plumage condition, although egg number was unaffected.

PRACTICAL FEEDING PROGRAMMES

For conventional poultry many nutritionists construct a feed programme by referring to tables of recommended nutrient concentrations such as those provided by breeding companies or public bodies such as NRC (1994). It is often forgotten that these recommendations are underpinned by assumptions about: (i) the desired growth or egg production curve, which is usually as close to genetic potential as economically feasible; (ii) the expected optimum nutrient intakes to achieve the level of production; (iii) assumed feed intakes; and (iv) costs of ingredients. Nutrient recommendation tables have general application for conventional production systems across many farms and countries because they use standardized controlled environments, common genotypes, consistent physical feed presentation and well-defined ingredients. Therefore, the assumptions listed above are applicable. However, for alternative meat systems in particular, with a longer growing period, these assumptions will vary considerably due to many adverse environmental feed intake factors. This severely reduces the suitability of general tables of nutrient recommendations for varied alternative systems.

Meat bird feed programmes

One approach to solving this dilemma for alternative broilers is to use an iterative framework to lead to an optimized local feed programme. A similar approach can be used for turkeys and consists of six steps.

1. Ensure the target weight and age at slaughter are defined and understood by all parties especially when this is constrained by the rules of the certification system being followed. The decision on the age and weight at slaughter may be determined by quality constraints such as meat quality or carcass finish.

2. Determine which genotype is being used and then compare the desired slaughter weight for age with the breed performance target. The objective is to assess whether the birds need to be grown near to their genetic potential or need to have growth slowed to meet the slaughter objective. This may occur in many organic systems when birds are slaughtered at ages over 70 days.

3. Determine intermediate target weights during growth and compare these with the growth objectives for the breed. This step should be a continuous process to monitor the impact of adverse feed intake factors constraining growth at particular ages. Common examples of adverse factors include:

- a sudden change of environment from brooding (warm) housing to outdoor or poorly insulated housing especially when feathering is poor, increasing maintenance requirements;
- the introduction of range leading to increased activity;
- stocking density, especially at the end of the ranging period;
- a change of physical feed form such as crumbles to pellets before chicks are old and heavy enough to consume large pellets (suggested limits are 300 g and 420 g body weight for introducing 3 and 4 mm pellets, respectively);
- a change in feeder type from chick to 'home made' adult feeders in ark housing;
- disease factors, especially infectious bursal disease virus or Marek's disease, which may occur more frequently in alternative systems where broilers are grown to older ages or because of ineffective vaccinal protection;
- the frequency and timing of coccidiosis and other protozoa infections, worms and digestive problems; and
- introducing whole wheat at high inclusion rates leading to feed spillage.

4. Determine the desired growth curve for the flock and monitor feed intakes. It is good practice to set key target body weights for producers, e.g. the age at which 750 g is achieved, especially when growth needs to be slowed down to meet the slaughter objectives. Only with this information is it possible, using amino acid response coefficients, to model target amino acid intakes and modify the existing feed programme correctly.

5. Does the quantity of alternative system feed justify special diets in a feed mill making conventional feed or will conventional diets have to be used in an adapted programme?

6. Where ingredient constraints are specified, such as minimum cereal contents or no synthetic amino acids, determine within these constraints by formulation exercises the ME concentrations that minimize cost per MJ paying close attention to amino acid balance.

The optimal ME concentrations may differ from those used for conventional broiler diets. One key question is whether the responses of fast (e.g. Saleh *et al.*, 2005) and slow growing genotypes to dietary available ME concentration are identical. Quentin *et al.* (2004) showed that slow and fast growing genotypes respond differently to physical form of the diet, with fast growing broilers reacting to a change from pellets to mash more dramatically than slow growing broilers.

Given the possible number of environmental factors, this decision process has to be continuous and iterative if nutrition is to be optimized. This means that cooperation between the producer and nutritionist is a key to success in all alternative systems. Some common scenarios include the following.

- Poor early growth due to management or feed intake factors. To attain the target slaughter weight requires more rapid growth in the pre-slaughter period. This scenario is more likely in brood-and-move systems with separate brooding and growing facilities. Increasing the nutrient density of the starter diet may not show economic benefits while feed intake is limited.
- Low growth rate in the immediate pre-slaughter period which requires that a target such as 750 g body weight is achieved as early as possible to ensure satisfactory final body weight.
- Programmed early slow growth relying on later compensatory growth. This is commonly practised in many organic systems slaughtering at ages over 70 days to avoid exceeding target weights.
- Optimal management requiring growth to be limited more by adjusting nutrient levels, either by reducing amino acid density before 28 days or diluting diets after 28 days with whole wheat to achieve the final growth target.

A further practical concern is maintaining amino acid balance when synthetic amino acids are not permitted, especially with regard to the potential impact on feather growth. Despite the reservations providing nutrient specifications listed above, Table 15.1 shows some typical total nutrient contents used in alternative systems in the UK and France.

Table 15.1. Example nutrient specifications for alternative systems using intermediate growth genotypes in the UK and France.

System	High cereal/low ME programme 49–56 days at slaughter			Conventional ME 49–56 days at slaughter			High stocking density – 49–56 days at slaughter			Organic ≥70 days at slaughter		
Age	0–21	21–40	40–56	0–21	21–40	40–56	0–21	21–40	40–56	0–28	28–56	56+
ME (MJ kg^{-1})	12.0	12.5	12.8	12.6	13.0	13.6	12.6	13.0	13.6	12.1	12.6	12.6
ME (kcal kg^{-1})	2875	2975	3050	3000	3100	3250	3000	3100	3250	2900	3000	3000
Lysine (g kg^{-1})	12.0	11.0	10.0	13.1	11.5	10.7	14.1	12.3	10.7	12.0	10.0	8.0
Methionine (g kg^{-1})	5.4	5.0	4.5	5.9	5.2	4.8	6.3	5.6	4.8	5.4	5.0	4.5
Methionine + cystine (g kg^{-1})	9.5	8.5	8.0	10.4	8.9	8.5	11.1	9.5	8.5	9.5	8.5	8.0
Threonine (g kg^{-1})	8.2	7.6	7.7	9.0	7.9	8.2	9.6	8.5	8.2	8.2	7.6	7.7
Tryptophan (g kg^{-1})	2.0	2.2	2.0	2.2	2.3	2.1	2.3	2.5	2.1	2.0	2.2	2.0
Calcium (g kg^{-1})	10.0	9.5	9.0	11.0	9.9	9.6	11.7	10.6	9.6	10.0	9.0	9.0
Available phosphorus (g kg^{-1})	4.8	4.2	3.8	5.3	4.4	4.0	5.6	4.7	4.0	4.8	4.2	3.8
Sodium (g kg^{-1})	1.7	1.7	1.7	1.9	1.8	1.8	2.0	1.9	1.8	1.7	1.7	1.7
Chloride (g kg^{-1})	1.8	1.8	1.8	2.0	1.9	1.9	2.1	2.0	1.9	1.8	1.8	1.8

ME, metabolizable energy.

Feeding programmes for egg layers

Primary breeders provide detailed optimum nutrient intakes for their breeds in a range of environments. In practice, feed intake factors such as increased levels of physical activity, variable environmental temperature and variable plumage condition must be considered for alternative layer systems. As a practical guide, versus a caged system, energy intake at peak production will increase by about 5 kJ day^{-1} (20 kcal day^{-1}) due to activity. A reduction of 1°C in environmental temperature will raise the energy requirement by approximately 1 kJ (+3.5 kcal), equivalent to 1.0–1.5% more feed in well-feathered hens in temperate climates. This temperature effect may be twice as great if feather condition is poor. Feed intake may therefore be 10–20% higher than in other indoor systems. It is important to stress that nutritionists need to know the feed intake in order to optimize the nutrient content of the feed.

Optimizing energy intake at peak production is a critical factor for free range production and, in practice, problems with low energy intake can often be associated with failure to ensure the correct body weight at sexual maturity, poor uniformity during rearing (both body weight and frame size), poor physical feed presentation and feeder management. It is important to remember that feed intake increases by over 50% from light stimulation to peak egg production so the management of feed intake in the late rearing period must not be ignored. This requires emphasis on feed particle size, avoiding dust in feed tracks and distributing feed after lighting in the morning and again 6 h before lights go out.

For free range production, the separate effects of dietary ME content on egg weight and plumage have also to be considered. For these reasons it is more important to use a phased feeding programme. This may utilize an increasing ME content programme through the late rearing period to the peak lay diet (by increasing the minimum inclusion of dietary fat) to ensure optimal ME intake and egg size in early lay. Fat addition also aids palatability and secures ME intake at peak production. The ME content during lay can then be progressively reduced by increasing the inclusion of ingredients with high dietary insoluble fibre content (high sources of acid detergent fibre). This may increase feeding time and gut fill and limit the desire to eat feathers that causes feather pulling and poor plumage condition.

ENVIRONMENTAL IMPACTS OF NUTRITION IN ALTERNATIVE SYSTEMS

The sustainability of poultry systems is discussed in other chapters in this volume, so we focus here only on matters directly concerning nutrition. Williams et al. (2006) calculated that organic poultry meat and egg production increase energy use by about 30% and 15%, respectively, compared with 'conventional' systems (Tables 15.2 and 15.3). Similar calculations have been performed by Bokkers and de Boer (2009). This is because the lower energy needs for producing organic feed crops is more than counterbalanced by lower poultry performance. Experimental approaches (e.g. Kratz *et al.*, 2004) have also confirmed that retention efficiency of nutrients was lower in free range or organic production. Among the reasons for the differences were duration of growth period, strain of birds and feeding strategy. Providing optimally balanced dietary protein is usually practicable only with supplemental amino acids and is well known to have large effects on nitrogen losses (Kim and MacLeod, 2001). There are further possible environmentally detrimental consequences of organic production, related to restrictions on the use of 'non-organic' raw materials. For example, an analysis by the Institute for Energy and Environmental

Table 15.2. Burdens of some alternative poultry meat systems, expressed per tonne of meat (from Williams *et al.*, 2006).

Impacts and resources used	Non-organic	Organic	Free-range (non-organic)
Primary energy used (MJ)	12,000	15,800	14,500
Global warming potential (GWP$_{100}$, kg 100 year CO_2 equivalent)	4,570	6,680	5,480
Eutrophication potential (kg PO_4^{3-} equivalent)	49	86	63
Acidification potential (kg SO_2 equivalent)	173	264	230
Pesticides used (dose ha^{-1})	7.7	0.6	8.8
Land use (ha)	0.64	1.40	0.73
Nitrogen losses (kg)			
NO_3^--N	30	75	37
NH_3-N	40	60	53
N_2O-N	0.3	0.3	1.6

Table 15.3. Environmental burdens of layer systems, expressed per 20,000 eggs (from Williams *et al.*, 2006).

Impacts and resources	Non-organic	Organic	Caged non-organic	Free range non-organic
Primary energy used (MJ)	14,100	16,100	13,600	15,400
Global warming potential (GWP$_{100}$, kg 100 year CO_2 equivalent)	5,530	7,000	5,250	6,180
Eutrophication potential (kg PO_4^{3-} equivalent)	77	102	75	80
Acidification potential (kg SO_2 equivalent)	306	344	300	312
Pesticides used (dose ha^{-1})	7.8	0.1	7.2	8.7
Land use (ha)	0.66	1.48	0.63	0.78
Nitrogen losses (kg)				
NO_3^--N	36	78	35	39
NH_3-N	79	88	77	81
N_2O-N	7.0	9.0	6.6	7.9

Technology GmbH (IFEU, 2002) shows that supplementation of 1 kg synthetic DL-methionine requires less than 16% of the energy needed to provide the equivalent amount of methionine from soybean or rapeseed meal.

Nutrition-related methods of reducing environmental impact

Nutrition is the most immediate and readily accessible route to reducing nitrogen, phosphorus and other losses. This can be achieved by:

1. optimizing nutrient balance where possible (amino acids are the prime example);

2. not including large excesses or safety margins of nutrients (e.g. metals, chlorides, phosphorus) – this is aided by reliable data on requirements; and

3. maximizing availability of nutrients so that total quantities added and then excreted are minimized – dietary enzymes, especially carbohydrases and phytases, have helped with this aim.

The contributions of the poultry industry to nitrogen pollution are determined by the feed, the bird and the interactions between the two. The best-known dietary method of reducing nitrogenous waste is to use a protein composition (amino acid blend) which is closely matched to the bird's requirements (Meluzzi *et al.*, 2001). This is often described as 'ideal protein'. Mack *et al.* (1999) studied the ideal ratio of the essential amino acids lysine, methionine, threonine, tryptophan, arginine, valine and isoleucine in broiler chickens. The ideal ratios relative to lysine were calculated to be 0.75 for methionine + cystine, 0.63 for threonine, 0.19 for tryptophan, 1.12 for arginine, 0.71 for isoleucine and 0.81 for valine on a true faecal digestible basis. There are many published examples showing the nitrogen-loss benefits of using a balanced ('ideal') protein diet, but one illustration is by Kim and MacLeod (2001) (Table 15.4). This experiment showed nitrogen retention efficiency falling from 0.66 on a near-ideal protein to 0.42 on an imbalanced diet. Nitrogen retention did not change significantly, because of a constant and limiting dietary lysine concentration, but there was a 2.5-fold increase in nitrogen excretion.

The degree to which the ideal protein concept is used in commercial practice is an economic, legislative or consumer matter because much is known about the relevant biology. As well as reducing nitrogen losses to the wider environment, it may improve bird welfare by reducing nitrogen excretion and therefore improving floor and litter conditions and may also reduce ammonia concentration in the poultry house environment.

The improvements which a science-supported industry has produced over the years in poultry mirror the mechanisms used to explain why poultry compare well with other farm animal species in life cycle analysis. Maintenance costs have been reduced as a proportion of productive output in both meat and egg breeds. Reproductive 'fitness' is high in that birds become sexually mature in a relatively short time and each bird then produces a large number of offspring. A dilemma for proponents of alternative systems is that breeding or feeding for lower growth rates will tend to reverse these benefits.

Table 15.4. Nitrogen retention and loss by broiler chickens on diets with the same lysine concentration but a wide range of crude protein content.

Diet	1	2	3	4	5	SED[a]	P[b]
Total ME (MJ kg^{-1})	13.4	13.4	13.4	13.4	13.4		
CP (g kg^{-1})	180	210	240	270	300		
Lysine (g kg^{-1})	11	11	11	11	11		
Lysine:CP ratio	0.061	0.052	0.046	0.041	0.037		
N intake (g day^{-1} per bird)	4.10	4.18	5.29	5.90	6.18	0.212	<0.001
N retention (g day^{-1} per bird)	2.68	2.43	2.60	2.61	2.60	0.147	NS
N loss (g day^{-1} per bird)	1.41	1.75	2.68	3.29	3.59	0.168	<0.001
Efficiency of N retention	0.66	0.58	0.49	0.44	0.42	0.022	<0.001

ME, metabolizable energy; CP, crude protein; NS, not significant.
[a]Standard error of a difference between two means.
[b]Statistical significance of differences between means.

CONCLUSIONS

The known rules of nutrition apply to alternative systems. However, because of the variables affecting alternative systems such as locomotor activity and thermal environment, the precise application of nutritional principles requires observation and recording of flock performance against defined targets, with iterative quantitative or qualitative adjustment of nutrition as required. Many of the practical challenges to the nutritionist arise because of the prohibition or selection of ingredients for dogmatic rather than fact-based reasons. There is also the paradox that concern about the global environment can often coexist, in the same person, with advocacy of nutritional ideas which are likely to have a detrimental impact on the environment. Sparks *et al.* (2008) surveyed organic pullet producers and found that the most frequent reason given for being involved in organic production was 'commercial', with 'environmental' and 'welfare' being the next most frequent categories. This order of motivational priorities should not be decried but van de Weerd *et al.* (2009) identified 'quality and availability of organic feed' as one of the main challenges of organic poultry production, which may not always sit comfortably with commercial imperatives.

REFERENCES

Al-Saffar, A.A and Rose, S.P. (2002) The response of laying hens to dietary amino acids. *World's Poultry Science Journal* 58, 209–234.

Antell, S. and Ciszuk, P. (2006) Forage consumption of laying hens – the crop as an indicator of feed intake and AME content of ingested feed. *Archiv für Geflügelkunde* 70, 154–160.

Bennett, C.D. and Classen, H.L. (2003) Performance of two strains of laying hens fed ground and whole barley with and without access to insoluble grit. *Poultry Science* 82, 147–149.

Bokkers, E.A.M. and de Boer, I.J.M. (2009) Economic, ecological, and social performance of conventional and organic broiler production in the Netherlands. *British Poultry Science* 50, 546–557.

Commission Regulation (2008) Commission Regulation (EC) No 889/2008 of 5 September 2008 laying down detailed rules for the implementation of Council Regulation (EC) No 834/2007 on organic production and labelling of organic products with regard to organic production, labelling and control. *Official Journal of the European Union* L 250, 18/09/2008, 1–84; available at: http://eur-lex.europa.eu/LexUriServ/LexUriServ.do?uri=CONSLEG:2007R0834:20081010:EN:PDF (accessed 14 October 2011).

Council Regulation (2007) Council Regulation (EC) No 834/2007 of 28 June 2007 on organic production and labelling of organic products and repealing Regulation (EEC) No 2092/91. *Official Journal of the European Union* L 189, 20/07/2007, 1–23; available at: http://eur-lex.europa.eu/LexUriServ/LexUriServ.do?uri=OJ:L:2007:189:0001:0023:EN:PDF (accessed 14 October 2011).

Curnow, R.N. (1973) A smooth population response curve based on an abrupt threshold and plateau model for individuals. *Biometrics* 29, 1–10.

Defra (2010) Guidance Document on European Union Organic Standards. http://www.defra.gov.uk/foodfarm/growing/organic/standards/pdf/guidance-document-jan2010.pdf (accessed 14 October 2011).

D'Mello, J.P.F. (1994) Amino acid imbalances, antagonisms and toxicities: In: D'Mello, J.P.F. (ed.) *Amino Acids in Farm Animal Nutrition*. CABI Publishing, Wallingford, UK, pp. 63–97.

Elwinger, K., Tufvesson, M., Lagerkvist, G. and Tauson, R. (2008) Feeding layers of different genotypes in organic feed environments. *British Poultry Science* 49, 654–665.

FAWC (1997) *Report on the Welfare of Laying Hens*. Farm Animal Welfare Council, Tolworth, UK.

Fisher, C., Morris, T.R. and Jennings, R.C. (1973) A model for the description and prediction of the response of laying hens to amino acid intake. *British Poultry Science* 14, 469–484.

Forbes, J.M. and Covasa, M. (1995) Application of diet selection by poultry with particular reference to whole cereals. *World's Poultry Science Journal* 51, 149–165.

France, J. and Thornley, J.H.M. (1984) *Mathematical Models in Agriculture*. Butterworths, London.

Hadorn, R., Gloor, A. and Wiedmer, H. (2000) Einfluss des Ausschlusses von reinen Aminosäuren bzw. Potentiellen GVO-Eiweissträgern aus Legehennenfutter auf pflanzlicher Basis (Effect of the exclusion of synthetic amino acids and potentially GMO-protein sources in vegetable diets for laying hens). *Archiv für Geflügelkunde* 64, 75–81.

Henuk, Y.L. and Dingle, J.G. (2002) Practical and economic advantages of choice feeding systems for laying poultry. *World's Poultry Science Journal* 58, 199–208.

Hetland, H. and Svihus, B. (2001) Effect of oat hulls on performance, gut capacity and feed passage time in broiler chickens. *British Poultry Science* 42, 354–361.

Hetland, H., Svihus, B. and Olaisen, V. (2002) Effect of feeding whole cereals on performance, starch digestibility and duodenal particle size distribution in broiler chickens. *British Poultry Science* 43, 416–423.

Horsted, K., Hermansen, J.E. and Ranvig, H. (2007) Crop content in nutrient-restricted versus non-restricted organic laying hens with access to different forage vegetations. *British Poultry Science* 48, 177–184.

Hughes, B.O. (1972) A circadian rhythm of calcium intake in the domestic fowl. *British Poultry Science* 13, 485–493.

Hughes, B.O. (1979) Appetites for specific nutrients. In: Boorman, K.N. and Freeman, B.M. (eds) *Food Intake Regulation in Poultry*. British Poultry Science Ltd, Edinburgh, pp. 141–169.

Hughes, B.O. (1984) The principles underlying choice feeding behaviour in fowls – with special reference to production experiments. *World's Poultry Science Journal* 40, 141–150.

IFEU (2002) *Ökolibalanz fur methionin in der Geflügelmast* (*Life Cycle Analysis for Methionine in Broiler Meat Production*). Institute for Energy and Environmental Technology GmbH, Heidelberg, Germany.

Kim, J.H. and MacLeod, M.G. (2001) In: Chwalibog, A. and Jakobsen, K. (eds) *Energy Metabolism in Animals, Proceedings of the 15th EAAP Symposium on Energy Metabolism in Animals*, Snekkersten, Denmark, 10–16 September 2000. EAAP Scientific Series No. 103. Wageningen Academic Publishers, Wageningen, The Netherlands, pp. 113–116.

Kratz, S., Halle, I., Rogasik, J. and Schnug, E. (2004) Nutrient balances as indicators for sustainability of broiler production systems. *British Poultry Science* 45, 149–157.

Larbier, M. and Leclercq, B. (1994) *Nutrition and Feeding of Poultry* (translated into English by J. Wiseman). Nottingham University Press, Nottingham, UK.

Leeson, S. and Summers, J.D. (2008) *Commercial Poultry Nutrition*. Nottingham University Press, Nottingham, UK.

Lemme, A. Wijtten, P.J.A., van Wichen, J., Petri, A. and Langhout, D.J. (2006) Responses of male growing broilers to increasing levels of balanced protein offered as coarse mash or pellets of varying quality. *Poultry Science* 85, 721–730.

Mack, S., Bercovici, D., de Groote, G., Leclercq, B., Lippens, M., Pack, M., Schutte, J.B. and van Cauwenberghe, S. (1999) Ideal amino acid profile and dietary lysine specification for broiler chickens of 20 to 40 days of age. *British Poultry Science* 40, 257–265.

MacLeod, M.G. and Jewitt, T.R. (1985) The energy cost of some behavioural patterns in laying domestic fowl: simultaneous calorimetric, Doppler-radar and visual observations. *Proceedings of the Nutrition Society* 44, 34A.

MacLeod, M.G., Jewitt, T.R., White, J., Verbrugge, M. and Mitchell, M.A. (1982) The contribution of locomotor activity to energy expenditure in the domestic fowl. In: Ekern A. and Sundstol, E. (eds) *Energy Metabolism of Farm Animals, Proceedings of the 9th EAAP Symposium*, Lillehammer, Norway, September 1982. Department of Animal Nutrition, Agricultural University of Norway, Aas, pp. 297–300.

McKeegan, D.E.F., Savory, C.J., MacLeod, M.G. and Mitchell, M.A. (2001) Development of pecking damage in layer pullets in relation to dietary protein source. *British Poultry Science* 42, 33–42.

McNab, J.M. and Boorman, K.N. (eds) (2002) *Poultry Feedstuffs: Supply, Composition and Nutritive Value*. CABI Publishing, Wallingford, UK.

Meluzzi, A., Sirri, F., Tallarico, N. and Franchini, A. (2001) Nitrogen retention and performance of brown laying hens on diets with different protein content and constant concentration of amino acids and energy. *British Poultry Science* 42, 213–217.

Meyer, G.B., Babcock, S.W. and Sunde, M.L. (1970) Decreased feed consumption and increased calcium intake associated with the pullet's first egg. *Poultry Science* 49, 1164–1169.

Mongin, P. and Sauveur, B. (1979) The specific calcium appetite of the domestic fowl. In: Boorman, K.N. and Freeman, B.M. (eds) *Food Intake Regulation in Poultry*. British Poultry Science Ltd, Edinburgh, pp. 171–189.

Morris, T.R. and Gous, R.M. (1988) Partitioning of the response to protein between egg number and egg weight. *British Poultry Science* 29, 93–99.

Morris, T.R., Gous, R.M. and Fisher, C. (1999) An analysis of the hypothesis that amino acid requirements for chicks should be stated as a proportion of dietary protein. *World's Poultry Science Journal* 55, 7–22.

NRC (National Research Council) (1994) *Nutrient Requirements of Poultry*, 9th revised edn. National Academy Press, Washington, DC.

Nys, Y. (2000) Dietary carotenoids and egg yolk coloration – a review. *Archiv für Geflügelkunde* 64, 45–54.

Park, K.J., Robertson, P.A., Campbell, S.T., Foster, R., Russell, Z.M., Newborn, D. and Hudson, P.J. (2001) The role of invertebrates in the diet, growth and survival of red grouse (*Lagopus lagopus scoticus*) chicks. *Journal of Zoology* 254, 137–145.

Pesti, G.M., Vedenov, D., Cason, J.A. and Billard, L. (2009) A comparison of methods to estimate nutritional requirements from experimental data. *British Poultry Science* 50, 16–32.

Picard, M., Melcion, J.P., Bertrand, D. and Faure, J.M. (2002) Visual and tactile cues perceived by chickens. In: McNab, J.M. and Boorman, K.N. (eds) *Poultry Feedstuffs: Supply, Composition and Nutritive Value*. CABI Publishing, Wallingford, UK, pp. 279–300.

Ponte, P.I.P., Rosado, C.M.C., Crespo, J.P., Crespo, D.G., Mourao, J.L., Chaveiro-Soares, M.A., Bras, J.L.A., Mendes, I., Gama, L.T., Prates, J.A.M., Ferreira, L.M.A. and Fontes, C.M.G.A. (2008a) Pasture intake improves the performance and meat sensory attributes of free-range broilers. *Poultry Science* 87, 71–79.

Ponte, P.I.P., Alves, S.P., Bessa, R.J.B., Ferreira, L.M.A., Gama, L.T., Bras, J.L.A., Fontes, C.M.G.A. and Prates, J.A.M. (2008b) Influence of pasture intake on the fatty acid composition, and cholesterol, tocopherols, and tocotrienols content in meat from free-range broilers. *Poultry Science* 87, 80–88.

Ponte, P.I.P., Prates, J.A.M., Crespo, J.P., Crespo, D.G., Mourao, J.L., Alves, S.P., Bessa, R.J.B., Chaveiro-Soares, M.A., Gama, L.T., Ferreira, L.M.A. and Fontes, C.M.G.A. (2008c) Restricting the intake of a cereal-based feed in free-range-pastured poultry: effects on performance and meat quality. *Poultry Science* 87, 2032–2042.

Quentin, M., Bouvarel, I. and Picard, M. (2005) Short- and long-term effects of feed form on fast- and slow-growing broilers. *Journal of Applied Poultry Research* 13, 540–548.

Rama Rao, S.V., Praharaj, N.K., Reddy, M.R. and Panda, A.K. (2003) Interaction between genotype and dietary concentrations of methionine for immune function in commercial broilers. *British Poultry Science* 44, 104–112.

Rose, S.P. and Kyriazakis, I. (1991) Diet selection of pigs and poultry. *Proceedings of the Nutrition Society* 50, 87–98.

Rose, S.P., Fielden, M., Foote, W.R. and Gardin, P. (1995) Sequential feeding of whole wheat to growing broiler chickens. *British Poultry Science* 36, 97–111.

Saleh, E.A., Watkins, S.E., Waldroup, A.L. and Waldroup, P.W. (2004) Effects of dietary nutrient density on performance and carcass quality of male broilers grown for further processing. *International Journal of Poultry Science* 3, 1–10.

Savory, C.J. and MacLeod, M.G. (1980) Effects of grouping and isolation on feeding, food conversion and energy expenditure of domestic chicks. *Behavioural Processes* 5, 187–200.

Sparks, N.H.C., Conroy, M.A. and Sandilands, V. (2008) Socio-economic drivers for UK organic pullet rearers and the implications for poultry health. *British Poultry Science* 49, 525–532.

Svihus, B. and Hetland, H. (2001) Ileal starch digestibility in growing broiler chickens fed a wheat-based diet is improved by mash feeding, dilution with cellulose or whole wheat inclusion. *British Poultry Science* 42, 633–637.

Tullett, S.G., MacLeod, M.G. and Jewitt, T.R. (1980) The effects of partial defeathering on energy metabolism in the laying fowl. *British Poultry Science* 21, 241–245.

Umar Faruk, M., Bouvarel, I., Même, N. and Roffidal, L. (2010) Adaptation of wheat and protein–mineral concentrate intakes by individual hens fed *ad libitum* in sequential or in loose-mix systems. *British Poultry Science* 51, 811–820.

Van de Weerd, H.A., Keatinge, R. and Roderick, S. (2009) A review of health-related welfare issues in organic poultry production. *World's Poultry Science Journal* 65, 649–684.

Van Kampen, M. (1976) Activity and energy expenditure in laying hens. 2. The energy cost of exercise. *Journal of Agricultural Science, Cambridge* 87, 81–84.

Ward, S. and MacLeod, M.G. (1991) Doubly labelled water measurements of the energy metabolism of an avian species under different ambient temperatures and reproductive states. In: Wenk, C. and Boessinger, M. (eds) *Proceedings of the 12th Symposium on Energy Metabolism of Farm Animals*, Zurich, Switzerland, 9–7 September 1991. ETH Zurich, pp. 230–233.

Ward, S. and MacLeod, M.G. (1992) Energy cost of egg formation in quail. *Proceedings of the Nutrition Society* 51, 41A.

Ward, S. and MacLeod, M.G. (1994) Free-range calorimetry? Energy metabolism measured by the doubly-labelled-water technique (Abstracts of WPSA UK Spring Conference) *British Poultry Science* 35, 831.

Williams, A.G., Audsley, E. and Sandars, D.L. (2006) Determining the environmental burdens and resource use in the production of agricultural and horticultural commodities. Main Report. Defra Research Project IS0205. Bedford: Cranfield University and Defra. http://randd.defra.gov.uk/Document.aspx?Document=IS0205_3958_EXE.doc (accessed 20 July 2011).

Zuberbuehler, C.A., Messikommer, R.E. and Wenk, C. (2002) Choice feeding of selenium-deficient laying hens affects diet selection, selenium intake and body weight. *Journal of Nutrition* 132, 3411–3417.

CHAPTER *16*

Genotype–Environment Interaction: Breeding Layers with Different Requirements for Varying Housing Systems

W. Icken, M. Schmutz and R. Preisinger

ABSTRACT

In addition to conventional selection criteria like egg production, feed conversion and egg quality, traits related to animal welfare have become more important in Europe and North America. To improve these traits and simultaneously capture performance data in non-cage environments hen-specifically, the Weihenstephan funnel nest box (FNB) was developed. The FNB captures egg production and egg quality data individually as well as nesting behaviour traits. A comparison of performance parameters from full siblings, tested in single bird cages and the FNB, leads back to potential genotype–environment interactions that will determine which testing system should be mainly used in the future for continuous improvement of egg production and egg quality. Low genetic correlations between full siblings, tested in varying housing systems, were estimated for egg number during the main laying periods. Otherwise, high genetic correlations and therefore no potential genotype–environment interactions could be assumed for the traits egg weight and egg number at the beginning of production. An additional breeding tool which has the potential to improve selection traits, regardless of the housing system, is genome-wide selection. Therefore, phenotypic performance recording must first be established for new traits before markers can be applied. Due to all these assumed effects, for a comprehensive performance testing with an evaluation of birds, consequent selection and reproduction of the best layers, layer breeding companies should implement hen-specific tests in non-cage systems in their breeding programmes.

INTRODUCTION

The majority of table eggs produced in the world today are produced in cage systems (IEC, 2007). However, a number of egg consumers dislike the idea of laying hens being confined in cages. For this reason, these consumers prefer having eggs from layers which are kept under less confined conditions, such as floor or barn housing and free range systems. Additionally, regulatory requirements such as the directive to ban unenriched cages in the European Union (EU) from 2012 onwards, or the total cage ban already implemented in several European countries, are among several key factors which determine breeding goals.

The advantages of egg production systems which are alternatives to cage systems are that the layers can literally avoid each other, both horizontally and vertically, and thus show their inherited behaviour patterns. These advantages are obtained at the expense of higher housing cost per hen, more feed per kilogram of egg mass, higher labour cost, more floor eggs, more dirty eggs, feather pecking and/or cannibalism. Furthermore, integrating these behavioural traits in the breeding process requires more comprehensive and hen-specific data recording. For several performance traits, the accurate recording of each individual bird is most practical in single bird cages. However for behaviour traits, hens have to be tested in a large group of layers in which it is more difficult to get individual hen data.

Controlled housing in single bird and group cages has historically been considered as the most favourable environment for testing birds and lines to be housed in commercial cage production units. On several occasions, it has been suggested that cage testing of pure line and cross-line stocks results in birds that are specifically adapted to cages and less capable of adapting to alternative systems. This view is reinforced by the fact that alternative systems may be far more stressful to laying hens as compared with cages. Feather pecking and cannibalism, together with increased bacterial infections, are the main explanations for reduced performance and higher mortality rate in these systems (Table 16.1).

DIFFERENCES IN PERFORMANCE BETWEEN CAGE AND FLOOR HOUSING

Strains found to be superior in a cage environment may not be able to maintain their superiority in the environment of floor management. Therefore, the magnitude of genotype–environment interactions has to be estimated to optimize testing systems for within-line selection and to select the most suitable line combination for cross-line breeding. In a similar pattern to field experiences with random sample tests using floor management systems, a significant

Table 16.1. Laying hen mortality in different housing systems (Kreienbrock *et al.*, 2004).

Observation	Floor		Aviary		Cage
	No range	Free range	No range	Free range	
No. of flocks	46	50	8	30	172
Mean mortality (%)	12.9	14.0	15.1	17.8	8.2
Best 10%	4.6	6.1	2.3	7.2	3.6

Table 16.2. Comparison of random sample test results in different environments.

	Mortality (%)		Egg number		Egg weight (g)		FCR[c]		IOFC ranking[d]	
Strain	Cage[a]	Floor[b]	Cage	Floor	Cage	Floor	Cage	Floor	Cage	Floor
A	2.0	0.8	322	300	66.0	66.6	1.94	2.25	1	1
B	5.0	10.3	327	294	63.7	64.2	2.06	2.40	3	2
C	2.1	9.2	312	281	63.4	63.2	2.21	2.56	4	3
D	1.0	18.3	323	260	63.8	62.6	1.99	2.51	2	4
Mean	2.5	9.7	321	284	64.2	64.2	2.05	2.43		

[a]Cage testing: Random Sample Test results from Haus Düsse 1998–2000.
[b]Floor testing: Random Sample Test results from Neu-Ulrichstein 1999–2000.
[c]Feed conversion rate.
[d]IOFC = 1.60 × egg mass – 0.40 × egg mass × FCR.

increase in mortality and a reduced efficiency of egg production for a floor system compared with a cage system can be observed (Table 16.2). Differences in ranking for income over feed cost between testing stations indicate that there is a variable degree of adaptability of strains to various environments. For both stations, the egg weight profile and feed efficiency were very similar on a strain level. The rate of lay and mortality show the biggest changes from one environment to another. On the other hand, the number of tested birds per strain is very limited (maximum 360 birds per strain) and the general liveability in the cage environment was very good.

Field and test station results indicate that data recording and selection have to be conducted in an environment that resembles the production environment as closely as possible to minimize the risk of selection errors due to genotype–environment interactions. Recording of mortality and the cause of death in pedigree stocks during production must be recorded as part of the selection procedure. If cannibalism is the major reason for mortality in group housing, then the separation of aggressors and victims would be necessary. However, this is extremely labour intensive in large populations. In commercial breeding programmes preference has been given to sire ranking for the rate of mortality from different locations and housing systems, with family-wise housing in groups to optimize within-line selection and to avoid the risk of genotype–environment interaction.

RECORDING INDIVIDUAL PERFORMANCE IN GROUP HOUSING

It is recognized that recording individual performances and behaviour in group housing systems serves as the basis for genetic improvement of laying hens in alternative production systems. Since 2005, the Institute for Agricultural Engineering and Animal Husbandry of the Bavarian State Research Centre for Agriculture and Lohmann Tierzucht GmbH have cooperated in developing nest boxes for investigating individual laying performance and nesting behaviour of laying hens at the Thalhausen experimental station (Fig. 16.1). With the aid

Fig. 16.1. Experimental unit with 48 single nest boxes for individual testing in floor housing systems.

of 48 Weihenstephan funnel nest boxes (FNBs), daily egg number, nest acceptance, exact oviposition time and the duration of stay in a nest box are automatically recorded for each hen in flocks of up to 360 layers in one group (Icken *et al.*, 2010). Transponder technology is used in combination with a specifically developed single nest box to allocate the hen's behaviour and performance. Based on this recording system, genetic parameters for behaviour and performance traits in group housing systems were estimated and families with desirable performance profile were selected for line improvement in the Lohmann Tierzucht primary breeding programme.

Functionality of the funnel nest box

To support individual data recording, an antenna is integrated in the funnel nest floor and each hen is tagged with a passive transponder (Texas Instruments, HDX, ISO, 23.10 mm × 3.85 mm; Texas Instruments Incorporated, Dallas, Texas, USA). This transponder is fixed to the leg of the hen using a leg band (Roxan iD, Selkirk, UK). This transponder–antenna system guarantees that each hen is identified as soon as she enters the single nest box and recorded until she leaves. Furthermore, a trap device at the entrance of the FNB (Figs 16.2 and 16.3) ensures that the single nest box is occupied by only one hen at any one time. The funnel nest floor locks the nest box while a hen is inside and helps to orientate the hen to place her head towards the nest entrance for a better reading accuracy as well as guarantee that every egg rolls out of the nest

(a)

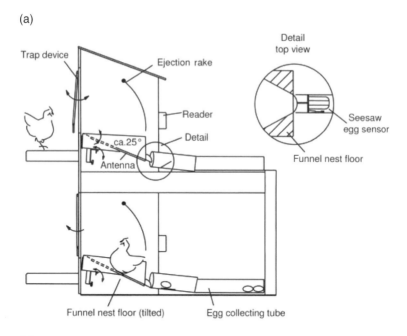

(b)

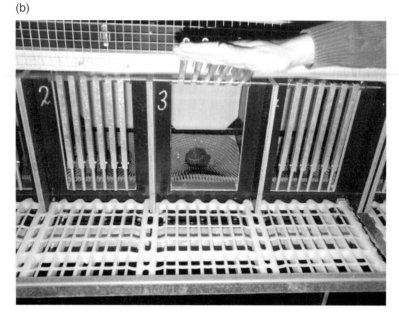

Fig. 16.2. Weihenstephan funnel nest box (FNB): (a) diagrammatic representation and (b) front view of unoccupied FNB.

(a)

(b)

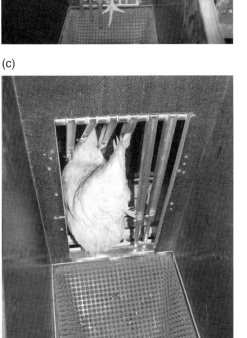

(c)

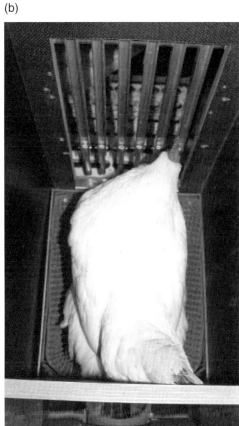

Fig. 16.3. A hen in three phases: (a) entering the nest, (b) staying inside and (c) leaving the nest.

immediately after being laid. The egg is registered at the seesaw egg sensor directly after leaving the nest box. All eggs in a nest box are collected in the order of lay in the egg collecting tube located behind the nest box. The combination of the transponder signal, the signal from the egg sensor and the position of the egg in the egg collecting tube enables the assignment of each egg to the individual hen.

The reliability of the correct assignment of each egg to each hen was tested with a small percentage of additionally housed hens which laid a different coloured egg to those being recorded in the flock. For these differently coloured eggs, the position of each egg in the egg collecting tubes was manually recorded on a daily basis. Afterwards, a check was made to find out whether a white or brown layer had been correctly assigned to a white or brown egg. The results show that 95.8% of these other coloured eggs were correctly assigned. The reasons for incorrect allocation were: (i) double nest occupation (<1% of eggs, e.g. an egg was laid while two hens were in the nest box thus a correct assignment was not possible); (ii) incorrect egg identification (approximately 2% of eggs, e.g. a soft-shelled egg blocked the seesaw egg sensor causing subsequently laid eggs to pile up and therefore they could not be registered); (iii) plausibility problems (<1% of eggs, e.g. two eggs were assigned to one hen on the same day); and (iv) non-specific errors (1%, e.g. the position of another coloured egg was not written down correctly). These results are similar to results observed from a video surveillance test resulting in approximately 98% of nest entrances and exits being correctly recorded.

Selection criteria for nesting behaviour

Nest acceptance

In analysing the components of nesting behaviour, the main focus is placed on nest acceptance which is defined as the number of 'saleable' eggs laid in the nest. At the experimental station at Thalhausen, distinctly different white egg and brown egg strains of Lohmann origin were performance tested in pens for 360 hens, equipped with FNBs. Daily nest visits were recorded along with an oviposition time for each hen during a period of up to 1 year. Based on the recorded number of nest eggs per hen, a breeding value can be estimated for egg production taking nest acceptance into account. This is then combined with traditional selection criteria in a selection index in order to perform pure line selection.

Oviposition time

Under a lighting regime of 16 h light and 8 h darkness, individual records of the exact oviposition time were taken. Based on this, we were able to compare the laying patterns of different lines of brown-egg and white-egg layers. It was found that most brown eggs were laid about 2 h after the lights were turned on, whereas a high percentage of the white-egg layers started looking for a nest 3 h after daylight had begun. The brown-egg layers had already reached the maximum rate of daily production at just 3 h after the lights came on, whereas the white-egg layers laid most of their eggs 6 h after the beginning of daylight. As shown in Fig. 16.4, the White Leghorn line concentrated the nest visits within a period of 2 h, whereas the brown egg line spread its nest visits over more than 4 h. Such a short time period in which most of the eggs were laid has also been observed by many other authors such as Lillpers (1993) and Zakaria et al. (2005), although the time of the day for this period of egg production activity differed between studies. While Zakaria et al. (2005) observed broiler breeder flocks with maximum egg production in the morning,

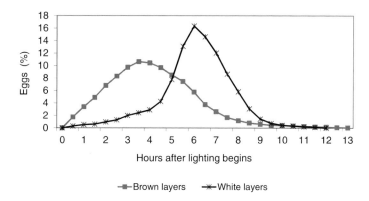

Fig. 16.4. Distribution of oviposition time during the day for two different strains.

Lillpers (1993) noticed that at between 31 and 51 weeks of age, the early afternoon was the main egg laying period for White Leghorn layers which were housed in individual cages.

Duration of time in the nest

The narrow time range of white-egg layers in terms of their oviposition time combined with longer nest visits means they require a longer nest occupation per egg laid (Table 16.3). Therefore, more nest space is needed as compared with brown-egg layers in order to avoid floor eggs. While the brown-egg layers occupied the nests for an average of 30 min, the white-egg layers spent 45 min in the nest for each oviposition. Shorter occupation times for white layers, which are in the same range as Icken *et al.* (2009) observed for brown layers, were investigated by Zupan *et al.* (2008). Nest visits without oviposition were mainly observed at the beginning of the laying period when hens habitually explore their new environment. Nest visits without oviposition lasted an average of 10 min for brown layers and nearly half an hour for white layers.

Reducing the duration of stay in the nest would be desirable if this can be achieved by selection. Apart from reduced investment for nests, faecal soiling of nests and eggs could also be decreased. However, possible negative correlations have to be kept in mind by reducing nest occupation, as there are hens that try hard to find a nest and therefore may also want to stay longer in it. The final objective must be to harvest the maximum number of saleable eggs from stress-free hens.

Table 16.3. Average oviposition time and duration of stay in the nest for brown and white layers.

Trait	Brown layer	White layer
Oviposition time (h:min)	08:00	09:45
Duration of time with oviposition (min)	30	45
Duration of time without oviposition (min)	10	28

Time intervals between ovipositions

As a component of total egg production, the variation in the time lag between two ovipositions could also be analysed from the detailed data (Table 16.4). Individual hens in four different flocks were classified by their mean time interval between ovipositions in a laying sequence. It was surprising to find that up to 22% of the hens in one flock laid two eggs with normal shells in less than 24 h. Conspicuously, these hens did not have the highest laying performance (Icken *et al.*, 2010). In the case of smaller clutches and more days off in between two clutches, they laid fewer eggs than the hens which belonged to the second category with an average time interval of between 24 h and 24 h and 15 min (Icken *et al.*, 2008a). The relationship between both traits is not linear and therefore our data only agree with the conclusion of Atwood (1929) for time intervals which take more than 24 h. Based on data for 172 laying hens, he stated that shorter time intervals between eggs is correlated with longer clutch lengths which indicates a higher overall laying performance. Later investigations of Yoo *et al.* (1988) on a White Leghorn strain, as well as of Bednarczyk *et al.* (2000) on more than 2000 Rhode Island White hens, confirmed these correlations and suggested that clutch traits may be used effectively in the selection index of laying hens.

Selection criteria for performance traits

Saleable nest eggs

The main breeding target for layers is still to maximize the number of saleable eggs that represents a high proportion of the economic efficiency of the egg producer in many markets. In alternative housing systems, this overall target has to be adjusted according to nest acceptance. This goal therefore changes to the number of saleable nest eggs. This trait is already recorded in the FNB when capturing the nest acceptance data described above. The simultaneous recording of the two important traits of nest acceptance and egg number is of additional benefit to the selection process. However, a larger number of

Table 16.4. Percentage of hens laying in different time interval categories and corresponding laying performance.

	Mean time interval of laying sequences (h:min)							
	<24:00		24:00–24:15		24:15–25:00		>25:00	
Flock	Hens (%)	Eggs (no. per 100 hen-days)	Hens (%)	Eggs (no. per 100 hen-days)	Hens (%)	Eggs (no. per 100 hen-days)	Hens (%)	Eggs (no. per 100 hen-days)
1	3	70	70	79	22	70	3	43
2	22	79	63	80	12	67	2	27
3	20	63	57	72	18	58	4	24
4	9	57	74	72	11	67	1	9

selection criteria may reduce the power of each single criterion in the selection index especially if the traits are negatively genetically correlated.

Over a period of 5 months, three flocks of Lohmann lines were tested in FNBs. Concurrently, full siblings of these layers were also tested in single bird cages. A comparison of their performance data, which is shown in Table 16.5, shows differences in the performance level of strains in different housing systems.

The hens of all three flocks were individually tested over five laying periods. One laying period always included 28 days and started when the hens were 20 weeks old. In accordance with the routine selection process, in which performance data of one testing year are divided into three different parts, the average laying performance was documented for the laying periods 1 to 2 (beginning of lay) and 3 to 5 (peak of production). For both parts of lay (beginning and peak), the hens tested in the FNB showed a reduced laying performance and therefore produced a lower number of saleable nest eggs than the hens in cages (Table 16.5). Next to some displaced floor eggs, the number of saleable nest eggs in the floor system was decreased by a small percentage of incorrectly allocated eggs which had to be discounted. However, in this study, the previously known higher performance potential of the white layer breed as compared with the brown layer breed was confirmed for both housing systems using both individual data recording techniques. On this basis, genetic parameters such as heritabilities and genetic correlations were estimated.

Table 16.6 shows that at the beginning of lay the estimated heritabilities for egg number were, but for one exception, higher than those for the laying periods 3 to 5. This is in accordance with many other studies which show lower heritabilities in terms of decreasing variance with increasing egg production (Savaş et al., 1998). Nurgiartiningsih et al. (2002) referred to heritability estimates of $h^2=0.02$ to 0.42 for the trait egg number, which are on a low to medium level. Therefore, the heritability of $h^2=0.63$ in Table 16.6 seems to be

Table 16.5. Average laying performance for brown and white layer sibling flocks, individually tested in a floor system or in single bird cages.

Flock	Housing system	Laying performance in periods[a] (no. per 100 hen-days)	
		1–2	3–5
1A[b]	Cage	43	95
1B[b]	Floor	47	84
2A[b]	Cage	61	95
2B[b]	Floor	43	81
3A[c]	Cage	70	96
3B[c]	Floor	61	89

[a]Laying performance in successive 28-day laying periods.
[b]Brown layer.
[c]White layer.

Table 16.6. Heritabilities (in bold) and genetic correlations (normal font) with their standard errors in parentheses for the number of saleable nest eggs in three flocks of white or brown layers housed in individual cages or housed on the floor with egg recording by the funnel nest box.

Flock	Housing	Egg number in periods 1–2		Egg number in periods 3–5	
		Cage	Floor	Cage	Floor
1A[a]	Cage	**0.26 (0.04)**	+0.97 (0.38)	**0.10 (0.04)**	+0.44 (0.23)
1B[a]	Floor		**0.15 (0.13)**		**0.63 (0.19)**
2A[a]	Cage	**0.29 (0.04)**	+0.56 (0.25)	**0.14 (0.03)**	+0.18 (0.24)
2B[a]	Floor		**0.31 (0.15)**		**0.29 (0.12)**
3A[b]	Cage	**0.39 (0.04)**	+0.94 (0.15)	**0.11 (0.03)**	+0.22 (0.41)
3B[b]	Floor		**0.38 (0.11)**		**0.12 (0.11)**

[a]Brown layer.
[b]White layer.

very high, i.e. at least for flock 1B. Additionally, Table 16.6 represents genetic correlations between full sibs, tested in the FNB or cages, respectively. With relatively high genetic correlations of r_g=+0.6 to +0.9 for the egg number at the beginning of lay, it can be assumed that there is nearly no genotype–environment interaction for this trait unlike at peak production, where the estimated correlations are much lower (r_g=+0.18 to +0.44) with relatively high standard errors. This suggests that genotype–environment interactions might exist for the egg number at peak production.

Egg quality

For consumers and the egg industry, an intact shell is the first and most important egg quality criterion. Unless the egg has an intact shell, it is downgraded and not saleable as an egg of good quality. Various sources of variation for shell quality have been reported in the literature, including strain, age of hen, nutrition, health, cage design and other mechanical stress factors from oviposition to the consumer (Carter, 1975; Cordts et al., 2001; Dunn et al., 2005).

   Primary breeders of egg laying chickens have always included shell strength in their breeding goals and probably improved shell strength at comparable ages. The problem of defective shells increasing towards the end of the laying period still remains and is often the main reason for depleting a flock when production is still above 80%. Data to predict egg breakage later in life are usually captured before the hens reach 1 year of age when the main selection on partial records is carried out. At this early age, most eggs have good shells and the accuracy of predicting the rate of breakage depends on the method used to evaluate shell quality.

   Carter (1971) concluded from pilot experiments that most cracks occurring in battery cages at oviposition are produced when the eggs drop on the cage floor. Variables affecting the probability of breakage at this point not only include intrinsic shell characteristics but also the material of the cage floor, egg m‹ the drop height, for which the author documented breed differences in a paper (Carter, 1975). In group housing systems, additional aspects wnich influence egg quality have to be considered as well. For example, management requirements typically provide one square metre nest space for up to 120 hens in floor housing (LTZ, 2010). On its own, this implies that many more hens can potentially damage the egg shell during oviposition than in conventional cages. Results of the international random sample test in Ustrasice 2008/2009 (Krekulová and Ripplová, 2009) show that the percentage of cracked eggs increases with increasing group size. Figure 16.5 shows the percentage of cracked eggs for all strains which is a little higher in enriched cages than in conventional cages (2.6% versus 2.2%) and much higher (7.8%) in alternative housing systems as compared with conventional cages.

Shell quality criteria

Direct selection against defective eggs cannot be very effective because the occurrence of shell defects is too low to exert significant selection pressure at the time of the main selection when the hens are less than 1 year of age. A simple way to support adequate shell quality is to only include eggs with apparently normal shells in the egg count, which can change the genetic correlation between egg production and shell strength from slightly negative to zero or even slightly positive.

All primary poultry breeders are still practising indirect selection for shell strength using a variety of destructive and non-destructive methods. The latter have the theoretical advantage that the eggs can still be used after measurement, but in view of the low prices per egg and EU food safety regulations, this argument carries less weight than speed and accuracy of measurement,

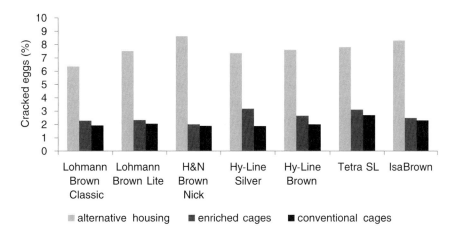

Fig. 16.5. Percentage of cracked eggs for different strains from the Ustrasice Random Sample Test 2008/2009.

heritability and genetic correlation with shell damage under commercial conditions (Dunn *et al.*, 2005). Indirect methods differentiate shell quality among eggs with apparently normal shells.

Shell breaking strength

To determine the egg's breaking strength, eggs are placed between two plates and subjected to increasing pressure until the shell breaks. The force necessary to break the shell is expressed in Newtons. Breaking strength may be measured between the poles or at the equator, which simulates different risks of breakage under field conditions. According to Cordts *et al.* (2001), the average breaking strength is somewhat lower when pressure is applied at the small pole but the variation is unaffected by the position during measurement. Shell breaking strength has been used in German random sample tests for many years as the main criterion of shell quality (Preisinger *et al.*, 1998) and is also being performed for all pure breed hens in the routine performance testing of Lohmann Tierzucht GmbH. Additionally, Lohmann Tierzucht is testing hen-specific eggshell breaking strength under floor housing conditions. This individual testing, however, is only possible with the FNB system described above which allows an assignment of the egg to the hen.

Eggshell quality was measured in three subsequent flocks of up to 280 white or brown layers, which were 25 to 30 weeks old. On the following eight to 15 days the eggs of each hen were collected and individually measured for their egg weight and breaking strength. Despite variation from flock to flock, the average breaking strength for all tested eggs was 49 N. Compared with their full sibs tested in single bird cages at an age of 39 weeks, the average value for each flock in floor housing was 2–3 N lower. The estimated heritabilities (Table 16.7) were independent of housing system and very constant in each flock. The values ranged from $h^2=0.32$ to $h^2=0.41$ on a medium level. Between full sib hens tested in cages or floor housing, respectively, the estimated genetic correlations for one brown (flock 2) and one white layer flock (flock 3) were

Table 16.7. Heritabilities (in bold) and genetic correlations (normal font) with their standard errors in parentheses for the egg quality traits of breaking strength and egg weight.

		Breaking strength		Egg weight	
Flock	Housing	Cage	Floor	Cage	Floor
1A[a]	Cage	**0.32 (0.05)**	+0.37 (0.20)	**0.72 (0.06)**	+0.78 (0.12)
1B[a]	Floor		**0.32 (0.16)**		**0.47 (0.15)**
2A[a]	Cage	**0.41 (0.04)**	+0.79 (0.16)	**0.67 (0.05)**	+0.99 (0.06)
2B[a]	Floor		**0.38 (0.13)**		**0.69 (0.11)**
3A[b]	Cage	**0.32 (0.04)**	+0.68 (0.16)	**0.71 (0.04)**	+1.00 (0.10)
3B[b]	Floor		**0.41 (0.12)**		**0.48 (0.03)**

[a]Brown layer.
[b]White layer.

high (r_g=+0.68 and r_g=+0.79). However, the genetic correlation in the breaking strength for the first flock was low (r_g=+0.37). Due to these inconsistent genetic correlations for breaking strength, it is not possible to give a clear conclusion regarding genotype–environment interactions for eggshell strength.

Egg weight

As shown in Table 16.7, the estimated genetic correlations for the relevant trait egg weight are very high. In the study genetic correlations that are close to +1.00 were emphasized, suggesting that the ranking of families is very similar for the highly heritable trait of egg weight. In the same study, a high genetic correlation was estimated for the body weight of hens reared and housed under different conditions. Therefore, no genotype–environment interaction is expected for different housing systems, too. Less costly data recording in single bird cages rather than in floor housing seems to be sufficient to select hens with an optimal egg weight for alternative housing systems.

Regardless of the data recording system, the difficulty of egg weight changing with hen age will always exist. Management factors such as lighting regime, body weight or feeding have an additional impact on the hen's age with the first egg as well as on the egg size but they do not compensate the effects of ageing at all. The egg producer demands a high percentage of eggs in the weight range preferred by local egg markets, which is usually achieved by selecting for a flat egg weight curve. This curve reaches the desired level soon after the start of lay begins, with moderate increases in egg weight.

As the egg weight in the first third of the entire production period increases considerably and this is above the preferred market optimum in the last third of the production, the breeder is challenged to select for rapidly increasing egg weight at the beginning of lay. If the optimum weight of about 60 g is achieved, all subsequent increases should continue on a minimum level.

As the egg weight in the first third of the production period is genetically closely correlated to the weight in the later periods (r_a≥0.85), the extent of the breeding influences on the egg weight curves is very restricted. Progress can only be achieved in small steps, based on continuous adjustments in each pure line generation (Ferrante *et al.*, 2009).

SELECTION CRITERIA FOR RANGING BEHAVIOUR

Additionally to the FNBs, the barn is also equipped with four electronic pop holes (EPHs). Through these EPHs, the hens have access to an adjacent winter garden (surface area about 40 m²) with a concrete floor that is littered with straw. The winter garden is covered with a waterproof plastic, permeable to light and surrounded by wired fences. Each EPH is equipped with two antennae which are integrated into the approaching boards that are attached to the pop holes on both sides to ease passage through the pop hole (Fig. 16.6). The direction of a bird entering the EPH can be determined with the order of the transponder readings at both antennae. A software package based on a specific evaluation routine determines the whereabouts of each hen (barn, pop hole or winter garden) throughout the day including the length of time spent at each location. These data make it possible to determine the time of the day when the hens go into the roofed winter garden and the total time spent in the outdoor area. Simultaneously, data on the laying performance of each individual hen are recorded in the FNBs in the same barn. Therefore, it is possible to estimate correlations between hen-specific ranging behaviour and laying performance.

The free range behaviour of the 272 laying hens was registered continuously 24 h per day, for 11.5 months from January until December. Despite the possibility to use the adjoined winter garden, around 35% of the hens did not visit the outdoor facilities with some variation observed between the laying periods. At the beginning of the observation period, only 26% of the hens used the opportunity to visit the roofed outside area. In the course of time and with increasing familiarity, the proportion of hens which used the winter garden rose to more than 60% from September onwards. From laying period 4 onwards, the proportions of hens per day which used the winter garden in each laying period were similar. If a hen went out once, she normally repeated this the next day. If not, there were many different environmental parameters which should be considered for the reduced acceptance. The outside stay per hen and day consisted on average of 11 single visits throughout all periods. The frequency of passages decreased from 13 passages per hen and day in the fifth period, to eight passages in laying period 12 (Icken *et al.*, 2008b). The

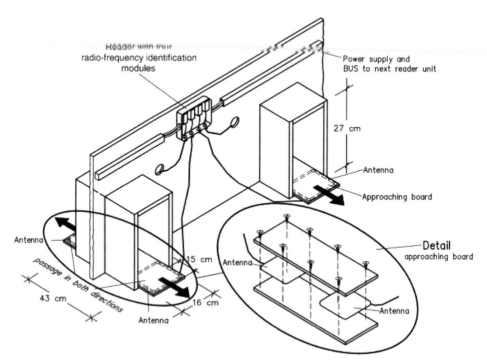

Fig. 16.6. Diagrammatic representation of the electronic pop hole.

length of each visit depended mainly on age and increased from the first to the 12th period from 13 min to 32 min (Fig. 16.7).

In order to analyse the genetic parameters, the following repeatability model was used for the traits frequency of passages and length of stay in the winter garden:

Model: $y_{ijk} = \mu + d_i + pe_j + a_k + e_{ijkl}$ (16.1)

where

y_{ijkl} = individual observation for the corresponding trait per day i and animal k within period j
μ = overall mean
d_i = fixed effect of day i
pe_j = permanent environmental effect for period j
a_k = random effect animal k
e_{ijkl} = random error.

The heritabilities were estimated for each 28-day laying period. Table 16.8 shows that the h^2 values were higher for the last five laying periods (8 to 12) than at the beginning of the observation time. During these periods, the heritabilities for the trait length of stay in the winter garden per day ranged from $h^2=0.21$ to 0.32 whereas the heritability for the trait frequency of passages per day ranged from $h^2=0.30$ to 0.49. The frequencies of passages for successive laying periods were highly positively correlated to each other ($r_g=+0.82$), showing that hens which often visited the winter garden in one laying period repeated this behaviour in other periods as well.

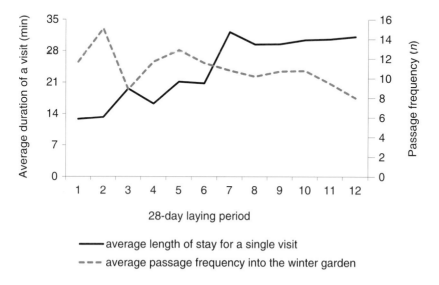

average length of stay for a single visit
average passage frequency into the winter garden

Fig. 16.7. Average length of stay for a single visit and average passage frequency into the winter garden per hen and day.

Table 16.8. Estimated heritabilities and standard errors (SE) for the traits length of stay in the winter garden per hen and day and frequency of passages into the winter garden.

| Laying period | Length of daily stay in winter garden | | Frequency of passages into the winter garden | |
	Heritability	SE	Heritability	SE
1	0.08	0.04	0.09	0.04
2	0.10	0.08	0.15	0.09
3	0.04	0.03	0.00	0.00
4	0.21	0.11	0.16	0.10
5	0.19	0.10	0.18	0.11
6	0.10	0.06	0.14	0.07
7	0.14	0.06	0.18	0.07
8	0.21	0.08	0.30	0.09
9	0.32	0.11	0.44	0.13
10	0.22	0.10	0.32	0.08
11	0.28	0.10	0.45	0.14
12	0.29	0.09	0.49	0.14

On basis of the FNB results, simultaneously recorded performance data make it possible to estimate the relationship between ranging behaviour and laying performance of one single hen. Table 16.9 shows genetic and phenotypic correlations between the traits laying performance, frequency of passages and length of stay in the winter garden. The corresponding heritability estimates can be found on the diagonal. Due to an early infection with the bacterium *Mannheimia haemolytica*, which strongly affected the behaviour and performance of the hens, genetic analyses were based on the captured data of laying periods 5 to 12. A heritability of $h^2=0.24$ was estimated for both traits (frequency of passages and length of stay in the winter garden). These estimates were higher than the estimated value for laying performance ($h^2=0.16$) during the same period of time. The genetic correlations were negative between both parameters for ranging behaviour and the laying performance. Only a slightly negative correlation between the traits frequency of passages and laying performance was observed, whereas a moderate negative correlation ($r_g=-0.34$) was detected between the traits length of stay in the winter garden and laying performance. The data suggest that hens which are often in the free range area appear to stay there for long periods and may therefore not return to the nest boxes in the barn for laying. Alternatively it has to be considered that because the laying performance was measured with the FNB, only eggs that were laid in the FNB were registered and the floor eggs were not assigned to the hen. The close genetic correlations between the frequency of passages and the length of stay in the winter garden was expected and validated by the high phenotypic correlations ($r_g=+0.82$ and $r_p=+0.86$). The phenotypic correlations of the ranging behaviour traits with laying performance were very low.

Table 16.9. Genetic correlations (above the diagonal), heritability estimates (bold, on the diagonal) and phenotypic correlations (below the diagonal), together with the corresponding standard errors (in parentheses), for the traits laying performance, frequency of passages and length of stay in the winter garden, for laying periods 5 to 12.

	Laying performance	Frequency of passages	Length of stay in winter garden
Laying performance	**0.16 (0.07)**	−0.08 (0.41)	−0.34 (0.14)
Frequency of passages	+0.08	**0.24 (0.13)**	+0.82 (0.14)
Length of stay in winter garden	+0.07	+0.86	**0.24 (0.12)**

The results for the heritability estimates of ranging behaviour traits suggest that the ranging activity of hens can be sustainably influenced through selection with specific selective breeding. With regard to the intensive technical effort to record individual ranging behaviour data and the negative genetic correlation of the ranging behaviour traits with laying performance, the question remains whether a selection for a better utilization of the ranging area might have negative effects on overall economic competitiveness. For special strains with high suitability for free range environments, the method described above may be an important approach. In order to meet generally reliable statements about the suitability of different genotypes for free range environments, tests on housing in different seasons need to be performed.

Selection criteria for non-nest related behaviour traits

Non-nest related behaviour traits are gaining more commercial interest even for birds specialized for alternative housing systems. In this context, the two traits, feather pecking and cannibalism, require specific consideration.

It has been suggested that cage testing of pure line and cross-line stocks results in birds that are specifically adapted to cages and less capable of adapting to alternative systems (Muir, 1996; Ellen *et al.*, 2010; Peeters *et al.*, 2010). This view has been ascribed to the fact that alternative systems and organic feeding methods are far more stressful. Feather pecking and cannibalism along with an increased risk of bacterial infection are the main reasons for reduced performance, higher mortality and far costlier disease prevention programmes in these systems (Kjaer, 2000; Klein *et al.*, 2000). If selection against feather pecking is part of a breeding programme, data recording and selection have to be done in an environment that resembles the production environment as closely as possible to minimize the risk of genotype–environment interaction. Group size has a significant effect on the social structure within an environment. If the group is small, the hierarchy among its members will be very stable. With increasing group size, the frequency of changes in ranking will be much greater, including the risk of fighting. Both feather pecking and cannibalism will become much more prevalent.

Scoring the quality of plumage on different parts of the body is a common tool for studying the genetics of feather pecking behaviour in laying hens. Individual scores of full and half sib daughters caged together are used as input for breeding value estimation and selection. With increasing age, a higher frequency of damage can be observed. Higher heritability estimates indicate that genetic variability is more visibly expressed in older birds. A disadvantage of this testing system is that feather pecking, aggressive pecking and cloacal cannibalism cannot be recorded as single traits.

Related to an aggressive behaviour of layers are different studies of feather pecking behaviour. Bilcik and Keeling (2000) and Kjaer (2000) have shown a genetic predisposition for feather pecking. Additionally, Kjaer (2000) reported differences in the tendency of feather pecking between and within strains. He showed that White Leghorn (LSL) strains of chickens have a considerably lower tendency for feather pecking than Lohmann Brown. Recently, in a genomic study Flisikowski et al. (2009) identified a locus with a large effect on the propensity for feather pecking. They also reported an association of the DRD4 gene with the exploratory behaviour in laying hens which can be used in genome-wide selection.

GENOME-WIDE SELECTION

Meuwissen et al. (2001), Goddard and Hayes (2007) and Calus (2009) state genome-wide selection as an important tool in the genetic improvement of livestock species in the prediction of breeding values. Genomic selection uses dense marker maps to predict the breeding value of animals with reported accuracies that are up to 0.31 higher than those of pedigree indices, without the need to phenotype the animals themselves or close relatives thereof. The basic principle is that because of high marker density, each quantitative trait locus is in linkage disequilibrium with at least one marker nearby. This approach has become feasible thanks to the large number of single nucleotide polymorphism (SNPs) discovered by genome sequencing and new methods to efficiently genotype a large number of SNPs.

Many theoretical advantages of genomic selection have to be weighed against the substantial expenditure. In addition to launching costs for the establishment of the method in each line or gene pool, there are also substantial costs for genotyping all candidates in each generation of selection. There is a theoretical potential for savings in performance testing (e.g. due to shorter testing periods and earlier selection decisions). However, in the learning process of the first several generations, there will be no possibility to economize on the costs of performance testing because the effective contribution of genomic selection depends on complex genotyping and the correlation between phenotypic parameters and markers.

The establishment of genomic selection in all lines and application to select between and within families requires that performance testing continues in future for all economically relevant characteristics in order to verify the linkage between the marker and the trait. Since commercial layers are a cross of different lines, genome-wide selection has to be established in all lines for pure and cross-line performance and continuously adjusted. The accuracy of this conventional phenotyping determines the success of subsequent genomic selection. The use of a broad calibration and genome-wide genotyping with SNPs will, it is hoped, identify regions of the genome associated with specific traits. Until there is a wealth of line repeatability studies, each line has to be analysed individually and the parameters estimated from one line cannot be simply extrapolated to another line. Commercial hybrids usually constitute four-way crosses which means the cost of genotyping four lines needs to be considered before results can be expressed in the field in the commercial generation. Furthermore, the four lines for a white-egg breeding programme have nothing in common with the four lines for a brown-egg breeding programme of the same primary breeder. The calibration process has to therefore be carried out at least eight times. After the first genomic selection and reproduction of cross-line offspring, it is possible to begin to measure the selection response in comparison with conventional selection. This comparison will provide information to assess the additional benefit of genomic selection. To reduce the cost for genotyping, the set of markers can be readjusted after this initial phase and reduced to the most informative regions. With a small line-specific SNP-Chip, the cost of routine genotyping can be substantially reduced.

For poultry, genomic selection is expected to contribute primarily to more accurate breeding value estimations and, for layers and (meat) breeders, to a shorter generation interval. These two factors will combine to speed up the annual rate of progress in selective breeding. The benefits of genomic selection should eventually become apparent in terms of lifetime productivity and lower susceptibility to disease. Furthermore, genome analysis can help to describe the current gene pool more accurately and to optimize effective population size without sacrificing selection intensity, while focusing on short-term breeding progress. Based on simulation studies, it has been calculated that breeding progress can be increased by 20–40% annually by extensive application of genomic selection (Avendaño *et al.*, 2010).

In layer breeding, the selection among full brothers at an early stage and the prediction of persistency of egg production and egg quality are of major interest for genomic selection. Males are selected traditionally based on the performance of their sisters and female relatives of previous generations. Therefore, full brothers have identical breeding values at point of selection although their real genetic potential varies greatly, as will be demonstrated by their progeny. If all males were reared and complete families of full brothers selected, the inbreeding would increase dramatically. Therefore, only a few sons per dam are presently being raised and the number of sons selected per sire is restricted. The objective with genomic selection would be to keep as many sons per family at hatch and to reduce them to suit the available rearing capacity by within-family selection on the basis of marker information.

First experiments using a DNA chip with 600,000 SNPs have begun. The calibration data consist of performance parameters from three or more generations. The first generation of offspring was reproduced from sires and dams selected exclusively on genomic breeding values in 2009, ignoring their phenotypic performance. The regression of offspring performance on parent performance will provide information on the accuracy of genomic selection and realized genetic gain (Wolc *et al.*, 2010).

As positive examples for effective marker-assisted selection in poultry, we can point to: (i) the elimination of fishy odour in brown-shelled eggs caused by a mutation on chromosome 8; (ii) reduced susceptibility to Marek's disease; and (iii) reduced susceptibility to infection by *Escherichia coli* (Honkatukia *et al.*, 2005; Cavero *et al.*, 2008). It is hoped that selection against feather pecking will also benefit from marker-assisted selection (Flisikowski *et al.*, 2009) in the near future.

CONCLUSIONS

Changing consumer preferences and regulatory requirements along with biological constraints are among the key factors which determine breeding goals. Egg production from conventional cages will inevitably be further reduced in the EU over the next few years. This will lead to a growing demand for hens that are specially adapted to floor and free range systems in order to secure income from egg production within Europe. As a consequence, breeding programmes and selection decisions will have to be based on a more complex selection index. Additional selection traits such as nesting and ranging behaviour will have to be included; but before including this as a selection objective, it is necessary to have a practicable and reliable data recording system that enables hen-specific data recording. Nesting behaviour traits such as the most important trait, nest acceptance, can be individually captured with the FNB. Furthermore, this newly developed single nest box allows the assignment of eggs to individual hens in floor housing systems. Therefore, it is also possible to get performance and egg quality data for each individual hen in alternative housing conditions. The balance between the most important economic traits – number of saleable nest eggs and egg quality parameters – may differ between various housing systems and markets. In terms of such potential genotype–environment interactions, full sibs in single bird cages were tested at the same time as their siblings in FNBs. Estimated genetic correlations between the data of sisters in single bird cages and nest boxes displayed a moderately close genetic correlation for egg number at the beginning of lay (r_g=+0.56 to +0.97) and a high correlation for egg weight (r_g=+0.78 to +1.00). The genetic correlations for the average egg number in the later production period were lower (r_g=+0.18 to +0.44). The low correlations for egg number during the main laying period, as well as the important trait of nest acceptance, enhance the importance of the FNB as a performance testing method in future layer breeding. Further investigations with a larger number of hens in a floor housing system have been planned with additional nest boxes. The newly recorded larger amount of data should lead to more accurate evaluations that will increase the value of the traits of interest – nesting behaviour and performance – in these systems.

Information from less costly performance tests in single and group cages should not, however, be ignored. They require a fraction of the effort in terms of labour, time and expense as the alternative recording system. Aside from self-made errors, data recording is 100% accurate. Also for management reasons, it is much easier to get reliable hen data from cages compared with floor housing systems even with sophisticated systems such as the FNB. Furthermore, group/family cages are the only currently feasible method for recording feather pecking at least on a family basis where a separation between victims and aggressors is possible and viable. Up to now other practicable solutions have been extremely time consuming, but combined with genome-wide selection may prove to be useful.

Genomic selection provides information which can already be used in growing animals of both sexes without performance testing. This increases the speed and accuracy of selection decisions. The prerequisite for the application is, however, upstream performance testing for all traits of commercial interest. Therefore, phenotypic performance recording must first be established for new traits before markers can be applied. Selection on molecular markers is no miracle to improve new traits directly. It is only an additional tool with the potential to increase the effectiveness of breeding without manipulating the genome of the birds. The breeding goals have to be defined and rates of progress predicted in order to offer the commercial poultry industry realistic expectations of future improvements. Short-term efforts to realize improvements in the areas of management and husbandry, hygiene and disease prevention, and last but not least to optimize nutrition, should not be reduced while expecting too much too soon from genomic selection.

For laying hens, there is still genetic variability to predict continuing progress in each generation in terms of persistency of production and egg quality, feed efficiency, health, behaviour and adaptability to different housing systems. Traits related to hen welfare will receive increasing attention in testing and selection. The high level of productivity already achieved, with rate of lay exceeding 90% for many months, is no reason to question whether further progress can be achieved. Testing in different housing systems and under field conditions will remain important while further developing genomic selection with company-specific DNA chips. With these tools, selection will become more complex and costly, but also help to tailor different strain crosses to specific needs of egg producers using different housing systems in different parts of the world market.

REFERENCES

Atwood, H. (1929) Observations concerning the time factor in egg production. *Poultry Science* 8, 137–140.

Avendaño, S., Watson, K.A. and Kranis, A. (2010) Genomics in poultry breeding – from utopias to deliverables. In: *Proceedings of the 9th World Congress on Genetics Applied to Livestock Production*, Leipzig, Germany, 1–6 August 2010. International Committee for World Congresses on Genetics Applied to Livestock Production and German Society for Animal Science, Bonn, Germany, p. 51.

Bednarczyk, M., Kiectzewski, K. and Swaczkowski, T. (2000) Genetic parameters of the traditional selection traits and some clutch traits in a commercial line of laying hens. *Archiv für Geflügelkunde* 64, 129–133.

Bilcik, B. and Keeling, L.J. (2000) Relationship between feather pecking and ground pecking in laying hens and the effect of group size. *Applied Animal Behaviour Science* 68, 55–66.

Calus, M.P.L. (2009) Genomic breeding value prediction: methods and procedures. *Animal* 4, 157–164.

Carter, T.C. (1971) The hen's egg: shell cracking at oviposition in battery cages and its inheritance. *British Poultry Science* 12, 259–278.

Carter, T.C. (1975) The hen's egg: relationships of seven characteristics of the strain of hen to the incidence of cracks and other shell defects. *British Poultry Science* 16, 289–296.

Cavero, D., Schmutz, M., Reinsch, N., Weigend, S., Voss, M. and Preisinger, R. (2008) Selection for Marek's resistance in White Leghorns. In: *Proceedings of the XXII World's Poultry Congress*, Brisbane, Australia, 30 June–4 July 2008. *World's Poultry Science Journal Supplement 2*. WSPA, Beekbergen, The Netherlands, p. 201.

Cordts, C., Schmutz, M. and Preisinger, R. (2001) Züchterische möglichkeiten zur verbesserung der schalenstabilität von eiern. *Lohmann Information* 3, 15–18.

Dunn, I.C., Bain, M., Edmond, A., Wilson, P.W., Joseph, N., Solomon, S., de Ketelaere, B., de Baerdemaeker, J., Schmutz, M., Preisinger, R. and Waddington, D. (2005) Heritability and genetic correlation of measurements derived from acoustic resonance frequency analysis; a novel method of determining eggshell quality in domestic hens. *British Poultry Science* 46, 280–286.

Ellen, E.D., Visscher, J., Rodenburg, T.B. and Bijma, P. (2010) Selection against mortality due to cannibalism in layers, does it work? In: *Proceedings of the 9th World Congress on Genetics Applied to Livestock Production*, Leipzig, Germany, 1–6 August 2010. International Committee for World Congresses on Genetics Applied to Livestock Production and German Society for Animal Science, Bonn, Germany, p. 399.

Ferrante, V., Lolli, S., Vezzoli, G. and Cavalchini, L.G. (2009) Effects of two different rearing systems (organic and barn) on production performance, animal welfare traits and egg quality characteristics in laying hens. *Italian Journal Animal Science* 8, 165–174.

Flisikowski, K., Schwarzenbacher, H., Wysocki, M., Weigend, S., Preisinger, R., Kjaer, J.B. and Fries, R. (2009) Variation in neighbouring genes of the dopaminergic and serotonergic systems affects feather pecking behaviour of laying hens. *Animal Genetics* 40, 192–199.

Goddard, M.E. and Hayes, B.J. (2007) Genomic selection. *Journal of Animal Breeding and Genetics* 124, 323–330.

Honkatukia, M., Reese, K., Preisinger, R., Tuiskula-Haavisto, M., Weigend, S., Roito, J., Mäki-Tanila, A. and Vilkki, J. (2005) Fishy taint in chicken eggs is associated with a substitution within a conserved motif of the *FMO3* gene. *Genetics* 86, 225–232.

Icken, W., Cavero, D., Schmutz, M., Thurner, S., Wendl, G. and Preisinger, R. (2008a) Analysis of the time interval within laying sequences in a transponder nest. In: *Proceedings of the XXII World's Poultry Congress*, Brisbane, Australia, 30 June–4 July 2008. *World's Poultry Science Journal Supplement 2*. WSPA, Beekbergen, The Netherlands, p. 231.

Icken, W., Cavero, D., Schmutz, M., Thurner, S., Wendl, G. and Preisinger, R. (2008b) Analysis of the free range behaviour of laying hens and the genetic and phenotypic relationships with laying performance. *British Poultry Science* 49, 533–541.

Icken, W., Thurner, S., Cavero, D., Schmutz, M., Wendl, G. and Preisinger, R. (2009) Analyse des nestverhaltens von legehennen in der bodenhaltung. *Archiv für Geflügelkunde* 73, 102–109.

Icken, W., Preisinger, R., Thurner, S. and Wendl, G. (2010) New selection traits from group housing systems. *Lohmann Information* 45(1), 22–26.

IEC (2007) *Comparison of International Country Data. International Egg Market. Annual Review 2007*. International Egg Commission, London.

Kjaer, J.B. (2000) Diurnal rhythm of feather pecking behaviour and condition of integument in four strains of loose housed laying hens. *Applied Animal Behaviour Science* 65, 331–347.

Klein, T., Zeltner, E. and Huber-Eicher, B. (2000) Are genetic differences in foraging behaviour of laying hen chicks paralleled by hybrid specific differences in feather pecking? *Applied Animal Behaviour Science* 70, 143–155.

Kreienbrock, L., Schäl, J., Beyerbach, M., Rohn, K., Glaser, S. and Schneider, B. (2004) EpiLeg-Orientierende epidemologische Untersuchungen zum Leistungsniveau und Gesundheitsstatus in Legehennenhaltungen verschiedener Haltungssysteme. *Abschlussbericht TiHo*. Stiftung Tierärztliche Hochschule Hannover, Hannover, Germany.

Krekulová, M. and Ripplová, E. (2009) *VII International performance test of laying type of hens – Final report 2008/2009*. Mezinárodní Testování Drůbeže státní podnik, Ústrašice, Czech Republic.

Lange, K. (1996) Untersuchungen zum Leistungsverhalten verschiedener Legehennenhybriden in alternative Haltungssystemen. *Lohmann Information* 4, 7–10.

Lillpers, K. (1993) Oviposition patterns and egg production in domestic hen. Dissertation, Swedish University of Agricultural Science, Uppsala, Sweden.

LTZ (2010) Management Guide. http://www.ltz.de/html/gb_page_121_123.htm (accessed 11 January 2011).

Meuwissen, T.H.E., Hayes, B.J. and Goddard, M.E. (2001) Prediction of total genetic value using genome-wide dense marker maps. *Genetics* 157, 1819–1829.

Muir, W.M. (1996) Group selection for adaptation to multiple-hen cages: selection program and direct responses. *Poultry Science* 75, 447–458.

Nurgiartiningsih, V.M.A., Mielenz, N., Preisinger, R., Schmutz, M. and Schüler, L. (2002) Genetic parameters for egg production and egg weight of laying hens housed in single and group cages. *Archiv für Tierzucht* 45, 501–508.

Peeters, K., Eppink, T.T., Ellen, E.D., Visscher, J. and Bijma, P. (2010) Survival in laying hens: genetic parameters for direct and associative effects in the reciprocal crosses of two purebred layer lines. In: *Proceedings of the 9th World Congress on Genetics Applied to Livestock Production*, Leipzig, Germany, 1–6 August 2010. International Committee for World Congresses on Genetics Applied to Livestock Production and German Society for Animal Science, Bonn, Germany, p. 575.

Preisinger, R., Flock, D.K. and Eek, R. (1998) Recent trends in shell strength of white and brown egg layers in German random sample tests. In: *Proceedings of the 6th Asian Pacific Poultry Congress*, Nagoya, Japan, 4–7 June 1998. Asian Pacific Poultry Association, Japan, pp. 260–261.

Savaş, T., Preisinger, R., Röhe, R. and Kalm, E. (1998) Genetische parameter und optimal prüfdauer für legeleistung anhand von teillegeleistungen bei legehennen. *Archiv für Tierzucht* 41, 421–432.

Wolc, A., Stricker, C., Arango, J., Settar, P., Fulton, J.E., O'Sullivan, N., Habier, D., Fernando, R., Garrick, D.J., Lamont, S.J. and Dekkers, J.C.M. (2010) Breeding value prediction for production traits in layers using pedigree and marker based methods. In: *Proceedings of the 9th World Congress on Genetics Applied to Livestock Production*, Leipzig, Germany, 1–6 August 2010. International Committee for World Congresses on Genetics Applied to Livestock Production and German Society for Animal Science, Bonn, Germany, p. 51.

Yoo, B.H., Sheldon, B.L. and Podger, R.N. (1988) Genetic parameters for oviposition time and time interval in a White Leghorn population of recent commercial origin. *British Poultry Science* 29, 627–637.

Zakaria, A.H., Plumstead, P.W., Romero-Sanchez, H., Leksrisompong, N., Osborne J. and Brake, J. (2005) Oviposition pattern, egg weight, fertility, and hatchability of young and old broiler breeders. *Poultry Science* 84, 1505–1509.

Zupan, M., Kruschwitz, A., Buchwalder, T., Huber-Eicher, B. and Štuhec, I. (2008) Comparison of the prelaying behaviour of nest layers and litter layers. *Poultry Science* 87, 399–404.

CHAPTER 17

Is There a Future for Alternative Production Systems?

V. Sandilands and P.M. Hocking

ABSTRACT

The trend towards alternative systems of production for poultry meat and eggs in the developed economies is paralleled by moves to increase intensification in the developing world. Alongside these changes is the clearly identified imperative to feed an increasingly large and affluent human population in a sustainable manner. Pressure to intensify from the economics of production continue to favour intensive systems but legislation to ban the most intensive systems of production for animal welfare concerns will have a major effect on the way poultry are kept. The least intensive of alternative systems may be associated with greater behavioural freedom for the animals, but can have a significantly greater environmental impact than intensive systems, higher mortality and possibly reduced product quality. In general there is a need for more evidence on all of the inputs and outputs from different systems and economic conditions. Such analyses will allow policy makers to identify areas that need to be changed or modified by appropriate action and by suitably targeted research.

INTRODUCTION

Whereas competition in a free market has and will lead to large-scale intensive production systems, the trend in the economically developed world towards alternative systems of production is driven largely by concerns about animal welfare, particularly with respect to laying hens kept in cages. In addition there has been unease, more recently, about the impact of intensive farming systems on the environment and biodiversity. These issues have been compounded by a general reaction against the pace of technology and a disengagement of most people from the production of food. In the developing world a move away from traditional extensive and backyard systems is driven by the overriding need to feed a growing population that is increasingly urbanized and, in countries with adequate resources, of developing export markets. How will the nations of the world reconcile these contrasting developments in a free market? What is the future of alternative production systems in the developed world? Will they be profitable? Are alternative systems of production environmentally sustainable?

A recent international report sponsored by the UK Government (Foresight, 2011) identified the major challenges facing the global food supply between 2010 and 2050. At the latter date the predicted world population is estimated to rise to more than 9 billion, thus placing increasing pressure on the world's finite resources. These pressures will be compounded by the fact that:

> many people will become wealthier creating demand for more varied, high quality diet requiring additional resources to produce. On the production side, competition for land, water and energy will intensify, while the effect of climate change will become increasingly apparent. (Foresight, 2011.)

The report goes on to state that many systems of food production are unsustainable as currently constituted (e.g. many livestock systems are dependent on large inputs of fossil energy in the form of fertilizers, herbicides and pesticides) and suggests that new technologies such as genetic modification must be adopted.

ECONOMICS AND LEGISLATION

One of the drivers of change is the economics (more accurately the profitability) of production. Genetic selection of poultry since the middle of the last century for feed efficiency, combined with increasing intensification, has led to poultry meat and eggs becoming a major, relatively cheap, protein source for many people in the world. Genetic selection for faster growth rates, for example, has more than halved the feed requirement for the same weight of product (McKay, 2009) and surely contributes to sustainability (except that far more poultry meat is consumed now compared with the 1950s). Nevertheless legislation to achieve significant improvements in poultry welfare (e.g. a ban on cages or reducing growth rates in boilers) can be achieved at relatively little costs to the consumer, albeit at substantial costs to the producer (McInerney, 1998), or at least with little impact on consumption (Sumner *et al.*, 2011). A similar outcome probably exists for legislation to protect the environment but to our knowledge there has been no analysis to date.

It is well known that the expressed wish for welfare-friendly products is generally not transferred to purchasing decisions, leading Webster (2001) to conclude that consumers should afford greater extrinsic value to farm animals. He suggested that welfare-based assurance schemes are a promising route to 'convert an expressed desire for higher welfare standards into effective demand' and underlined the importance of ensuring that the outcomes of these schemes are in fact good animal welfare. Currently there are a number of legal and policy instruments that may be used to protect and enhance animal welfare (see Pritchard, Chapter 2, this volume) that directly or indirectly address the problem of ascribing 'costs' to animal suffering and animal welfare. In addition to these mechanisms different retailers may adopt voluntary welfare codes to differentiate their market and to protect their reputation from adverse publicity (see also Appleby, Chapter 3, this volume).

The environmental costs of production, as for those of animal welfare, have traditionally been hidden but this situation is no longer tenable and there have recently been attempts to assess and quantify the sustainability of production systems for poultry meat and eggs.

ENVIRONMENTAL IMPACT

Sustainability was defined by Foresight (2011) as 'the use of resources that do not exceed the capacity of the earth to replace them'. To what extent does the move to alternative production systems for poultry and eggs, particularly in Europe, now mitigate against this overriding priority to defend global food supplies?

There are a limited number of studies comparing the sustainability of alternative systems of production for poultry meat and eggs using different techniques including environmental impact assessment, energy balance, life cycle analysis (LCA), economic viability and animal welfare singly or in combination. The difficulty of integrating economic, ecological and social indicators is well illustrated by the study of Bokkers and de Boer (2009): an organic broiler system scored highly on social factors (animal welfare, food safety and quality) and poorly for environmental aspects but economic performance was superior, based on the very high price of organic meat (and organic feed) at the time of the study. Comparisons of this nature should evaluate the sensitivity of the analysis to variation in the inputs, particularly the price received for the product, and, as indeed these authors point out, take account of the longer term effect of increased competition from similar or cheaper systems of production on the cost of inputs and the price of the product.

Intensive production systems rely heavily on a variety of inputs that are provided by the environment without monetary cost ('ecosystem services'): for example, the costs of environmental harm through nitrogen pollution, or the contributions to greenhouse gas emissions, are not usually borne by the production systems (but note also that positive effects are also possible, e.g. through the fertilizing value of excreta). However, several studies have shown that the extra feed required to support less-intensive poultry egg and meat production has a major effect on the environmental impact of alternative systems. The main environmental issues relate to ammonia emission, nitrogen, phosphorus, carbon dioxide and dust; and to chemical residues from the manufacture of vaccines, detergents, disinfectants and pesticides. An early attempt to quantify the effect of these variables for alternative egg production systems by de Boer and Cornelissen (2002) indicated that the conventional cage system made the least negative contribution to sustainable egg production compared with deep litter and aviary systems in the Netherlands. Mollenhorst et al. (2006) conducted a comprehensive analysis of data from 13–17 farms on each of four systems in the Netherlands (conventional cage, deep litter, deep litter with outdoor run, aviary with outdoor run) based on animal welfare, economics, environmental impact, ergonomics (farm labour perspective) and product quality criteria. They concluded that the aviary with outdoor run was a 'good alternative' to the conventional cage system on the basis of animal welfare and economics, but with worse scores on environmental impact.

'Emergy' is an estimate of the total amount of solar energy directly or indirectly required to make a certain quantity of product (poultry meat or eggs). Castellini *et al.* (2006) used this method to categorize the energy efficiency of conventional and organic broiler production based on data from a research farm. Their results showed that all the emergy-based indicators (the ratio of total and locally produced emergy and non-renewable to renewable resources) were in favour of the organic system whereas the overall solar transformity (lower emergy for the same output) of the two systems was not greatly different.

LCA is a standardized environmental accounting system used to catalogue all the material and energy inputs and emissions from resource extraction to product disposal and to relate these to specific environmental impacts. Pelletier (2008) used LCA to study broiler performance in the US broiler industry: by far the largest impact on the environment (80–97%) was associated with the production of feed ingredients in arable farms. In principle the use of organic feeds that do not rely on synthetic fertilizers could reduce this environmental impact but of course far more land would be required and, as currently formulated, organic standards would not make efficient use of this more environmentally benign system of growing cereals (see MacLeod and Bentley, Chapter 15, this volume). Boggia *et al.* (2010) compared a conventional broiler with two organic systems by LCA and demonstrated that the extra land required by organic systems contributed about 10% more as a proportion of the total environmental impact of the system than conventional broilers. The environmental effects of poultry rearing in all three systems affected acidification, eutrophication, respiratory inorganics and climate change. However, their analysis suggested that conventional broilers fed organic feed would have slightly less overall impact on the environment than conventional broilers, but that organic production using slow growing birds would have a much larger negative impact on the environment than conventional systems. The authors suggest that partial substitution of soybeans with alternative protein sources such as field peas or beans would improve the sustainability of organic systems but it would likely also benefit conventional broiler production, once more illustrating the difficulty of synthesizing the different aspects of sustainability into a single recommendation that is generally true. Whereas we agree with Xin *et al.* (2011) that far more needs to be done to quantify more precisely the economic efficiency, inputs, outputs and environmental footprint of alternative production systems for better LCA assessment of sustainability in different economic areas, the problem of integrating different aspects of sustainability (e.g. Mollenhorst *et al.*, 2006) remains as an unsolved problem.

WELFARE

The main premise for moving poultry housing away from conventional and into alternative systems is to improve bird welfare, in order to meet their needs through suitable environments. Laying hens have been the farm animal species at the forefront of these changes, because conventional cage housing was seen to thwart the meeting of those needs. Thus the earliest European Union (EU)-wide directive concerning poultry (1999/74/EC) was necessarily based on their housing and management (European Commission, 1999) in an attempt to improve their welfare. Concerns over the inability to perform many motivated behaviours in conventional housing has been the main driver for these changes in consumer demand and legislation. In a relatively short period of time, egg production in the most common housing system types has fluctuated hugely in the UK (Table 17.1), depending on market demands, and yet egg production continues to be a profitable business. Birds can indeed show a greater repertoire of behaviours in systems that are more environmentally complex. With time, these systems are becoming better suited to the needs of the birds while also satisfying the producer.

The ability to perform most natural behaviours is just one of the Five Freedoms that were originally formulated by the Brambell Committee (Brambell, 1965) and later adopted by the UK's Farm Animal Welfare Council as a means of ensuring that the welfare of animals is met. The other freedoms (freedom from hunger, thirst and malnutrition; freedom from pain, injury and disease; freedom from thermal and physical discomfort; and freedom from fear and distress) are just as important, and should not be overlooked in any system. It is an ongoing dilemma with housing systems that a system may fulfil some freedoms well, while others less adequately.

Disease risk is a serious concern when it comes to any farming system, but the greater incidence of several diseases and higher levels of parasite burdens seen in many loose-housed egg production sites is a cause for concern (see Lister and van Nijhuis, Chapter 4, this volume). Even without identifying any specific disease, mortality levels in laying hen extensive systems are generally higher than that seen in cage systems (see Rodenburg *et al.*, Chapter 12, this volume). Alternative systems require a greater degree of management, and therefore time and money, in order to keep these in check, but this is achievable. Whether such a level of management can be achieved in large-scale systems, on a long-term basis, remains to be seen.

Although they are not 'alternative' *per se*, it would have been an oversight not to mention enriched, or furnished, cages. With the ban of conventional cages in the EU from 2012, many hens are likely to be housed in these cages as an improvement to bird welfare over the conventional cage, but with many

Table 17.1. Changes in the percentage of eggs from different systems of production in the UK.

System	1951[a]	1966[a]	1980[b]	2000[c]	2010[c]
Cage	8	67	95	72	50
Barn	12	25	4	8	5
Free range (includes organic)	80	8	1	20	45

[a]Systems of production (Anonymous, 2010).
[b]Proportion of hens by system (Hewson, 1986).
[c]Eggs produced by system (Defra, 2011).

of the advantages that conventional cages have over floor systems, namely small groups and separation from faeces. However, consumers may find them indistinguishable from conventional cages ('a cage is still a cage') and welfare groups generally find them unacceptable due to the lack of continual access to a foraging substrate (given that the feed dispensed on to the scratch mat is infrequent and quickly depleted) or to the outdoors. Alternative laying hen systems provide greater environmental complexity but can jeopardize performance, health and hygiene (Rodenburg *et al.*, Chapter 12, this volume). Hens in these systems need careful management if outbreaks of feather pecking and cannibalism are to be avoided (although systems with range access may in fact help reduce these). There are acknowledged problems with all systems, and efforts should be made to improve housing systems to suit the needs of hens: a novel housing system has been developed in the Netherlands which is showing promising results (Koerkamp and Bos, 2008).

Turkey and broiler housing systems across Europe have gone through less radical changes than hen housing, partly because there is less to change in their environment to meet their needs: birds are already housed on litter, immature birds do not require nesting facilities and breeding stock are generally provided with both of these. However, alternative systems for chicken have increased (particularly alternative indoor methods) and free range is particularly popular in France, possibly due to the perceived improvement in meat taste and quality with this system, although that does not appear to be borne out in some taste tests (see Jones and Berk, Chapter 14, this volume). The new meat chicken directive (2007/43/EC), which came into force in 2010 (European Commission, 2007), is likely to have an effect on the quality of life of standard reared broilers, with longer continuous dark periods, limitations on stocking density and monitoring of foot pad health. Its effects on broiler welfare remain to be seen. The market share of alternative production for turkeys is generally small but the Traditional Farm Fresh turkey has a major share of the Christmas market in the UK. With both types of meat birds, ability to utilize range where provided is limited if fast growing strains are used: thus bird genotype is very important. As with hens, unwanted pecking behaviour can be more problematic in alternative systems, which require careful management control if producers are not going to resort to beak trimming. Ranging behaviour and feather pecking, pulling and cannibalism may be changed by genetic selection, if methods are developed for measuring the relevant trait (Icken *et al.*, Chapter 16, this volume), and genetic lines adapted to alternative systems may make a substantial contribution to the widespread adoption of alternative systems for laying hens.

Like meat birds, housing for breeding birds of all types has largely escaped the move to 'alternative' methods of housing, because they are already housed on the floor, in social groups, often with natural mating (de Jong and Swalander, Chapter 13, this volume). Few birds are given access to range, however. In contrast, game bird rearing, which has been more traditional (i.e. extensive) in the past, is showing signs of moving towards more intensive housing for parent stock in raised floor, as opposed to grass, pens. Rearing stock are often given access to grass runs, but not always, and housing types vary greatly from farm to farm. New welfare codes in the UK may unify and improve how these birds are housed, but there is still much scope for research in game bird housing, health, productivity and welfare (Pennycott et al., Chapter 9, this volume).

Waterfowl housing differs considerably across Europe and Asia, and also with waterfowl type (duck, goose, Muscovy) and use (meat, eggs, foie gras, feathers) (Guémené et al., Chapter 8, this volume). Clearly in Asia, where most waterfowl production takes place, 'conventional' farming is often still 'traditional', i.e. mixed systems (fish and birds) and/or small groups, with access to water (depending on geography), but these systems are modernizing towards the large scale and intensive. By comparison, in the West some alternative systems are incorporating range and water access. However, these make up a very small proportion of waterfowl production, which may be a reflection of the small market and lack of public demand for alternatively reared products.

THE FUTURE OF ALTERNATIVE SYSTEMS

There will always be a demand for some proportion of alternatively reared meat or eggs, although the proportion will undoubtedly fluctuate over the years. Housing system and poultry productivity, health and welfare will most likely continue to be a complex balancing act between conflicting motivations, both for the animals housed and for ourselves as animal keepers and consumers. Alternative systems have the potential for great good but also for great harm. Our responsibility is to strike a balance between producing animal protein economically, which may drive the proportion of poultry housed in alternative systems down, and to give animals 'a life worth living' (FAWC, 2009), which may drive the proportion up. Sometimes, however, the life worth living in an animal's eyes may not always mean that welfare is optimized. The LAYWEL report (2006) put it very well:

> the question arises, how the risk of diseases, damages and mortality is related to welfare. In many cases the hens do not perceive risk-bearing conditions as adverse experiences. They may even show preferences for these conditions. If there are established relationships between the risk factors and the occurrence of welfare problems, the potential hazard of the birds' welfare has to be balanced against the strength of preference.

So, although hens may for example enjoy drinking out of muddy puddles found on the range, the risk to health may be considered greater than the need to fulfil the desire to drink there.

We must strike this balance, which is not easy. Are we bound to housing systems that result in either a few birds suffering greatly, or many birds suffering somewhat? With time, we may manage to move to improved housing systems for poultry, so that most farmed birds have a life worth living.

CONCLUSIONS

The adoption of alternative production systems for poultry during the past few decades has been, and will continue to be, an ongoing process of adaptation and evolution and no one system may be appropriate for all conditions or for all time. As specific problems or issues are identified by methods such as LCA, so these will be addressed by research and development in universities and institutes or by the industry itself, and the relative advantages of different systems may change. Currently the biggest impact on the adoption of alternative systems in Europe is legislation and other countries are likely to follow suit (e.g. the USA; Mench *et al.*, 2011). In the longer term (at least by 2050) a burgeoning and more affluent world population will impact on these developments. Furthermore an increasing emphasis on the environmental sustainability of food production will become apparent, possibly though legislation to inject a cost to environmental pollution, but also in the increasingly scarce resources, particularly of oil for the manufacture of fertilizers and other chemical inputs, the availability of water and land for crop production. However, the authors would hope that there would not be a return to the more extreme forms of intensive production of poultry meat and eggs.

ACKNOWLEDGEMENTS

The Scottish Agricultural College (where the first author works) is supported by the Scottish Government. The work of the second author is supported by a core strategic grant from the Biotechnology and Biological Sciences Research Council to the Roslin Institute.

REFERENCES

Anonymous (2010) Animal Farm Life. http://www.animalfarmlife.eu/index.html (accessed 3 August 2011).

Brambell, F.W.R. (1965) *Report of the Technical Committee to Enquire into the Welfare of Animals kept under Intensive Livestock Husbandry Systems.* Command 2836. HMSO, London.

Boggia, A., Paolotti, L. and Castellini, C. (2010) Environmental impact evaluation of conventional, organic and organic-plus poultry production systems using life cycle assessment. *World's Poultry Science Journal* 66, 95–114.

Bokkers, E.A.M. and de Boer, I.J.M. (2009) Economic, ecological and social performance of conventional and organic broiler production in the Netherlands. *British Poultry Science* 50, 546–557.

Castellini, C., Bastianoni, S., Granai, C., Dal Bosco, A. and Brunetti, M. (2006) Sustainability of poultry production using the emergy approach: comparison of conventional and organic rearing systems. *Agriculture Ecosystems and Environment* 114, 343–350.

De Boer, I.J.M. and Cornelissen, A.M.G. (2002) A method using sustainability indicators to compare conventional and animal-friendly egg production systems. *Poultry Science* 8, 173–181.

Defra (2011) Defra egg statistics. http://www.defra.gov.uk/statistics/foodfarm/food/eggs/. UK egg packing station throughput and prices dataset (accessed 23 May 2011).

European Commission (1999) Council Directive 1999/74/EC of 19 July 1999 laying down minimum standards for the protection of laying hens. *Official Journal of the European Communities* L 203, 03/08/1999, 53–57; available at: http://eur-lex.europa.eu/LexUriServ/LexUriServ.do?uri=OJ:L:1999:203:0053:0057:EN:PDF (accessed July 2011).

European Commission (2007) Council Directive 2007/43/EC of 28 June 2007 laying down minimum rules for the protection of chickens kept for meat production. *Official Journal of the European Union* L 182, 12/07/2007, 19–28; available at: http://eur-lex.europa.eu/LexUriServ/LexUriServ.do?uri=OJ:L:2007:182:0019:0028:EN:PDF (accessed 7 July 2011).

FAWC (2009) *Farm Animal Welfare in Great Britain: Past, Present and Future.* Farm Animal Welfare Council, London; available at: http://www.fawc.org.uk/pdf/ppf-report091012.pdf (accessed 20 June 2011).

Foresight (2011) *The Future of Food and Farming (2011) Final Project Report.* The Government Office for Science, London; available at: http://www.bis.gov.uk/assets/bispartners/foresight/docs/food-and-farming/11-546-future-of-food-and-farming-report.pdf (accessed 30 January 2011).

Hewson, P. (1986) Origin and development of the British poultry industry: The first hundred years. *British Poultry Science* 27, 525–539.

Koerkamp, P.W.G.G. and Bos, A.P. (2008). Designing complex and sustainable agricultural production systems: an integrated and reflexive approach for the case of table egg production in the Netherlands. *NJAS – Wageningen Journal of Life Sciences* 54, 133–145.

LAYWEL (2006) Welfare implications of changes in production systems for laying hens: D 1.2 Report with consensual version of welfare definition and welfare indicators. http://www.laywel.eu/web/pdf/deliverable%2012.pdf (accessed 7 July 2011).

McInerney, J.P. (1998) The economics of welfare. In: Mitchell, A.R. and Ewbank, R. (eds) *Ethics, Welfare, Law and Market Forces: The Veterinary Interface.* Universities Federation for Animal Welfare, Wheathampstead, UK, pp. 115–132.

McKay, J. (2009) The genetics of modern commercial poultry. In: Hocking, P.M. (ed.) *Biology of Breeding Poultry.* CABI Publishing, Wallingford, UK, pp. 3–9.

Mench, J.A., Sumner, D.A. and Rosen-Molina, J.T. (2011) Sustainability of egg production in the United States – the policy and market context. *Poultry Science* 90, 229–240.

Mollenhorst, H., Berentsen, P.B.M. and de Boer, I.J.M. (2006) On-farm quantification of sustainability indicators: an application to egg production systems. *British Poultry Science* 47, 405–417.

Pelletier, N. (2008) Environmental performance in the US broiler poultry sector: life cycle energy use and greenhouse gas, ozone depleting, acidifying and eutrophying emissions. *Agricultural Systems* 98, 67–73.

Sumner, D.A., Gow., H., Hayes, D., Mathews, W., Norwood, B., Rosen-Molina, J.T. and Thurman, W. (2011) Economic and market issues on the sustainability of egg production in the United States: analysis of alternative production systems. *Poultry Science* 90, 241–250.

Webster, A.J.F. (2001) Farm animal welfare: the five freedoms and the free market. *The Veterinary Journal* 161, 229–237.

Xin, H., Gates, R.S., Green, A.R., Mitloehner, F.M., Moore, P.A. and Wathes, C.M. (2011) Environmental impacts and sustainability of egg production systems. *Poultry Science* 90, 263–277.

INDEX